CATALYSIS –
Science and Technology

Volume 1

Akademie-Verlag · Berlin

Die
Originalausgabe erscheint
im Springer-Verlag Berlin · Heidelberg
New York

Vertrieb
ausschließlich für alle Staaten mit Ausnahme
der sozialistischen Länder:
Springer-Verlag Berlin · Heidelberg
New York

Vertrieb
für die sozialistischen Länder:
Akademie-Verlag Berlin

Erschienen im
Akademie-Verlag, DDR-1086 Berlin, Leipziger Straße 3–4

Lizenznummer: 202 · 100/523/82
Gesamtherstellung: VEB Druckerei „Thomas Müntzer", 5820 Bad Langensalza
Umschlaggestaltung: Eckhard Steiner
Bestellnummer: 763 111 7 (6706) · LSV 1215
Printed in GDR
DDR 142,— M

General Preface to Series

In one form or another catalytic science reaches across almost the entire field of reaction chemistry, while catalytic technology is a cornerstone of much of modern chemical industry. The field of catalysis is now so wide and detailed, and its ramifications are so numerous, that the production of a thorough treatment of the entire subject is well beyond the capability of any single author. Nevertheless, the need is obvious for a comprehensive reference work on catalysis which is thoroughly up-to-date, and which covers the subject in depth at both a scientific and at a technological level. In these circumstances, a multi-author approach, despite its well-known drawbacks, seems to be the only one available.

In general terms, the scope of *Catalysis: Science and Technology* is limited to topics which are, to some extent at least, relevant to industrial processes. The whole of heterogeneous catalysis falls within its scope, but only biocatalytic processes which have significance outside of biology are included. Ancillary subjects such as surface science, materials properties, and other fields of catalysis are given adequate treatment, but not to the extent of obscuring the central theme.

Catalysis: Science and Technology thus has a rather different emphasis from normal review publications in the field of catalysis: here we concentrate more on important established material, although at the same time providing a systematic presentation of relevant data. The opportunity is also taken, where possible, to relate specific details of a particular topic in catalysis to established principles in chemistry, physics, and engineering, and to place some of the more important features into a historical perspective.

Because the field of catalysis is one where current activity is enormous and because various topics in catalysis reach a degree of maturity at different points in time, it is not expedient to impose a preconceived ordered structure upon *Catalysis: Science and Technology* with each volume devoted to a particular subject area. Instead, each topic is dealt with when it is most appropriate to do so. It will be sufficient if the entire subject has been properly covered by the time the last volume in the series appears. Nevertheless, the Editors will try to organize the subject matter so as to minimize unnecessary duplication between chapters, and to impose a reasonable uniformity of style and approach. Ultimately, these aspects of the presentation of this work must remain the responsibility of the Editors, rather than of individual authors.

The Editors would like to take this opportunity to give their sincere thanks to all the authors whose labors make this reference work possible. However, we all stand in debt to the numerous scientists and engineers whose efforts have built the discipline of catalysts into what it is today: we can do no more than dedicate these volumes to them.

Preface

Catalysis has made major contributions to many areas of chemical industry. Before embarking upon detailed considerations of catalytic science and technology, it is very helpful to look first at the nature of industrial catalysis, and how it has evolved and grown to meet demands imposed by changing industrial needs. Dr. H. Heinemann is uniquely qualified to place industrial catalysis in a historical perspective: in his distinguished industrial career he has been closely involved with many of the major innovations in industrial catalysis.

Before a catalytic process is commercialized, the supporting research and development work is carried out in chemical reactors. It is obviously imperative that the behaviour of such reactor systems should be thoroughly understood by those who use them, and by those who may have to interpret their results, yet all too often this basic need is not met. Professor J. C. R. Turner provides a straightforward yet thorough account of catalytic reactor theory which should make it impossible for any catalytic practitioner to plead ignorance.

The catalytic hydrogenation of dinitrogen to ammonia is one of the world's great industrial processes and the catalytic activation of molecular dinitrogen is a key step in that process. The chapter by Professors A. Ozaki and K. Aika deals with the chemistry of dinitrogen activation at the catalyst surface, and shows how this relates to the synthesis of ammonia. The chapter also deals with the activation of dinitrogen by molecular complexes in homogeneous systems.

The Fischer-Tropsch synthesis is now again seen as an important possible route for the production of synthetic liquid fuel and other industrial chemicals. The process has a long history and it has been the subject of considerable development since it was first discovered. Dr. M. E. Dry has occupied a central position in optimizing the process at the only Fischer-Tropsch plant which is currently operating on a commercial scale. Dr. Dry's chapter deals mainly with the processes using "classical" Fischer-Tropsch catalysts, but he also relates this to other catalytic systems which have, as yet, only been studied on a laboratory scale.

Catalytic reforming is a key process for the production of high-octane gasoline. The chemistry of catalytic reforming has been intensively studied, as have the properties of the multifunctional catalysts which are used. Dr. J. H. Sinfelt has played an important role in elucidating a number of aspects of catalytic reforming chemistry, and he has also been responsible for the introduction of one of the major industrial innovations in the field in recent years. His chapter gives an account of the main chemical features of catalytic reforming, and he also indicates how the chemistry of the process relates to the technology.

Contents

List of Contributors

Professor K. Aika
Research Laboratory of Resources Utilization,
Tokyo Institute of Technology,
Nagatsuta 4259, Midori-ku, Yokohama, Japan

Dr. M. E. Dry
Research Department,
SASOL, Sasolburg, South Africa

Dr. H. Heinemann
Lawrence Berkeley Laboratory,
Materials and Molecular Research Division,
University of California,
Berkeley, CA 94720, USA

Professor A. Ozaki
Research Laboratory of Resources Utilization,
Tokyo Institute of Technology,
Nagatsuta 4259, Midori-ku, Yokohama, Japan

Dr. J. H. Sinfelt
Corporate Research Laboratories,
Exxon Research and Engineering Co.,
Linden, N.J. 07036, USA

Professor J. C. R. Turner
Department of Chemical Engineering,
University of Exeter,
North Park Road, Exeter EX4 4QF, UK

Chapter 1

A Brief History of Industrial Catalysis

H. Heinemann

Lawrence Berkeley Laboratory,
Materials and Molecular Research Division,
University of California,
Berkeley, CA 94720, USA

The use of catalytic processes has grown almost exponentially during the last few decades. This chapter reviews industrial catalytic developments, which have been commercialized during the last forty years. Emphasis is put on heterogeneous catalytic processes, largely in the petroleum, petrochemical and automotive industries, where the largest scale applications have occurred. Homogeneous catalytic processes are briefly treated and polymerization catalysis is mentioned.

Contents

1. Introduction

Industrial catalysis is an old art. Wine and soap makers have employed catalytic agents for thousands of years, though without knowledge or understanding of their workings. Large-scale conscious use of industrial catalysts originated in the mid-18th century with the introduction of the lead chamber process for manufacture of sulfuric acid, in which nitric acid was used to oxidize SO_2 to SO_3 in the presence of water. The lower oxides of nitrogen formed are in turn oxidized with air to form nitric acid. While the need for a catalyst was recognized, the scientific basis for its chemical and kinetic action came only much later. This is a trend that persists to the present. In spite of great advances in the science of catalysis, major industrial applications and novel uses have almost always been based on empirical findings. Scientific explanations followed later and frequently led to process improvements and refinements. The goal of catalytic scientists to be able to predict industrial catalytic behavior of substances and processes has thus far remained elusive. There is hope, however, that with the rapidly increasing sophistication of tools and observational means catalytic science may in the future replace catalytic art.

Mills and Cusumano [1] have pointed out that the use of catalytic processes has grown almost exponentially from the early 18th century to the present. It has been estimated that at present over 20 percent of all industrial products have underlying catalytic steps in their manufacture. Early catalytic processes were used mostly for the production of inorganic chemicals (sulfuric acid, nitric acid, chlorine, ammonia), with catalytic processes involving organic reactions becoming prominent only in the 20th century, but rapidly dominating the industry, mostly because of the widespread application of catalysis in fuels production.[1]

[1] Since this Chapter deals with catalysis in an industrial situation, a number of units have been retained which are relevent to industrial rather than scientific usage. In particular, it is convenient here to summarise the following:
1 barrel (bbl), equal to 0.159 m^3,
1 ton (U.S. ton), equal to 2000 lb,
1 pound (lb), equal to 0.454 kg,
1 standard cubic foot of gas (SCF), equal to 0.0832 m^3

The present chapter is mostly limited to the rapid growth of industrial catalysis since the second World War. A few brief excursions into earlier history have been found necessary. The author found it difficult to ascertain exact dates for many innovations. References can in most instances be found for the time of a first commercial operation of a process, but it is much harder and in some cases impossible to determine the time of conception. In fact, the research and development leading to new technology often involves so many people and ideas that the resulting process cannot always be attributed to specific individuals. In addition, publications and even patent application dates frequently lag considerably behind conception.

The great majority of catalytic processes are still based on heterogeneous catalysis. Homogeneously catalysed processes however, have assumed much more importance in recent years and their impact is often underestimated because much of the volume and value of catalytic processes is concentrated in the petroleum refining industry which uses predominantly heterogeneous catalysts. The relative growth of homogeneous catalytic process technology is far greater in chemical and petrochemical applications than in all other industrial applications, including those of the petroleum industry.

While there has been a large number of process developments during the period under consideration, the majority are of an evolutionary type and there are relatively few process ideas that have opened up new chemistry and engineering and/or started new catalytic industries. The list presented in Table 1 gives the author's admittedly subjective impression of what might be called "breakthroughs" in catalytic technology during the last 35–40 years.

Table 1. Major catalytic innovations, 1935–1978

Year of first commercialization	Event	Area of industry
1936	Catalytic cracking	Petroleum
1941	Fluid-bed technology	Petroleum-petrochemicals
1942	Thermofor catalytic cracking	Petroleum
1942	Paraffin alkylation	Petroleum
1950	Catalytic naphtha reforming (Pt-catalysts)	Petroleum
1955	Ziegler-Natta polymerization	Polymers
1960	Acetaldehyde from ethylene (Wacker Chemistry)	Chemicals
1963	Low-pressure ammonia synthesis	Fertilizer
1963	Ammoxidation	Chemicals
1964	Zeolite catalysts	Petroleum-petrochemicals
1964	Oxychlorination	Monomers
1966	Olefin disproportionation	Petrochemicals
1967	Bimetallic reforming catalysts	Petroleum
1968	Shape selective catalysis	Petroleum-petrochemicals
1976	Emission control catalysts	Automotive

All of these will be discussed in this chapter along with many other developments of importance. No claim can be made for completeness.

During the early years of industrial catalysis development described in this chapter, there were several new technologies requiring extensive engineering as well as catalyst developments. Fluid catalytic cracking, catalytic reforming, and low-pressure ammonia synthesis are examples. In the last 15 to 20 years there has been more emphasis on novel catalysts that produced better products and product yields, and which could be used in existing or slightly modified equipment. Examples of this type are zeolite and bimetallic reforming catalysts. A major reason for this trend lies in the spiraling construction costs of industrial plants, with the concomitant increase in the

Table 2. Major applications of heterogeneous catalysis in U.S. petroleum industry

	Capacity		U.S. Catalyst sales/10^6 Ib per year in 1978	Catalyst value/ 10^6 \$ per year in 1979
	barrel per day	metric tonne per day		
Catalytic Cracking	5,000,000	635,000	286	143
Catalytic Hydrocracking	900,000	114,000	~2	20
Naphtha Reforming	2,000,000	222,000	5	27
Catalytic Alkylation	890,000	85,000	3,700	128
Catalytic Hydrotreating	2,000,000	260,000	22	45

Table 3. Major applications of heterogeneous catalysis in U.S. petrochemical industry

	Approximate 1979 production/ 10^6 tons per year	Product value/ 10^6 \$ per year in 1979
Ammonia Synthesis	15	1,500
Methanol	4.1	693
Steam Reforming	(6.3×10^9 SCFD)	535
Oxidation		
— ethylene oxide	2.5	1,600
— formaldehyde	3.5	490
— phthalic anhydride	3.5	2,870
— maleic anhydride	0.2	200
Acrylonitrile (ammoxidation)	1.0	560
Styrene (dehydrogenation)	3.8	2,660
Hydrogenation		
— aniline	0.3	274
— cyclohexane	1.2	600
Vinyl Chloride Monomer (oxychlorination)	3.5	1,050
Vinyl Acetate Monomer (oxychlorination)	0.9	522
Butadiene	1.7	884

Table 4. Major applications of heterogeneous catalysis in commodity chemicals

	Approximate 1979 production/ 10^6 tons per year	Product value/ 10^6 \$ per year in 1979
Sulfuric Acid	40	1,800
Nitric Acid	8.5	2,040[a]

[a] pure HNO_3

financial risk of failure or protracted break-in periods of novel facilities. A catalyst failure at worst may require a change back to a previously used catalyst, with a loss of some days in down-time, while major equipment changes may require weeks and months during which costly facilities are nonproductive. In addition, there has been a trend to ever larger unit operations. New engineering technology is best tried in relatively small units, which however, are no longer competitive with large production facilities. One can expect the trend to improve older catalysts, and introduce novel catalysts, in existing equipment to continue for some time.

To support the importance of heterogeneous catalysis to industrial production, three tables illustrate catalytic uses; for the Petroleum Industry in the U.S.A. (Table 2), giving capacities, catalyst sales and values; for the Petrochemical Industry (Table 3)[1] giving product volume and value; and for the Commodity Chemicals Industry (Table 4)[1] giving product volume and value. A similar table on industrial homogeneous catalytic uses is contained in section XI (Table 8).

2. Catalytic Cracking and other Acid Catalysed Reactions

Acid (and base) catalysis are involved in some of the oldest industrial reactions, such as hydrolysis of esters for soap manufacture and inversion of sugar cane. Friedel-Crafts reactions were discovered in 1877–1878 and aluminum chloride — a typical Friedel-Crafts type catalyst — was the first commercial catalyst used in converting heavier petroleum hydrocarbons to lighter fragments, particularly in the gasoline boiling range.

Gasoline (boiling between 38 °C and 210 °C and having a C_4—C_{13} range) comprises only about 15–25 percent of natural petroleum. Most of this "straight run" gasoline consists of normal or slightly branched paraffins, some naphthenes, and a few aromatics, most of these components having low octane numbers. "Cracking" of heavier petroleum fractions over selective catalysts enhances the obtainable yield of gasoline from a barrel of

[1] Data for these tables courtesy of Catalytica Associates, Inc.

crude oil and results in the formation of larger quantities of highly branched paraffins, olefins, and aromatics, all of which are high octane number components. The McAffee $AlCl_3$ process found limited application in the years following 1915, and was operated as a batch process with a severe problem in disposing of the sludges consisting of spent aluminium chloride dissolved in hydrocarbons. Gurwitsch [2] and Herbst [3] observed and detailed the catalytic activity of certain activated clays as early as 1912 and 1926, respectively. A major breakthrough occured in 1936 after Eugene J. Houdry had solved a series of problems involving catalyst deactivation, regeneration, and stability, and overcame formidable engineering problems. It is interesting to note that Houdry was a mechanical engineer who was also an automobile race driver, and as such recognized that the limitations of the internal combustion engine at that time were not of a mechanical nature but lay in the constraints imposed by the low-octane number characteristics of gasoline then available. In searching for a better gasoline, he studied the chemistry of hydrocarbons and the synthesis of branched chain paraffins and olefins, and of aromatics, by catalytic cracking of gas oils. Houdry devised a system of cyclic reaction and regeneration which maintained the cracking unit in heat balance, and which could be practiced commercially in a continuous mode. Cyclic operation of fixed-bed cracking units, utilizing the exothermic heat of regeneration to provide the required cracking temperatures (cracking is endothermic) became a reality when Houdry Process Corporation, together with Socony-Vacuum Oil Company and Sun Oil Company, built the first commercial units at plants of the two oil companies in 1936–1938, shortly before the outbreak of the Second World War. The catalyst was contained in numerous parallel tubes that were suspended in a molten salt heat exchanger, as shown in Figure 1 [4, 5]. Figure 2 shows a Houdry unit. In spite of the rapid refinements that followed, some of the original units were still in operation in the early 1960s. About 90 percent of the aviation gasoline base stock used in the battle of Britain came from Houdry Units.

The sudden demand for large quantities of aviation gasoline during World War II accelerated the rapid expansion of the cracking process, and numerous units were built during the period from 1938 to 1950. At the same time, major improvements were made in the mechanical design of the cracking units and in the cracking catalyst.

The cyclic operation of the fixed bed units was replaced by designs which moved the catalyst continously from a reactor through a purge zone to a regenerator, and from there through another inert gas-purge zone back to the reactor. This was accomplished by two quite different methods: In the moving-bed type of operation first introduced by Socony-Vacuum Oil Company in 1942 [5], the pelleted or extruded catalyst moved by gravity through reaction and regeneration zones, and was lifted from the bottom of one vessel to the top of the other by a bucket elevator (Figure 3). In the early 1950s, this design was further refined by replacing the elevator with a lift pipe in which the catalyst was blown by a high-velocity gas stream to the top vessel [6]. Similar designs were commercialized by Socony-Vacuum

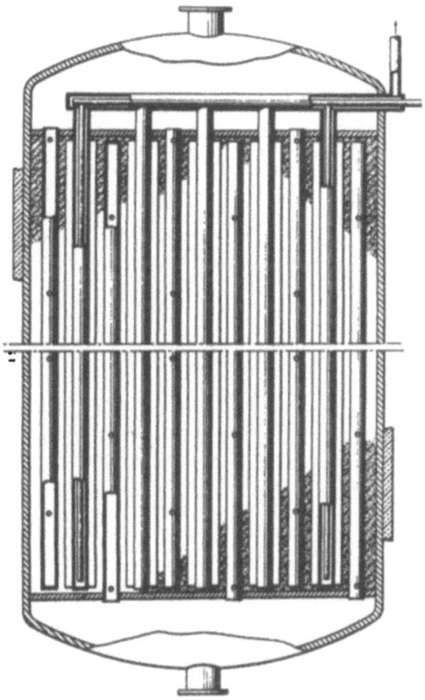

Figure 1. Houdry catalytic cracking unit reactor. Tubular catalyst containers in molten salt medium. Figure taken from reference 4

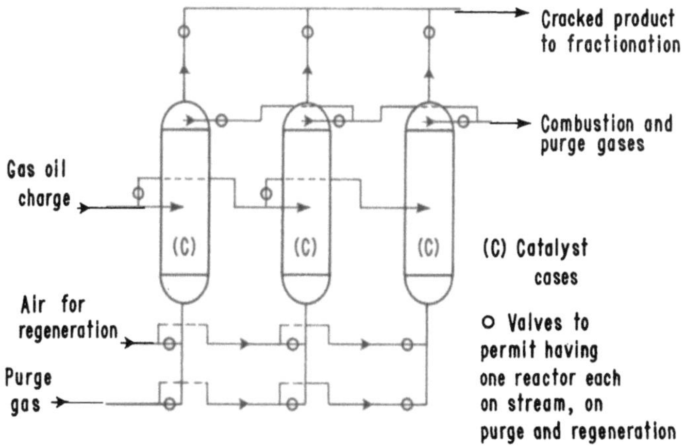

Figure 2. Schematic drawing of Houdry fixed bed catalytic cracking unit

Oil Company under the name of "Thermofor Catalytic Cracking" (TCC), and by Houdry Process Corporation under the name of "Houdriflow". These units operated satisfactorily for many years, but are now gradually disappearing because their capacity is limited by heat-flow constraints. Units of larger than 20,000 bbl/day have not been built.

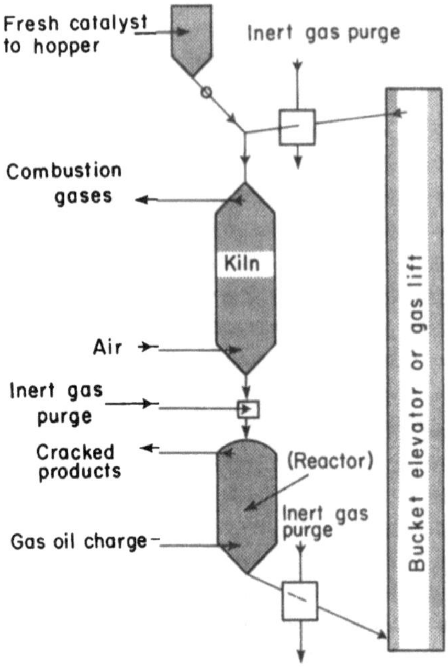

Figure 3. Schematic drawing of TCC cracking unit (moving bed)

In 1941 a group of companies under the leadership of Standard Oil Company of New Jersey introduced the first "Fluid bed Catalytic Cracking Unit" (FCC) [5. 7]. In this revolutionary design, based largely on work by Lewis and Gilleland at M.I.T., the catalyst in the form of fine particles in the 30–200 mesh range was maintained in suspension in a stream of vaporized hydrocarbons, blown through the reactor and collected in a separator and in cyclones, passed through a stand-pipe in which it was purged to the bottom

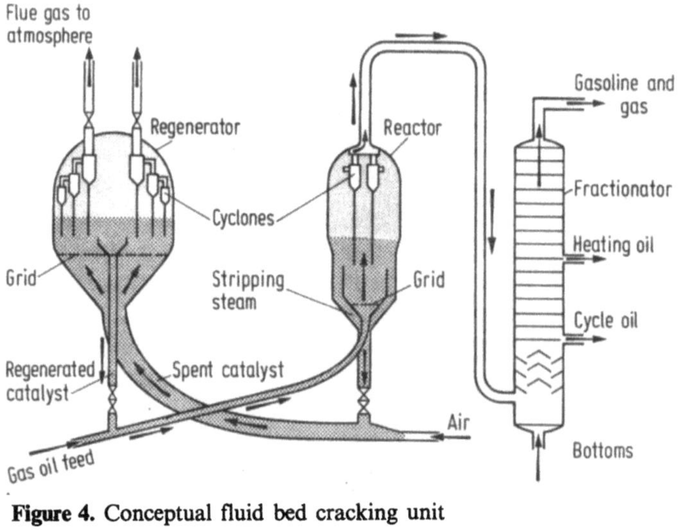

Figure 4. Conceptual fluid bed cracking unit

of the regenerator, and blown through the regenerator by an oxygen-containing gas stream and finally returned to the reactor (Figure 4). Over the years, many improvements — such as short-contact time and riser cracking — have been made in the engineering design of FCC units [8]. Riser cracking was made possible also by improvements in the physical strength of cracking catalysts. A modern short-contact time fluid cracker is shown in Figure 5 [8]. The FCC design could be scaled to very large units because of the rapid heat exchange between carrier gas and catalyst, which permits very close temperature control. The vast majority of present cracking units are of the FCC type, and the fluid-bed technology has been applied to other processes, particularly those of a highly exothermic or endothermic nature. In 1978 the U.S. catalytic cracking capacity was about 5,000,000 bbl/day.

While these design changes were taking place, improvements were made in the catalyst type and composition. The original cracking catalysts were acid-treated clays of the montmorillonite type. They permitted larger yields of gasoline of higher octane number than had previously been obtainable by thermal cracking of gas oils. The clays were gradually replaced by amorphous synthetic silica-alumina catalysts which were more stable under regeneration conditions and also gave a better product distribution [9]. Gasoline yields obtainable from gas oil increased from about 20 percent by thermal cracking to over 40 percent with silica-alumina catalysts. The importance of the catalyst shape and pore distribution was recognized about 1945. Bead catalysts were invented by Marisic at Socony-Vacuum and resulted in lower attrition losses than pelleted or extruded catalysts in TCC-type units. The attrition and activity advances of bead versus extruded-clay catalysts is shown in Figure 6 [4]. Open structure beads, produced by incorporating crystalline alumina during gelling, increased activity and reduced diffusion limitations. Variations of the silica/alumina ratio (normally 65/35) permitted fine tuning of product yields. Silica-magnesia catalysts were introduced in 1952 and resulted in better gasoline yields (but of slightly lower octane number). However, these catalysts never reached large-scale use because of regeneration problems [10].

Silica-alumina catalysts were used in FCC units as well as in TCC units, but because of the small particle size used in FCC did not require the attention to diffusional problems encountered in TCC-type units. The whole area of FCC cracking has been reviewed in detail by Venuto and Habib [8].

A major revolution in cracking catalysts occurred in the early 1960s, and this will be described in section 3 dealing with zeolite catalysis.

Leaving catalytic cracking temporarily, there are a series of other acid catalysed reactions which have become important, mostly in the fuel area. These include polymerization of olefins to dimers, trimers, and tetramers; alkylation; and isomerizations of paraffins and aromatics. All of these have been previously described in the literature [11], and no breakthroughs have occured in polymerization of C_3 and C_4 hydrocarbons to fuels in the last 30 years although a number of refinements have been introduced.

In alkylation to high-octane gasoline, both the sulfuric-acid and hydrofluoric-acid processes continue to dominate the field [12].

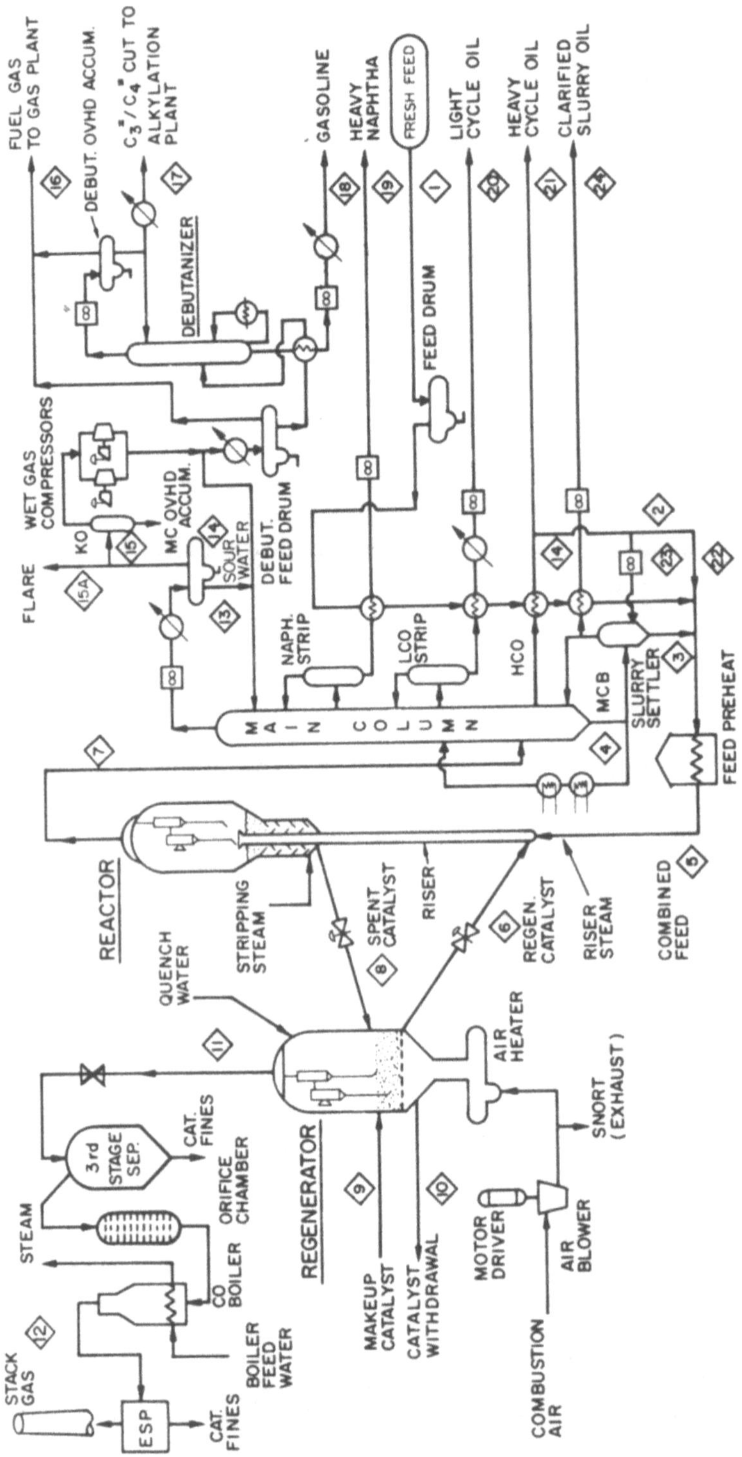

Figure 5. Modern F.C.C. unit. Figure taken from reference 8

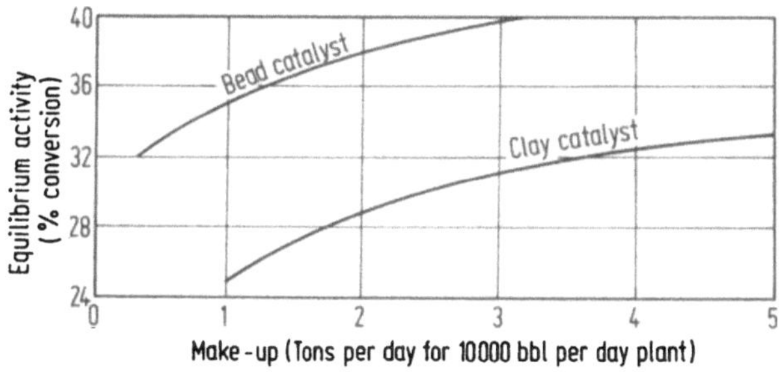

Figure 6. Activity and make-up rate of clay and silica-alumina bead catalysts. Figure taken from reference 4

Addition of an olefinic hydrocarbon to another molecule is being practiced in two major areas: (1) in the alkylation of isobutane with butenes or propene to produce highly branched C_7 and C_8 hydrocarbons as high octane number gasoline components; (2) in the alkylation of aromatics with ethylene or propene to produce alkyl-aromatics. Paraffin alkylation was discovered by V. N. Ipatieff in 1935 and commercialized in 1942. The two catalysts in commercial use are sulfuric acid and hydrofluoric acid. Alkylation supplied large volumes of aviation gasoline in World War II. The subject has been reviewed by R. N. Kennedy [11] and others and improvements in the process technology since then do not involve major inventions.

Alkylation of aromatics with olefins is used primarily in the production of ethylbenzene and of cumene. Anhydrous $AlCl_3$ + HCl catalysts have been used since the early 1940's. A major improvement in catalyst technology was introduced in 1977 by the use of zeolites and will be discussed in the section on zeolites. It has eliminated the problems of acids sludge corrosion.

In aromatics alkylation and aromatics isomerization novel process technology has emerged which will also be described in the section on zeolites.

The need for paraffin isomerization arose during World War II. Alkylation was one of the few routes to high-octane-number aviation gasoline. While sufficient amounts of C_4 olefins were available from catalytic and thermal cracking, there was a shortage of isobutane. On the other hand, there were supplies of n–butane which could be isomerized. Two routes, commercialized by Shell Oil Company and Texaco in 1941 were used for isomerization of normal to isobutane. Both were based on aluminium chloride as a catalyst and gaseous HCl as a promoter [13]. Ober 40 units were built. In one process, a sludge of $AlCl_3$ in aromatic hydrocarbons served as a catalytic liquid through which the n-butane gas was passed; in the other, anhydrous $AlCl_3$ was deposited on alumina or on bauxite, and the process was operated in a fixed-bed configuration. The major problem with both processes was the highly corrosive nature of the sludge or of the sludge drippings from the solid catalyst. Frequent reactor replacement was required. Paraffin isomerization, previously dependent on $AlCl_3$-type catalysis, has since then made progress by the introduction of dual functional catalysts, described in section 4. The newer processes have concentrated on C_5 rather than C_4 hydrocarbons.

3. Zeolite Catalysis

Crystalline alumino silicates possessing base exchange properties have been known for well over 100 years and occur quite frequently in nature. They have found early application in ion–exchange chemistry, but their catalytic usefulness was discovered only in the late 1950s. Early attempts to use them as a base for catalytic cracking failed, and for a long time it was believed that the regular and uniform pore structure of a crystalline material was inferior to the pore size distribution of amorphous catalysts. In the mid–1950s Union Carbide Corporation first commercially produced synthetic zeolites of the x and y type (faujasites) as adsorbents (Figure 7); they later became ingredients of zeolite catalysis. While Rabo et al. pointed out in 1960 [14] that these materials possessed activity for such reactions as isomerization, it remained for Plank and Rosinsky at Socony–Mobil Oil Corporation to stabilize zeolites x and y so that they could withstand regeneration temperatures and steam partial pressures occuring in cracking without sintering and losing crystallinity. They achieved this by ion exchanging rare earth metals for alkali metals, and using a matrix of silica-alimina [15, 16] to separate zeolite crystallites.

The new zeolite cracking catalysts exhibited greater activity and selectivity than all previous catalysts (Tables 5 and 6). While their initial introduction was for use in TCC units in late 1961, fluid-bed zeolite catalysts were soon manufactured and used. The great selectivity of these catalysts sharply reduced the amount of gas oil required to produce gasoline. Figure 8 shows the U.S. catalytic cracking capacity as a function of time. The sudden change of slope in the years between 1964 and 1970 is due to the fact that the capacity of existing units was increased by the use of zeolite catalysts to such an extent that the steadily increasing demand for gasoline could be met for several years without new capacity. It has been estimated that

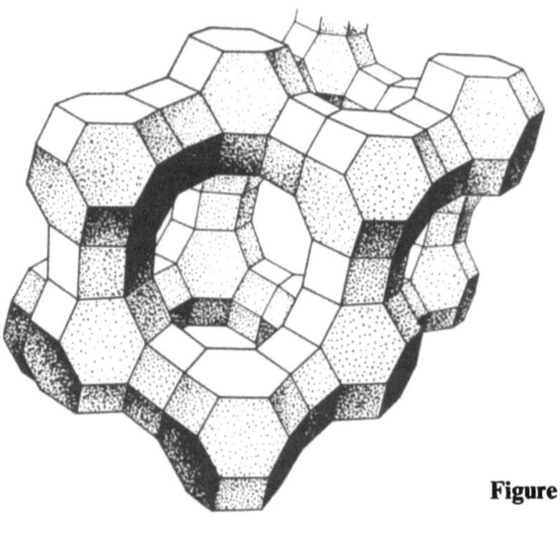

Figure 7. Faujasite zeolite

Table 5. Comparison of gasoline compositions from gas oil cracking catalysed by silica-alumina and zeolite [15]

Feed:	Calif. virgin gas-oil		Calif. coker gas-oil		Gachsaran gas-oil	
Catalyst, Durabead 1 or Durabead 5	5	1	5	1	5	1
% Paraffins	21.0	8.7	21.8	12.0	31.9	21.2
% Cycloparaffins	19.3	10.4	13.4	9.5	14.3	15.7
% Olefins	14.6	43.7	19.0	42.8	16.3	30.2
% Aromatics	45.0	37.3	45.9	35.8	37.4	33.1

Durabead 1 = silica-alumina
Durabead 5 = early generation zeolite (REHX)

Table 6. Yields of products from cycle stocks cracked over Durabead 5 and Durabead 7 compared with silica-alumina [15]

	Si/Al	Durabead 5[a]		Durabead 7[b]	
	Yields	Yields	diff.[e]	Yields	diff.[e]
	Augusta catalytic light fuel oil[c]				
Conversion, vol %	35.6	35.6	0	35.6	0
C_5 + gasoline, vol %	22.1	25.9	+3.8	29.2	+7.1
Total C_4's, vol %	8.7	7.9	−0.8	6.2	−2.5
Dry gas, wt %	5.2	4.1	−1.1	3.5	−1.8
Coke, wt %	4.3	2.2	−2.1	1.4	−2.9
	Beaumont heavy catalytic fuel oil[d]				
Conversion, vol %	42.5	42.5	0	42.5	0
C_5 + gasoline, vol %	24.5	26.3	+1.8	30.6	+6.1
Total C_4's, vol %	9.4	9.4	0	8.2	−1.2
Dry gas, wt %	6.5	5.2	−1.0	4.7	−1.5
Coke, wt. %	8.7	7.8	−0.9	4.9	−3.8

[a] Contains REHX in silica-alumina.
[b] Contains REHY in silica-alumina.
[c] Properties – 27.3° API, Aniline No. = 139.5 °F, (59.8 °C), ASTM boiling range = 516°–666 °F. (269–353 °C).
[d] Properties – 19.5° API, Aniline No. = 157.5 °F, (69.8 °C) ASTM boiling range = 410°–760 °F. (210–405 °C).
[e] Difference between Durabead and Si/Al yields.

savings of crude oil alone by this improved selectivity amounted to about $ 200 million per year, prior to the quadrupling of oil prices in 1974. No history of industrial catalysis would be complete without mention of the patent litigation that ensued for more than ten years and ended with upholding the validity of the Plank and Rosinsky patents. About 90 percent of all catalytic cracking today employs zeolite catalysts. About 290 million pounds of cracking catalyst with a value of $ 145 million were produced in 1978 [17].

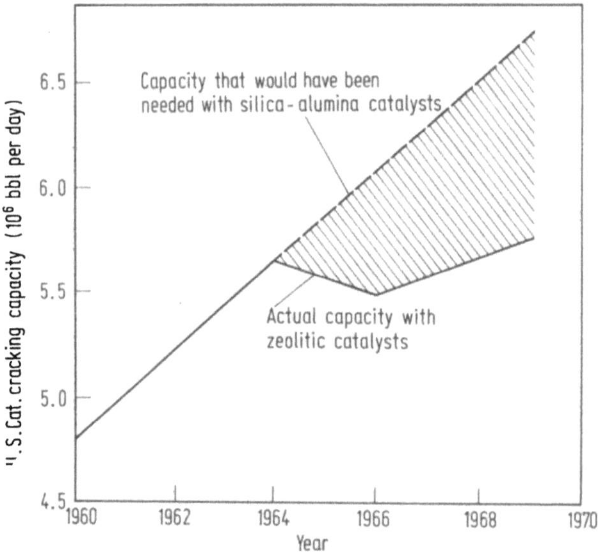

Figure 8. Effect of zeolitic catalysts on catalytic cracking capacity

A series of catalyst improvements followed the initial introduction of zeolites, each adding to stability or selectivity of the catalyst [18]. In 1977, a new series of catalysts came on the market which contained, in addition to the zeolite, a combustion promoter. This permitted catalyst regeneration to very low residual coke levels at relatively low regeneration temperatures, and also permitted oxidation of CO to CO_2, reducing pollution and heat-loss problems. These catalysts, again developed by Mobil Oil, contain extremely small amounts (0.01–50 ppm) of platinum impregnated on the catalyst or introduced with the feed stock [17]. The role of oxidation promoters was well-known. But many previous attempts to incorporate them into cracking catalysts failed, because the oxidation component acted also as a dehydrogenation component during cracking, and resulted in undesirably large yields of hydrogen. No such effect has been observed with the new catalysts. One must marvel, however, at the turnover rates these tiny amounts of precious metal must achieve during regeneration. It raises the old question of how much of the surface of catalysts is active in a catalytic reaction.

The introduction of a zeolite-cracking component into hydrocracking catalysts will be discussed in section 4 on dual functional catalysis. This was commercialized by Chevron and Union Oil Company of California about 1970.

Almost simultaneously with the introduction of zeolite-cracking catalysts came the discovery of catalytic shape selectivity by P. B. Weisz and his co-workers at Mobil [19]. They stipulated and demonstrated that diffusional constraints prevented the entry of molecules above certain dimensions into the pores of certain zeolites, and introduced the concept of "molecular engineering". The first process based ion this concept was disclosed in 1968 and was called "Selectoforming" [20]. The catalyst used was a naturally occuring zeolite with about 5 Å pore openings (contrasting the 8–10 Å openings of faujasites and the 3 Å of zeolite A). When a catalytic reformate was passed over this catalyst containing a small amount of nickel as a hydrogenation component on the presence of hydrogen, a selective cracking of n–paraffins occurred while branched chain hydrocarbons and aromatics passed undisturbed. This resulted in the removal of the lowest octane number components of gasoline, converting them mostly to LPG hydrocarbons.

Since 1974, a series of novel processes was introduced by Mobil Oil based on the unique properties of a synthetic zeolite called ZSM–5 [21] (Figure 9). This material has pore openings of 5–7 Å and exhibits shape selectivity, acid activity, and an unusual resistance to coking. It extended the range of "Selectoforming" by cracking both normal and singly branched paraffins, but not the high-octane number hydrocarbons possessing a quaternary C atom. In addition, it permitted simultaneous alkylation of simple aromatics like benzene with the olefinic fragments from paraffin cracking, resulting in higher liquid yields [22]. The "M-forming" process was first operated in Germany in 1976.

Shape selective cracking also is the basic reaction of the "Mobil Distillate Dewaxing" (MDDW) process [23]. This process was first tested in a commercial installation in France in 1974 and has been operating in a number of refineries in various parts of the world since 1978. It serves two purposes: one is lowering the pour point of middle distillate fractions such as diesel and heating oils to make them suitable for cold-weather operations, the other is permitting the inclusion of higher boiling material in these fractions to increase their yield-per-barrel of crude oil. In both cases, the pour point reduction is achieved by cracking waxy normal paraffins selectively with a relatively small yield loss. The cracked product consists largely of gasoline of good octane number. The reaction is carried out over ZSM-5-type zeolite in the presence of hydrogen, but there is essentially no hydrogen consumption. Catalyst deactivation is gradual and can be reversed by a hydrogen purge, indicating that it occurs probably by sorption of nitrogen compounds on the catalyst.

Figure 9. Pore arrangement of ZSM-5 zeolite [21]

In section 4 (dual functional catalysis), mention will be made of xylene isomerization. ZSM-5-type catalysts have largely replaced Pt—Al_2O_3 in this application. They isomerize the three xylenes to equilibrium. Ethylbenzene is largely disproportionated, eliminating the need for fractionation to remove ethylbenzene from the feed. Catalyst life in this application has exceeded two years. Recent patents indicate that chemical modification of ZSM-5 with phosphorous or carbon can further increase selectivity and result in p-xylene yields far exceeding equilibrium.

The same type of catalyst can replace $AlCl_3$ in the alkylation of benzene with ethylene to produce ethylbenzene [21]. This eliminates catalyst disposal problems and substitutes a fixed-bed reactor system for a sparged tower.

Toluene disproportionation to benzene and xylenes is another commercial process operated with the ZSM-5 class of catalysts.

The latest application of this type of catalyst — which has not yet become commercial although it has been in operation in a sizeable pilot plant — is conversion of methanol to high-octane gasoline [24]. This new chemical reaction involves an internal dehydration and polymerization with simultaneous isomerization and hydrogen transfer. Methanol goes via dimethylether to an olefinic entity, which then forms isoparaffins and aromatics. The catalyst is ZSM-5 and operation can be in either fixed of fluid bed at quite mild conditions. The importance of the reaction lies in the possibility of converting either coal or natural gas via methanol (a well-established old technology) to gasoline. Several reaction mechanisms have been suggested for this reaction that had not been previously observed. It is not limited to methanol, but works with higher alcohols, ethers, and other oxygenates. The yields are stoichiometric in the case of methanol giving 44% hydrocarbons and 56 percent water. The reaction is highly exothermic and heat removal is the major engineering process problem. Variations in operating conditions permit changes in the aromatics/paraffin ratio, and allow relatively high yields of ethylene and propylene.

4. Dual Function Catalysis

A. Naphtha Reforming

Catalytic reforming of naphthas serves to improve the octane number of gasoline by isomerizing paraffins, dehydrogenating cyclohexanes, dehydroisomerizing methylcyclopentanes, aromatizing some paraffins and also hydrocracking some paraffins.

The reforming of straight-run naphthas to achieve production of high-octane number gasolines developed slowly from thermal reforming [25] to conversion over molybdena-alumina catalysts at elevated pressures and in the presence of hydrogen. This process was used during World War II to produce toluene from methylcyclohexane. A continous fluid bed process commercialized by Standard Oil of Indiana never achieved broad application after an explosion destroyed the first commercial plant in 1947, demonstrating the hazards of operating regenerative high-pressure hydrogenative processes in a continuous mode. About 1950 a new generation of reforming processes were introduced. "Platforming", the most successful, was developed by Universal Oil Products Company (UOP) and was also first on the market, closely followed by "Catforming" (Atlantic Refining Company) and "Houdriforming" (Houdry Process Corporation). All three processes employed a platinum catalyst on an acidic base. In "Platforming" and "Houdriforming", 0.3–0.8 percent Pt was supported on γ-alumina, and high activity levels were maintained by adding very small amounts of a hydrogen halide or hydrogen halide precursor to the feed. In "Catforming", the catalyst support consisted of a silica-alumina gel. All three processes operated at 400–600 psig (27–40 atm) in the presence of hydrogen, and at 800–950 °F (430–510 °C). Life cycles were long and activity was maintained by gradually raising the temperature of operation to balance catalyst deactivation. After six to twelve months of operation, the catalyst was replaced by a fresh batch and the platinum of the spent catalyst was recovered by solution chemistry. Somewhat later it was learned that these catalysts could be regenerated by careful oxidation. The major improvements achieved by these catalysts were the ability to isomerize paraffins to highly branched entities, dehydrogenate naphthenes to aromatics, dehydroisomerize methylcyclopentanes to aromatics, and convert some paraffins to aromatics by dehydrocyclization. In the course of naphtha reforming, hydrogen is produced as another product, and this reformer hydrogen supplies a large percentage of refinery demand for hydrogen. The dual functional nature of reforming catalysts, possessing hydrogenative-dehydrogenative function as well as acidic properties, was described in 1953 by Mills, Heinemann, Milliken, and Oblad [26] and is illustrated in Figure 10. Their stipulation of the intermediary role of olefinic entities in the reaction mechanism was confirmed by P. B. Weisz and C. D. Prater in 1956 (27], who showed the presence of these olefins in the small amounts permitted by equililibrium considerations.

Since aromatics have very high octane numbers ($>$ 100) and can be tolerated in gasoline up to almost 50 volume percent, it was soon recognized that a high degree of aromatization was desirable and would proceed best at lower pressures. Figure 11 indicates the yield at various octane numbers at different pressures. Since catalyst deactivation increases as pressure is lowered, utilization of this concept required more stable and regenerative catalysts. It was found in 1953 that eta-alumina was a more stable and active base; and in 1954 Standard Oil of Indiana introduced "Ultraforming", followed in 1956 by Esso's "Powerforming". These latter processes employ a cyclic mode of regeneration. Because of the relative ease of regeneration, the reforming can be operated at lower pressures; this permits better aromatization and higher octane number at the expense of more rapid catalyst deactivation.

All reforming processes use a multireactor system (usually three reactors, see Figure 12 [28]. The first two reactors are endothermic because the major

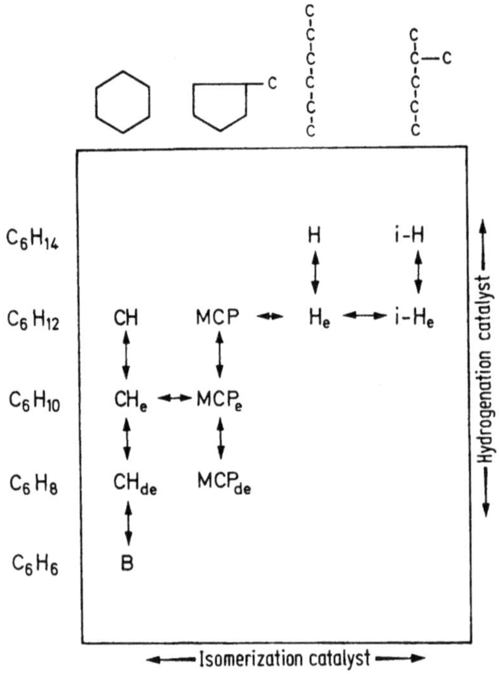

Figure 10. Dual functional reforming. Figure taken from reference 26

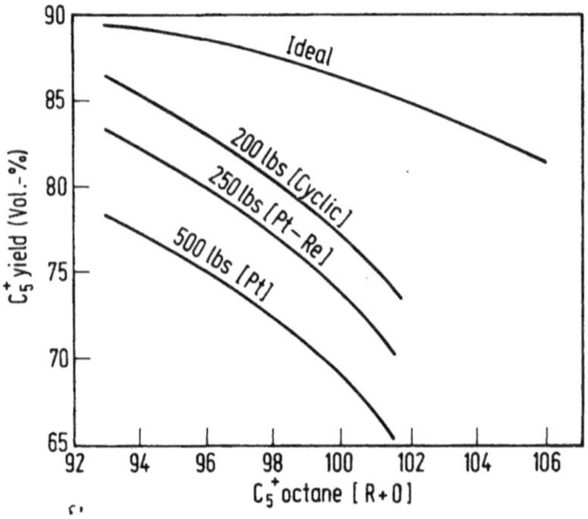

Figure 11. Yield vs. octane for reforming of C_6-360 mid-continent naphtha

reaction in these is dehydrogenation; and the third reactor is exothermic because of hydrocracking and hydrogenation. Interstage heaters are employed to permit the same temperature at each reactor inlet. The last reactor operates at the highest average temperature and shows the most rapid aging. In the cyclic processes there is a spare reactor which undergoes regeneration, after

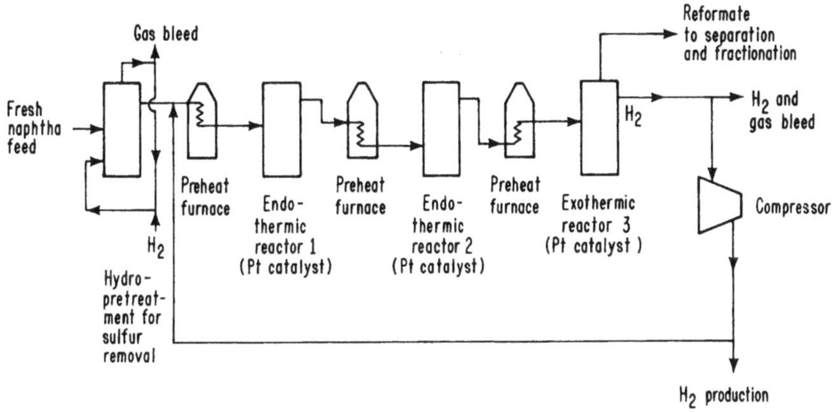

Figure 12. Catalytic reforming unit

which it is substituted for the first reactor, which becomes Number 2. The second reactor becomes Number 3, and the third is withdrawn and regenerated.

In the case of naphtha reforming, just as in catalytic cracking, the original process development involved the engineering design of a new unit concept as well as catalyst development. In both cracking and reforming, further impressive improvements were achieved by catalyst modifications that permitted use of the new catalysts in existing units. In a period of high investment-and-construction costs, this reduced the risk of introducing new process technology — a risk that might otherwise have been unacceptable. A major step forward in the art of naphtha reforming occurred in 1967 with the introduction of catalysts containing, in addition to platinum, another metal (or metal oxide) on an alumina base. Chevron's "Rheniforming" process [29] was first on the market. This catalyst, containing rhenium in addition to platinum, provides greater stability. In 1975 Exxon commercialized operation with another catalyst, said to contain iridium and platinum and providing stable operation at quite low pressure. Sulfur sensitivity is greater than that for Pt catalysts. About 5 million pounds of reforming catalysts worth $ 28 million were sold in 1978 [17].

Bimetallic catalysis is based on research on the concept of bimetallic cluster catalysts. Such catalysts consist of metallic clusters composed of atoms of two or more different metals in a state of high dispersion on a carrier [30]. The impact of this research, while already apparent in catalytic reforming as outlined above, is likely to increase over the next few years. Isomerization, hydrocracking, and hydrogenation will be affected. Vinylacetate manufacture with palladium-gold catalysts [31], and olefin partial oxidation (see section 9 on oxidation) with silver-gold or copper-cold catalysts [32], are in the offing.

B. Isomerization

Catalysts used for isomerization are almost identical with those used in reforming. This is natural since isomerization is one of the important functions of reforming. Of the several processes developed, the Shell and UOP C_5 isomerization processes have, since the late 1950s, replaced $AlCl_3$ catalysis. In the case of aromatics isomerization, one is primarily concerned with xylene isomerization. In this case, *p*-xylene is removed (e.g., by fractional crystallization) from a mixture of xylenes. The remaining o- and m-xylene rich mixture is isomerized to equilibrium and the p-xylene again removed, recycling the remainder. Atlantic Refining Company's "Octafining" process using $Pt–Al_2O_3$ or $Pt–SiO_2–Al_2O_3$ catalyst was introduced in 1960 by Mitsui and has found broad application as *p*-xylene demand for dacron fabrics mushroomed [33]. It operates in the presence of hydrogen at about 400 to 500 °C (205–260 °C) and 150–350 psig (10–23 atm), and the catalytic mechanism is similar to the one involved in reforming. Imperial Chemical Industries (ICI) has commercialized another process that employs silica-alumina cracking catalysts at 400–500 °C and atmospheric pressure in the absence of hydrogen. This process requires frequent catalyst regeneration (every second or third day), while the "Octafining" process operates on the same catalyst charge for six to twelve months, without regenration.

In 1976 Mobil Oil Corporation introduced the "Mobil Vapor Phase Isomeration Process" (MVPI) and in 1978 the "Mobil Low Pressure Isomerization Process" (MLPI), each employing zeolite catalysts. These have been discussed in the section on zeolites. The MVPI process has conquered about three-fourths of the free-world's C_8 isomerization capacity. It is not a dual functional catalyst, nor is the ICI catalyst.

C. Hydrocracking

Hydrocracking has assumed increasing importance as feedstocks for fuels became heavier and supplies of gas oil for catalytic cracking and of naphtha for reforming became insufficient to supply the gasoline market. The relatively high carbon/hydrogen ratio of high-boiling fuels and their aromaticity made them unsuitable as cracking feedstocks, at least without prior hydrogenation. Conversion of heavier gas oils by hydrogenative processing dates back to pre-World War II technology developed by M. Pier and his associates at I.G. Farben Industrie in Germany between 1925 and 1930, in collaboration with Standard Oil of New Jersey, operating at pressures of 3,000 to 5,000 psig (200–333 atm). The combination of hydrogenative and acid functions in catalysts to convert heavy petroleum fractions to gasoline and dry gas under high hydrogen partial pressure, however, did not become a large-scale commercial reality until the 1960s. This was primarily because long catalyst life was required for a process operating at high pressure (with the time-consuming and costly requirement to depressure, purge, and after regeneration to purge and repressure again), and also because nitrogen compounds in the feed tended to rapidly deactivate the cracking function. About 1960, catalysts and systems were found that operated at lower pressures (2,000 psig).

Union Oil Company (together with Esso) and Chevron Oil Company pioneered hydrocracking processes using nickel or nickel-tungsten on silica-alumina as catalysts [34]. In later versions, zeolites of the faujasite type were substituted for the silica-alumina base in about 1966–1967. With feeds having high nitrogen compound content, a two-reactor system was used in which the nitrogen compounds were converted in a first hydrogenation reactor to ammonia over nickel-tungsten or tungsten sulfide on alumina or silica with, of course, concomitant hydrogenation of some aromatics. Ammonia at high hydrogen partial pressure is less of a poison to acid catalysts than organic nitrogen and would be tolerated in the second hydrocracking reactor. In one version of the process, separate hydrogen circulation systems are used for the two reactors, with an acid wash to remove ammonia and a carbonate wash to remove H_2S from the hydrogen cycle.

Hydrocracking has become an integral part of refinery operations, with a U.S. capacity of about 900,000 bbl/day, but has lately encountered competition from hydrogenation (see section 5 on metallic and multimetallic catalysis) and subsequent catalytic cracking, and capacity has remained fairly constant during the last five years.

5. Hydrogenation Catalysis and Hydrogen Production

One of the oldest of catalytic reactions is the saturation of fats over nickel catalysts. In general, metallic catalysts have found their most widespread use in the activation of the hydrogen molecule.

A. Desulfurization and Denitrification

The geatest new industrial uses during the last 30 years have again occurred in the petroleum industry. With increasing supply of high-sulfur crude oils and the need to remove sulfur and nitrogen compounds for pollution abatement as well as to produce streams that can be subsequently treated over sulfur or nitrogen sensitive catalysts (e.g., in catalytic reforming, methanation, or hydrocracking), large hydrogenation units have been built, mostly to convert organic sulfur and nitrogen compounds to H_2S and NH_3, respectively, which in turn can then be removed by washes or adsorption. The most common catalysts used are cobalt-molybdena on alumina, molybdenum sulfide on alumina, and tungsten and/or nickel sulfide on supports. For the desulfurization of distillates, the operation is carried out at 500–700 psig (33–47 atm) pressure, and temperature of 600–800 °F (315–430 °C) space velocities of 1–5 v/v/hr. The catalyst can be regenerated periodically by air burning, and catalyst life between regeneration is long — usually more than six months. This technology is derived from old German work of the 1930s, but found general use only in the 1960s.

More recently — in the late 1960s and early '70s — desulfurization of residual materials became important. This is more difficult, and catalyst poisoning is irreversible because of the deposition of heavy metals (V, Ni, Fe) from the oil on the catalyst. Conditions of this type of operation, pioneered largely by Gulf Oil and Esso, are more severe; pressures range up to 2,000 psig (130 atm) sand space velocities are as low as 0.3 v/v/hr. It was found that the pore characteristics of the catalyst support are of great importance. Large pores are required to allow some of the very large hydrocarbon molecules containing hetero atoms or metals to diffuse into the catalyst, while small pores are needed to provide the overall high surface area. The technology has developed in two directions: 1) Catalysts with a very wide pore distribution have been used, ranging from 10 Å to 1,000 Å; 2) Catalyst beds have been graded with large pore catalysts first contacting the oil, and a decreasing pore-size catalyst treating partially converted feed. Hydrodesulfurization and hydrodenitrogenation catalysts are not selective to S and N compounds, and in all operations saturation of aromatics occurs as a side reaction, using up much more than the stoichiometric amounts of hydrogen required to convert sulfur or nitrogen compounds.

B. Selective Hydrogenation

Since 1975, considerable thought has been given to the hydrogenation of recycle stocks. These highly aromatic fractions from the catalytic cracking of heavy gas oils can be saturated, essentially in the type of operation described above, and are then suitable as catalytic cracking feeds. In many cases, this eliminates the need for hydrocracking.

Another application of hydrogenation is a selective one. In the steam cracking of ethane or naphtha to produce ethylene and propylene, a byproduct called pyrolysis gasoline is obtained which contains aromatics (benzene, toluene, xylenes) as well as olefins and diolefins. Before the aromatics can be extracted, it is necessary to saturate the olefins and diolefins without saturating the aromatics. Several commercial plants were built in the period between 1955 and 1965 using mostly nickel–sulfide catalysts for this treatment at relatively mild conditions. In a variant of the process, only the diolefins are saturated and the aromatic-olefinic product is blended as a high-octane component into gasoline.

Olefins produced from naphtha by steam cracking usually contain small but bothersome amounts of diolefins and acetylenes, such as butadiene, isoprene, and methylacetylene. These must be removed prior to polymerization of the olefins. This is ususally done by selective hydrogenation over supported palladium catalysts under conditions that will not cause hydrogenation of mono-olefins [36]. Operating conditions are 35–100 °C and 3–30 atm pressure. Nickel-cobalt-chromium catalysts have also been used for this purpose. In 1963, Farbenfabriken Bayer announced a liquid-phase selective hydrogenation process using a palladium catalyst which operates at very low temperatures. This has since been commercialized.

C. Hydrogen Production

Hydrogen for hydrogenation reactions comes mostly from two sources: reformer hydrogen, and the product from the steam-hydrocarbon (or steam-carbon) reaction. The steam-hydrocarbon reaction is called "steam reforming". It produces "synthesis gas", a mixture of hydrogen and carbon monoxide. The carbon monoxide is reacted over a water-gas shift catalyst with water to form more hydrogen and carbon dioxide, which can be crubbed from the gas mixture leaving relatively pure hydrogen.

Hydrogen as a by-product from naphtha reforming has already been mentioned in an earlier section. Production of hydrogen by steam reforming from methane, naphtha, heavy oil, and coal has achieved increasing importance in recent years. This is because of the increased demand for hydrogen, both for petroleum refining processes such as hydrotreating and hydrocracking, and for petrochemical use in the production of ammonia and methanol, among others. Steam reforming has recently been reviewed [37].

While the reaction between carbonaceous material or hydrocarbons and steam over catalysts such as nickel on supports has long been known and used, a major breakthrough was achieved in 1962 [38]. Until then, the reaction of hydrocarbons and steam was limited to relatively low pressures. The reason for this is that, in addition to the reaction $C + H_20 = CO + H_2$ and similar reactions for hydrocarbons, there is also a decomposition of hydrocarbons to produce carbon and hydrogen. The deposition of carbon on the catalyst rapidly deactivates the catalyst. At increasing pressures, this decomposition becomes faster at the temperatures involved than the reaction of steam and water. It was therefore necessary to accelerate the carbon-steam reaction so that it equaled or became faster than the hydrocarbon decomposition. This was achieved almost simultaneously by the M. W. Kellogg Company in the United States and the Imperial Chemical Industries (ICI) in Great Britain [39]. It was found that the carbon-steam reaction could be greatly accelerated by the presence of alkali or earth-alkali ions, and that catalysts containing sodium or potassium oxide in additions to the customary nickel [38] could perform at relatively high pressures up to 500 or 600 psig (33 or 40 atm). Without this discovery, the new generation of ammonia plants (see following paragraphs) which operate at pressures of about 2,000 psig (133 atm) would not have been possible. By using a liquid feed such as naphtha or heavier hydrocarbons at pressures up to 500 pounds (33 atm) which can be achieved by pumps, the remaining pressure difference of about 1,500 psig (100 atm) can be obtained by a single-stage compressor. The discovery of alkali-promoted nickel catalysts for the steam-carbon reaction made possible a revolution in the ammonia industry and the creation of large-scale plants to produce ammonia at about one-half the cost of older generation plants.

Alkali catalysts will also be important for the production of hydrogen and of methane from coal or char and water.

As already mentioned, if the production of hydrogen alone rather than sŷnthesis gas is desired, it is necessary to react carbon monoxide with water to produce CO_2 and hydrogen [40]. The exothermic water-gas shift process is used at 310–490 °C with an excess of water vapor; chromia-iron oxide catalysts are used with many improvements in detailed catalyst composition having occured during the past 20 years. Copper oxide-zinc oxide-alumina catalysts have also found application. Shift catalysts are available from several catalyst manufacturs.

D. Ammonia Synthesis

The synthesis of ammonia ($N_2 + 3\,H_2 \rightarrow 2\,NH_3$) was discovered in the early part of the 20th century and has been widely described. The first plant was built in 1913 by BASF in Germany. While improvements have been made in the catalyst used [41], these are of relatively minor importance and can be neglected in a history of industrial catalysis. Most catalysts are based on magnetite containing some alumina and CaO with alkali promoters. A major breakthrough, however, was achieved in the late 1950s when it became possible to build very large (1,000 ton per day and above) ammonia plants of a simplified nature, and with greatly reduced investment and operating costs. The resultant reduction in the cost of ammonia increased the demand for ammonia as a fertilizer and has greatly contributed to avoiding famine in many parts of the world. This breakthrough is based on a com-

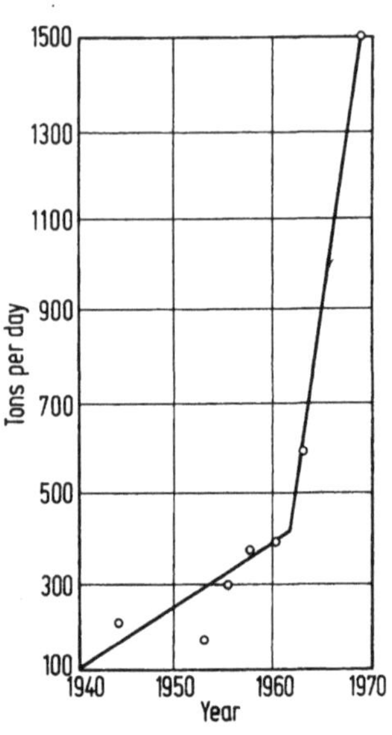

Figure 13. Capacity of individual ammonia units built

bination of chemical and mechanical inventions. The chemical part of this invention, namely the production of hydrogen at relatively high pressures, has been detailed in a previous paragraph; the mechanical part of the invention consists of the introduction of single-train large units with one-stage centrifugal compressor rather than two or more stages of reciprocal compressors. Operation is being carried out at somewhat lower pressures than previously used (current operating pressures are in the order of 2,000 psig (113 atm)). even though equilibrium conditions are somewhat less favorable there than at higher pressure. However, this is greatly overcome by savings in operating and utility costs. Other mechanical improvements in ammonia synthesis include new reactor designs that permit greater efficiency. The first company to introduce the new design of ammonia plants was M. W. Kellogg.

Increased demand for ammonia as a result of these improvements is shown in Figure 13, which presents the size of individual plants as a function of time, and illustrates the rapid increase in capacity after the new generation of plants came into existence.

E. Methanol Synthesis ($CO + 2 H_2 \rightarrow CH_3OH$)

The synthesis of ammonia and of methanol parallel each other in many ways. In fact, the unit design for the two processes is almost identical. Like ammonia, methanol synthesis dates back many years (it was first commercialized in 1924). Mechanical improvements parallel those of ammonia synthesis. Chemical improvements were achieved around 1966 by Imperial Chemical Industries' discovery of a low-pressure methanol catalyst. This catalyst is comprised of zinc, copper and alumina, and permits operation at significantly lower pressure (750 psig (50 atm) vs. 2,000 psig (133 atm)) and lower temperature than the old historic zinc–alumina–chromia catalyst allowed, and has therefore contributed to a more economical production of methanol [42]. The advantages of a high-pressure production of synthesis gas as outlined above are as applicable to the methanol synthesis as they are to the ammonia synthesis.

In addition to new catalyst developments, there have also been new reactor designs for methanol synthesis. The Lurgi reactor consists of a large number of relatively small-diameter catalyst-containing tubes. A bundle of these tubes is surrounded by a jacket filled with pressurized water. Control of the steam pressure in the jacket controls the temperature of the water, and provides an excellent heat exchange medium for the exothermic methanol synthesis reaction.

These developments in ammonia and methanol synthese are a clear demonstration of the fact that even technologies considered mature can undergo revolutionary developments that will change the demand and supply picture because of greatly reduced product prices.

6. Catalytic Hydrocarbon Dehydrogenation

Commercial dehydrogenation processes relate essentially to two types of operation involving different catalysts: 1) butane dehydrogenation to butenes and/or butadiene, 2) ethylbenzene dehydrogenation to styrene monomer.

Butane dehydrogenation was pioneered by Eugene J. Houdry and Houdry Process Corporation during World War II and has been reviewed in detail [43]. It is carried out to yield either butenes or butadiene, or both. Chromia-alumina is the preferred catalyst for this reaction. The catalyst is easily deactivated by steam, the presence of which during reaction must be avoided. Improvements made since installation of the first commercial plants in the mid-1940s are mostly of a mechanical nature and relate to operational time and conditions, and to catalyst regeneration time. Since the reaction is highly endothermic, heat generation for maintaining reaction temperature is dependent on burning coke deposits on the catalyst. A fine balance is required between coke lay-down and regeneration and operating time.

Oxidative dehydrogenation of butanes or butenes to butadiene is practiced over bismuth molybdate catalysts at 400–500 °C. The exothermicity of oxidation supplies most of the heat requirements of the endothermic dehydrogenation.

The impetus for commercial production of styrene in the United States lay in the critical need for a substitute for natural rubber during World War II. As a result of an industry-wide cooperative effort, an annual production capacity of over 400 million pounds was installed in only two years. This effort has been surveyed by Boundy and Boyer [44]. The best catalysts, still in use after more than 30 years, are comprised of alkali-promoted iron oxide and the reaction is carried out in the presence of steam to reduce hydrocarbon partial pressure. The best catalysts are unsupported. The major by-products from the dehydrogenation are benzene and toluene. A discussion of the effect of promoters on iron oxide can be found in E. H. Lee's review [45].

7. Catalytic Dealkylation

Hydrodealkylation is a process used for the production of benzene from toluene. Thermal as well as catalytic processes have been developed. Demethylation must be selective and hydrogenation of the aromatic ring must be avoided. High purity, low sodium chromia-alumina catalysts are used at 550–650 °C and 35–80 atm pressure [12]. Because of the exothermicity of the reaction, close temperature control is required.

8. Catalytic Coal Liquefaction and Gasification

A. Liquefaction

Production of liquid fuels from coal is based on two technologies discovered in Germany in the 1920s: the Bergius process for direct catalytic hydrogenation in a solvent, and the Fischer-Tropsch process involving the reaction of coal and steam to produce synthesis gas ($CO + H_2$) and subsequent hydrocarbon synthesis from this gas. Both technologies have been repeatedly reviewed [46, 47, 48]. Both processes found large-scale commercial application in Germany during World War II, and a Fischer-Tropsch plant was built in South Africa in the mid-1940s and has been operating since, with a second larger facility to start operating in 1981. A third one was announced in 1979. While there was much active research on both the direct (hydrogenation) and the indirect (Fischer-Tropsch) liquefaction routes in the United States and in England during the 1940s and early '50s, interest lapsed when large volumes of inexpensive petroleum were discovered in the Middle East. Not until the 1970s was research intensified again, especially after the Arab oil embargo of 1974. No new commercial facilities other than the South African SASOL plant have been built as yet, but several new processes are being developed.

The new direct hydrogenation processes are directed toward operating at lower-pressure and lower hydrogen consumption than the Bergius process, as shown in Figure 14. Two of the three major processes being developed employ catalysts. In the H-coal process, the coal is dissolved in a recycle solvent containing largely aromatic and hydroaromatic hydrocarbons, and the resulting solution and/or slurry is contacted with a cobalt-molybdena-alumina catalyst in an ebbulating bed reactor. Catalyst usage is said to

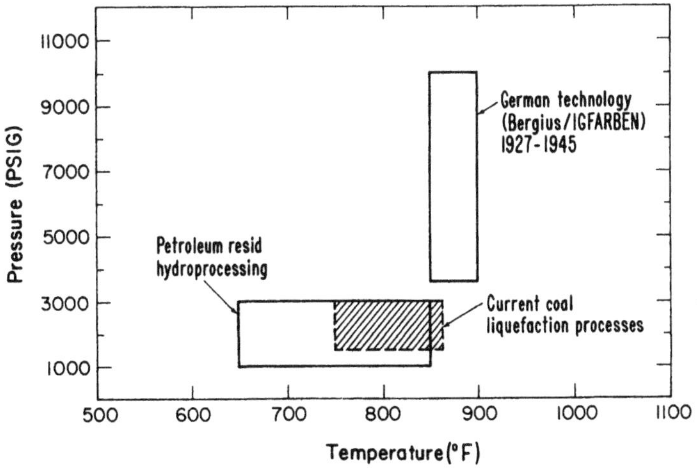

Figure 14. State of the art Petroleum-Coal hydroprocessing. Figure by courtesy of R. H. Fischer, U.S. Department of Energy

be about one pound per ton of coal charged, or per 3 bbl of product. The operation is carried out at pressures in the range of 2,000–3000 psig (133 to 200 atm) and at temperatures of 750–850 °F (400–450 °C). After separation of the ash, products are fractionated and the solvent is recycled. Depending upon the severity of the operation, the process operates in either a "syn-crude" or "fuel oil" mode. Hydrogen consumption is higher in the "syn-crude" mode (5000 cft/bbl) than in the "fuel oil" mode (3500 cft/bbl).

The "Exxon Donor Solvent" process distinguishes from the H-coal process by carrying out the hydrogenation by means of a donor solvent rich in hydroaromatics. This solvent is then rehydrogenated in a separate loop, using a conventional cobalt-molybdena catalyst. In this case, the catalyst is not subjected to potential poisons in the dissolved coal and slurried ash, and its life is therefore appreciably longer. The U.S. Department of Energy, jointly with industry, is currently (1980) sponsoring the erection and operation of large pilot plants for both processes.

The third major process (SRC) does not employ a catalytic reactor but it appears that the ash in coal possesses catalytic properties, which enhance thermal hydrogenation.

Indirect liquefaction also has been dormant until the mid-1970's, except for the SASOL plant. The major innovation at SASOL had been a fluid-bed operation of the synthesis reaction that permitted better temperature control and better yields of gasoline. H. Kölbel developed a liquid phase synthesis in a large pilot plant on Germany in 1948–1950 [49, 50]. Conventional iron-type Fischer-Tropsch catalysts are suspended in a heavy oil. High per-pass conversions ($\geq$ 95 percent) are obtained and it is possible to operate with relatively low hydrogen/CO ratios such as 0.6/1. Internal shift supplies the additional hydrogen required. The low H_2/CO ratio could be of increasing importance as new and less expensive gasifiers supplying low H_2/CO ratio gas become commercial.

A new development announced in 1976 is the Mobil methanol-to-gasoline process [21] mentioned in the section on zeolites. This process, for which a large pilot plant is currently (1980) under construction, will provide a new route from coal to high-octane gasoline, and appears to be competitive with other coal liquefaction routes.

B. Gasification

In coal gasification there are two worthwhile developments. One involves the reaction of coal with steam in the presence of alkali carbonate. In the early and mid-1960's, the M. W. Kellogg Company did pilot plant work on a process in which coal dissolved in molten carbonate was reacted with steam to produce synthesis gas at high temperature and pressure [51]. Sulfur compounds in the coal react with alkali carbonate to form hydrosulfides which can then be reconverted to carbonate under high CO_2 particle pressure. In 1977, Exxon announced work on a process to produce methane directly from coal and water at relatively low temperatures by impregnating the coal with 7–10 percent of alkali carbonate. Complete recovery of the alkali carbonate has not yet been demonstrated.

Many coal gasification plants are in operation and many new designs are being developed. None of these employs added catalysts and the catalytic role of coal ash in gasification, if any, is not well defined.

C. Methanation

Several methanation processes have been announced. The major catalytic step is the methanation of synthesis gas which takes place over a nickel catalyst. An interesting case is the use of a Raney nickel catalyst with which the reactor walls have been coated [52]. This is a novel technique and provides an alternative to supported nickel catalysts for the $CO + 2\,H_2 \rightarrow CH_4$ reaction.

9. Heterogeneous Oxidation, Ammoxidation, Chlorination, and Oxychlorination Catalysis

A. Oxidation

Oxidation of naphthalene over molybdena or vanadia catalysts is an old

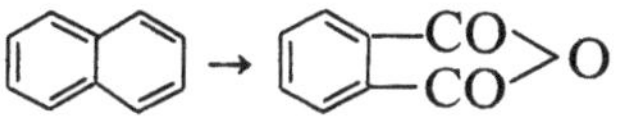

art used for many years for the production of phthalic anhydride. In 1946, Oronite Company (now Chevron) initiated the production of phthalic anhydride from *o*-xylene over vanadium oxide catalysts. The feed can be impure because other C_8 aromatics and paraffins are oxidized to CO_2. Considerable progress was achieved when the first fluid-bed oxidation plants for naphthalene or *o*-xylene charges were built in 1952–1953 by Imperial Chemical Industries and American Cyanamid Company. Since then, the Badger-Sherwin Williams fluid-bed process has found application in several large plants. The better temperature control of the fluid bed has permitted better selectivities. Similar processes are used for oxidation of benzene to maleic anhydride, and of toluene to benzoic acid.

The conversion of light hydrocarbons into products containing oxygen or other heteroatoms is of great importance to the chemical industry. Ethylene oxide and ethylene glycol, propylene oxide, acrolein, acrylic acid, and acrylonitrile are all large-volume chemicals that are produced today from ethylene or propylene by catalytic oxidation or ammoxidation, respectively.

Ethylene oxide [53] is a very large-scale commercial product. Early attempts to oxidize ethylene directly to the oxide failed. In 1931, Lefort first reported success using silver as a catalyst. Union Carbide first commercialized a direct oxidation process in 1937; and Scientific Design Company and Shell Oil Company have licensed many installations since then. Silver is still the major active catalyst ingredient. Promoters are used with the silver, such as oxides of alkali and alkaline earth metals. Organic halides minimize the formation of CO_2. Ethylene oxidation processes are operated under recycle conditions in fixed-bed equipment at 250–325 °C and 10 to 50 atm pressure. Conversion to ethylene oxide is about 70 percent, with much of the rest being lost to CO_2.

A new catalytic process for production of propylene oxide from propylene was commercialized by Oxirane Company in 1969 [54, 69]. Since it involves homogeneous catalysis, it is described in the section on Homogeneous Catalysis, as are other oxidation processes.

Hearne and Adams [55] described in 1948 the production of acrolein from propylene and oxygen ($C_3H_6 + O_2 \rightarrow CH_2 = CH{-}CHO + H_2O$) over cuprous oxide catalysts. Yields were about 50 percent. Further improvements by Shell involved a copper-oxide/silicon-carbide catalyst promoted by iodine.

A most important discovery was patended in 1959 by Idol and in 1962 by Callahan [56, 57], who reported yields of acrolein much higher than those obtained in the old cuprous oxide system. The improvement was due to the selective action of a new type of catalyst — bismuth molybdate. Based on this discovery, Standard Oil Company of Ohio (SOHIO) commercialized vapor phase propylene oxidation and ammoxidation processes.

B. Ammoxidation

Propylene, ammonia, and air are reacted in a fluid bed to give acrylonitrile and water ($C_3H_6 + NH_3 + 1^1/_2\, O_2 \rightarrow CH_2 = CH{-}CN + 3\, H_2O$) Operation is at 400–500 °C and 3–30 psig (0.2–2 atm) pressure. A once-through operation without recycle is feasible and the process has found worldwide application. Even more selective catalysts of the novel uranium antimonate system were disclosed in 1965, and in 1970 SOHIO introduced "multi-component" catalysts containing several elements of the group consisting of nickel, cobalt, iron, bismuth, molybdenum, potassium, manganese, and silicon. Production of acrylnitrile by this type of catalysis was 2.5 million tons per year in 1977 [17].

An alternative propylene oxidation route using tin plus antimony oxides as catalysts was also developed by SOHIO [54, 58].

A process which is somewhat related to ammoxidation involves the synthesis of methylpyridines from ammonia and acetaldelyde. It is carried out over silica-alumina catalysts modified by thorium, zinc, or cadmium, at 750–950 °F (400–510 °C). It gives yields of 40–60 percent.

C. Hydrohalogenation and Oxychlorination

Catalysts comprising inorganic metal chlorides have found application in hydrohalogenation reactions. Thus, bismuth and antimony trichloride are used to add HCl to ethylene or propylene and mercuric chloride on carbon is a common catalyst for reacting acetylene with HCl to produce vinyl chloride.

Oxidation of HCl to chlorine over cupric chloride was first described by Deacon about 100 years ago. Attempts to operate a process on this reaction failed because of major corrosion problems. In 1969, M. W. Kellogg Company announced a successful process (which is discussed in the section on Homogeneous Catalysis).

In 1964, Goodrich, Dow, and Monsanto commercialized oxychlorination processes [58, 59]. Most of these are fixed- or fluid-bed operations over copper chloride catalysts, reacting ethylene and HCl and oxygen to dichloroethane and water. The copper salts are molten and sorbed in the alumina support at operating conditions; HCl produced in the direct chlorination of ethylene can thus be converted into the desired vinyl chloride monomer. This development followed an earlier operation of the Raschig-Hooker process, in which benzene is chlorinated to chlorobenzene by the reaction of benzene with HCl and oxygen over copper chloride on an inert support.

Chlorides of rare earths and alkali metals often serve as promoters for $CuCl_2$ in oxychlorination. Workers at the M. W. Kellogg Company [60] have described a homogeneous version of the oxychlorination process, using an aqueous solution of copper salts. This has the advantage of easy heat removal by water evaporation.

D. Hydrogen Cyanide

The formation of hydrogen cyanide from ammonia, methane, and oxygen over rhodium or iridium-promoted platinum was disclosed by Andrussow some time ago ($CH_4 + NH_3 + 1^1/_2\ O_2 \rightarrow HCN + 3\ H_2O$) and was commercialized using promoted platinum gauze, similar to the process used to make nitric acid from ammonia. An improvement of the Andrussow process was commercialized in the 1950s by Degussa. In this process, small diameter refractory tubes are coated with the catalyst and the reaction proceeds on the reactor walls.

10. Olefin Disproportionation Catalysis

A new catalytic reaction was disclosed by Banks and Bailey in 1964 [61, 62]. Called "olefin disproportionation", it converted linear olefins into homologs of shorter and longer chains in a highly specific and efficient manner. The total moles of product olefins heavier than the feed equaled the total moles of lighter olefins. Propylene for example could be converted to ethylene and butenes ($2\ C_3H_6 \rightarrow C_2H_4 + C_4H_8$). While the original discovery involved the use of molybdenum or tungsten hexacarbonyl catalysts supported on alumina, many other catalysts — both heterogeneous and homogeneous —

Table 7. Olefin disporportionation catalyst supports and active constituents/promoters [63]

	Supports			Active constituents/Promoters		
Oxides	Al	Si	Fe	Mo	W	Re
	Ni	Zr	Sn	Nb	Rh	Sn
	W	Th	SiAl	Te	La	Ta
	AlTi	AlTh	MgSi	Os	Ir	
	MgTi					
Phosphates	Al	Ti	Ca			
	Zr	Mg				
Hexacarbonyls				Mo	W	Re
Sulfides				Mo	W	

have since been reported for this reaction [63]. Table 7 shows some of the heterogeneous support and active constituen promoter combinations used. Typical homogeneous catalysts described in the literature are those obtained by interaction of tungsten hexachloride, ethanol, and ethylaluminiumdichloride, or by reaction of nitrosyl complexes of halides.

The first commercial application of the new process chemistry occurred in 1966. Shawinigan Chemicals Company in Canada installed a Phillipps Petroleum Company "Triolefin" process unit to convert propylene into polymerization-grade ethylene and high purity butenes. Other facilities have been installed since then.

11. Industrial Homogeneous Catalysis

Applications of homogeneous catalysis have greatly increased in recent years. Where sulfuric acid catalysed alkylation and the cobalt carbonyl catalysed Oxo reaction ($CH_3CH{=}CH_2 + CO + H_2 \rightarrow C_3H_7CHO$) were almost alone in this field 25 years ago, there are now over 20 industrial processes. Most employ soluble metal compounds as catalysts. Monomers and polymers are the major products. G. W. Parshall has recently published a review of homogeneous catalytic processes [64] and Table 8 summarizes some major applications.

Olefin polymerization is probably the largest scale application of organometallic catalysed processes. Many of the catalysts used in the process technology initiated by Ziegler and Natta [72] are not soluble and therefore are discussed separately. However, a significance amount of linear polyethylene is produced with soluble titanium catalysts. Ethylene-propylene-diene elastomers and stereoregular polybutadiene are produced with analogous catalyst mixtures based on other transition metal compounds.

In 1938–1946 Roelen in Germany developed the Oxo process, still one of the largest applications of homogeneous catalysis. It involves hydroformylation, the reaction of an olefin with carbon monoxide and hydrogen to produce aldelydes (e.g., butyraldehyde from propylene or propionaldehyde from ethylene) and fatty alcohols from higher olefines. Cobalt carbonyls were the usual catalysts [65]. They were generated in situ from cobalt salts and synthesis gas ($CO + H_2$) in the presence of a base. They are used at

Table 8. Major applications of homogeneous catalysis in the U.S. chemical industry [64]

	Approximate 1975 capacity or production (thousands of metric tonnes)
Carbonylations	
$CH_3CH{=}CH_2 + CO + H_2 \rightarrow C_3H_7CHO$ (includes other oxo products)	650
$RCH{=}CH_2 + CO + 2\,H_2 \rightarrow RCH_2CH_2CH_2OH$	170
$CH_3OH + CO \rightarrow CH_3COOH$	190
Monoolefin Reactions	
$CH_2{=}CH_2 + O_2 \rightarrow CH_3CHO$	410
$CH_3CH{=}CH_2 + ROOH \longrightarrow CH_3\underset{\diagdown O \diagup}{CH{-}CH_2} + ROH$	250
$CH_2{=}CH_2 \rightarrow$ Polyethylene (excludes oxide supported catalysts)	150
$CH_2{=}CH_2 + CH_3CH{=}CH_2 +$ diene $\rightarrow$ EPDM rubber	85
Diene Reactions	
$3\,CH_2{=}CHCH{=}CH_2 \rightarrow$ cyclodecatriene	10
$C_4H_6 + CH_2{=}CH_2 \rightarrow$ 1,4-hexadiene	2
$C_4H_6 + 2\,HCN \rightarrow NC(CH_2)_4CN$	70
$C_4H_6 \rightarrow$ *cis*-1,4-polybutadiene	290
Oxidations	
$c\text{-}C_6H_{12} \xrightarrow{O_2} c\text{-}C_6H_{11}OH + c\text{-}C_6H_{10}{=}O \xrightarrow[HNO_3]{O_2\ or}$ adipic acid	610
$c\text{-}C_{12}H_{24} \xrightarrow{O_2} c\text{-}C_{12}H_{23}OH + c\text{-}C_{12}H_{22}{=}O \xrightarrow{HNO_3}$ dodecanedioic acid	10
$CH_3{-}C_6H_4{-}CH_3 \xrightarrow{O_2}$ terephthalic acid and esters	2,100
$n\text{-}C_4H_{10} \xrightarrow{O_2} CH_3COOH$	470
$CH_3CHO \xrightarrow{O_2} CH_3COOH$	335
Other Reactions	
$CH_2{=}CHCHClCH_2Cl \rightleftharpoons ClCH_2CH{=}CHCH_2Cl$	270
$ClCH_2CH{=}CHCH_2Cl + C\ NaCN \rightarrow NCCH_2CH{=}CHCH_2CN$	125
$ROOC{-}C_6H_4{-}COOR + HOCH_2CH_2OH \rightarrow$ polyester	1,900

200–300 atm pressure and 100–150 °C. A major improvement in catalyst technology was made when Union Carbide commercialized in 1976 the production of butyraldehyde from xylene employing homogeneous rhodium catalysts containing a phosphine ligand. The new catalysts operate at much lower pressure, e.g., 10–25 atm and at low temperature (~ 100 °C).

In 1960, a major advance in metallorganic catalysis occurred when Wacker Chemie, a subsidiary of Höchst, introduced a simple high-yield process for producing acetaldehyde from ethylene [66]. This process practiced in the United States by companies such as Celanese and Texas Eastman has largely displaced syntheses based on ethanol or acetylene. The oxidation of ethylene is carried out either as a two-stage process using air, or as a one-stage process using oxygen. The oxidizing catalyst is an aqueous solution of cupric chloride and palladium chloride and proceeds in three steps:

$$CH_2{=}CH_2 + H_2O + PdCl_2 \rightarrow CH_3CHO + Pd + 2\,HCl$$

$$Pd + 2\,CuCl_2 \rightarrow PdCl_2 + 2\,CuCl$$

$$2\,CuCl + {}^1/_2\,O_2 + 2\,HCl \rightarrow 2\,CuCl_2 + H_2O$$

Ethylene is oxidized by water and palladium chloride. The resulting palladium metal is reconverted to palladium chloride by cupric chloride. The cuprous chloride is reoxidized by air or oxygen. Farbenfabriken Bayer has introduced a heterogeneous version of this catalyst. The same type of homogeneous catalysis is also used to produce vinyl acetate from ethylene and acetic acid.

Acetic acid is produced by oxidation of acetaldehyde with soluble metal catalysts. The largest type of process is however, based on the carbonylation of methanol [67]. Badische Anilin and Sodafabrik (BASF) pioneered a process that has been used in the U.S. by Borden Chemical Company since 1968. It uses a cobalt carbonyl catalyst promoted by iodine. More recently, Monsanto commercialized in 1977 a process employing rhodium complexes and methy iodide as cocatalysts [68]. The Monsanto process uses milder process conditions and exhibits very high selectivity to acetic acid but must, to be economical, recover almost all of the rhodium.

According to Parshall [64], one of the largest applications of homogeneous catalysis is the production of terephthalate esters. For this, *p*-xylene is oxidized by air using soluble cobalt and manganese salts as catalysts. Amoco Chemicals Company is a major user of this process.

An interesting new process was discovered and commercialized in 1969 by Oxirane Corporation [69]. Propylene is oxidized to propylene oxide by *t*-butylhydroperoxide, catalysed by soluble molybdenum compounds such as molybdenum carbonyl. The t-butylhydroperoxide is obtained by air oxidation of isobutane. T-butyl alcohol is a coproduct with propylene oxide. This process is finding application in various parts of the world and is replacing older methods of propylene oxide manufacture.

Also, in 1969 the M. W. Kellogg Company announced a different type of homogeneously catalysed process, a new version of the old Deacon process to convert HCl to chlorine. The "Kelchlor" process [70] has been commercialized on a large scale by E. I. duPont de Nemours & Company. It uses oxides of nitrogen as the catalyst in combination with a powerful dehydrating agent, H_2SO_4. The latter drives the equilibrium towards high yields, since it effectively removes water of reaction from the reaction product. The process ist of particular use in conjunction with chlorination reactions, since it permits the by-product HCl to be reconverted to chlorine.

A very large-scale application of dissolved catalysts is the production of dichloroethane from ethylene and chlorine, employing $FeCl_3$, $CuCl_2$, or $SbCl_3$ catalysts. The dichloroethane is then thermally dehydrochlorinated to vinyl chloride.

12. Catalytic Polymerization

Polymerization catalysis is an important and relatively old field. Before and during the second World War, efforts were largely concentrated on dimerizing and trimerizing ethylene and propylene to high-octane gasoline [71]. Work was also done to obtain higher molecular weight products in the lubricating oil boiling range. With the need for synthetic rubber spiraling during World War II, technology was developed for butadiene polymers and butadiene-styrene as well as isobutylene-isoprene copolymers. Numerous improvements have been made in the polymerization of these and other monomers, such as vinyl chloride, which cannot be enumerated here.

After World War II the polymer industry began to develop rapidly, and polyethylene — and to a somewhat lesser extent polypropylene — became articles of large scale commerce. The high pressure polyethylene process commercialized by Imperial Chemical Industries in England, BASF in Germany, and others, involved a thermal free radical process at pressures of 1000 atm and above, and it dominated the field until major discoveries were made by Karl Ziegler in Germany and G. Natta in Italy. These men later shared a Nobel Prize in 1964 for their contributions.

Ziegler disclosed in 1955 a catalyst system [72] that would polymerize ethylene to polyethylene at low pressures, e.g., 10 atm (The "low-pressure" polyethylene is more linear, has a higher melting point, and is more rigid than "high-pressure" polyethylene). Catalysts spanning the homogeneous-heterogeneous interface were used (see also the section on Homogeneous Catalysis). One such catalyst is prepared from aluminum alkyl, e.g., $AlEt_3$ and titanium tetrachloride, using an excess of the alkyl [73]. The preparation is carried out in a dry hydrocarbon solvent, usually hexane. The precipitate plus the aluminum alkyl still in solution is "the catalyst". The reaction is carried out at 50–75 °C and about 10 atm pressure, and the polymer forms as a white powder suspended in the solvent. Many catalyst modifications have been developed by various manufacturers. In the Amoco Chemical Company process [74], which operates at 240–300 °C and 35–100 atm pressure, the catalyst is molybdic acid on alumina, plus reducing agents such as aluminum alkyl, lithium borohydride, or metallic sodium as co-catalysts. The Phillips Petroleum Company catalyst [75] comprises chromium trioxide on silica or alumina in cyclohexane as the solvent. Promoters are used, as shown in Table 9 [76].

Table 9. Some oxide promoters for supported chromia catalysts [72]

Supports	SiO_2	Al_2O_3	ZrO_2	ThO_2
Promoters	CuO			
	ZnO	ZnO		
	SrO	SrO		
	WO_3			
	Mn_2O_3	Mn_2O_3		
	Co_2O_3			
	Fe_2O_3			
		MgO		
		BaO	BaO	
		B_2O_3		

G. Natta in 1969 described the polymerization of propylene and higher α-olefin to highly ordered, crystalline polymers. "Isotactic" polypropylene has the carbon atoms in each succeding propylene unit ordered in the same identical way. C. L. Thomas refers to a statement by Natta that the term "isotactic" was suggested by Mrs. Natta because the structure reminded her of the ordered lines and columns of marching soldiers.

Natta requires a crystalline or microcrystalline catalyst for the production of sterospecific products. Crystalline α-titanium trichloride plus aluminum alkyl constitutes such a catalyst system. Operating conditions of this type of catalyst for propylene polymerization are similar to Ziegler's, except for slightly higher temperatures.

The Ziegler and Natta discoveries have resulted in applications of great technological importance, and over one-half of all olefin polymers produced is based on this technology.

Urethane foams constitute an important item of plastics manufacture. They are obtained from isocynates and alcohols, particularly diols in the presence of some water. The reaction is catalytic and most of the product is manufactured using a catalyst first introduced in 1959 by Houdry Process Corporation, called "Dabco" (triethylene diamine) [77]. This is often promoted by organic stannous compounds. The commercial use of this catalyst combination appears to have been the first introduction of an organic compound as a catalyst. Dabco has such high activity that it permits a one-step process.

13. Catalysis for Motor Vehicle Emission Control [78]

It is a rare event when a whole new field of catalytic applications opens up. This happened when the State of California enacted laws on air quality and motor-vehicle emissions standards in 1959 and 1960. The laws were to go into effect when at least two devices had been developed that could meet thes standards. While these laws stimulated a considerable amount of research, one should not omit mention of the fact that Eugene J. Houdry, the pioneer of so much industrial catalysis (see sections on Catalytic Cracking and Dehydrogenation) foresaw the need for emission control — at least in confined spaces — as early as 1949, and developed catalytic mufflers for indoor vehicles, marketed by Oxy-Catalyst, Inc. One of these early catalysts consisted of monolith-porcelain rods covered with alumina upon which platinum was deposited. This anticipated later developments.

Following the California law enactment, three groups of catalyst and muffler manufacturers were certified by the California Motor Vehicle Control Board in 1964 and 1965. These were W. R. Grace-Norris Thermador, Universal Oil Products-Arvin, and American Cyanamid-Walker. However, engine modifications by the car manufacturers enabled them to meet specifications without the use of catalysts.

The U.S. Federal Clean Air Act of 1970 set requirements that could not be met by existing technology, and this spurred an intensive research effort. Although enforcement of the law was later delayed by one year, from 1975 to 1976, and the law was modified by setting interim standards, catalytic mufflers have been installed on all new cars in the U.S. since 1976. The only exception has been Honda cars which used an improved mechanical system.

Among the numerous organization doing research and development in this area during this period was a combination of companies doing joint research in the "Interindustry Emission Control Program". The group consisted of Ford Motor Company, Mobil Oil Corporation, Volkswagen, Toyota, Fiat, and others.

The initial objective of emission control research was to reduce carbon monoxide and hydrocarbon emissions to specified levels, 90 percent or more below those previously emitted. The reduction of nitrogen oxide emissions was undertaken somewhat later. A totally new problem for catalytic reactor design was immediately recognized. While most industrial catalytic operations involve continuous and steady-state conditions, automobile operations are intermittent and transient, with wide fluctuations of flow rate and temperature that depend on driving conditions and cold or warm engine starts.

As shown in J. Wei's review of the field [78], exhaust gas-flow rates can vary from 20 to more than 200 SCF/min, and exhaust temperatures from 600 to 1800 °F (391–980 °C) and above. Many materials such as base-metal oxides and alloys oxidize CO and hydrocarbons to CO_2 and water at quite high conversions. Requirements are, however, that more than 90 percent of CO and hydrocarbons be removed at space velocites of up to 200,000 v/v/hr within a temperature range of about 500–2000 °F (250–1100 °C). In addition, the oxidizing atmosphere contains 15 percent water and not much more than a stoichiometric amount of oxygen. The catalyst, further, must last for at least 25,000 vehicle miles.

Only precious metal catalysts have thus far been used commercially to meet the many difficult specifications. In most cases, the metal is dispersed as small crystallites on thermally stable, chemically inactive supports such as α-alumina. Three major support shapes comprise spheroids, wire mesh, and monolithic honeycombs. A "washcoat" of high surface area alumina is ususally placed on the ceramic surface and acts as substrate for the precious metal.

A major requirement for the use of platimum-type emission catalysts has been the elimination of potential poisons from the gasoline, most prominently the lead compounds. All cars having catalytic mufflers must use lead-free gasoline. This has resulted in the inability to boost octane number by addition of tetraethyl lead to gasoline, and in the need to produce relatively high octane (91–93 Research) gasoline by catalytic reforming or cracking; it has increased the amount of crude oil processed, since yield of gasoline decreases with increasing octane requirement.

The amount of precious metal used in emission control devices is about 0.05 troy oz/car. Reactor design is complex, since high pressure drop must be avoided at very high space velocities. Manufacturers of commercial emission control catalysts in the U.S. include Engelhard, W. R. Grace & Company, Universal Oil Products Company, The Catalyst Company, and the Houdry Process Division of Air Products & Chemicals, Inc.

In addition to CO and hydrocarbon emission standards, specifications for NO_x emissions have been set for 1977 and later cars in the United States. Three-way precious metal "Redox" catalysts can take advantage of a "window" at an air/fuel ratio of exactly 14.7:1, at which NO_x conversion is still high and CO and hydrocarbon conversion is very substantial (all conversions > 80 percent) (Figure 15). However, the window is so narrow that maintenance of this air/fuel ratio is difficult. Proposals for a separate NO_x catalyst bed have thus far not been commercialised. For the total catalytic system, two converters or a converter containing a dual-bed catalyst will probably be necessary. The first bed will catalyse the reduction of nitrogen oxides to nitrogen, while the second oxidizes hydrocarbons and CO to CO_2 and water. Three-way converters will be required to meet the 1981 standards.

Development of catalysts for automotive emission control has been and is a great challenge because of the extreme mechanical and operating con-

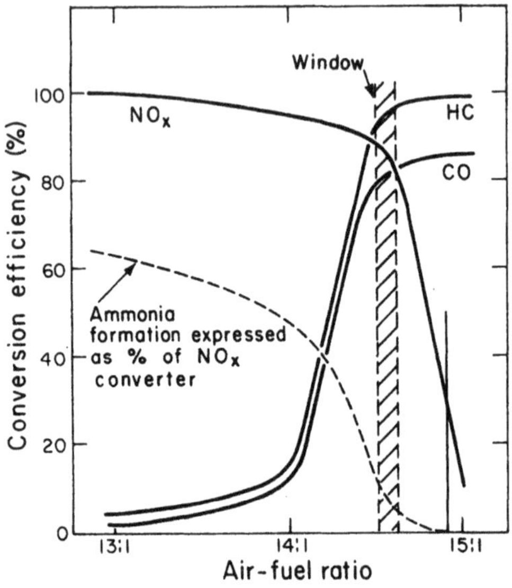

Figure 15. "Three-way" redox catalyst conversion characteristics. Figure taken from reference 78

dition variations encountered. The use of catalysts in cars in 1978 is shown in Table 10 [17].

Table 10. Use of catalysts in cars in 1978 [17]

Material	Estimated use[a]	Estimated value[a]/10^6 \$
Alumina pellets	37.6 million lb	\$71.0
Cordierite monolith	8.9 million lb	34.0
Alumina wash coat	1.6 million lb	1.4
Platinum	395,000 troy oz	93.6
Palladium	130,000 troy oz	9.2
Total		\$ 209.2

[a] For installation on cars manufactured in the U.S. last year. Actual use would be at least 10 percent higher because of converters for Canadian cars and spares.

14. Fuel Cell Catalysis

Catalysts are used at both the anode and cathode of a fuel cell. Hydrogen has been the most common fuel. It is oxidized at the anode and the oxidant is reduced at the cathode. While much research has been undertaken on fuel-cell catalysis, industrial applications are thus far very limited. Fuel cells were used in the Gemini and Apollo space crafts using noble metal on carbon catalysts. A large electric utility demonstration is planned by Consolidated Edison in New York for 1979. This unit will generate about 5 megawatts [79].

References

1. Mills, G. A.; Cusumano, J. A.: Catalysis. In Kirk-Othmer: Encyclopedia of Chemical Technology (3rd Ed., Vol. 5). New York: Wiley, 1979
2. Gurwitsch, L.: Kolloid-Z. *11*, 17 (1912)
3. Herbst, H.: Erdöl und Teer *2*, 265, 411 (1926)
4. Sachanen, A. N.: Conversion of petroleum (2nd Ed.). New York: Reinhold, 1948
5. Sittig, M.: Pet. Refiner *31*, 9, 263 (1952)
6. Ardern, D. B.; Dart, J. C., Lassiat, R. C.: Adv. Chem. *5*, 13 (1951)
7. Murphree, E. V.: Adv. Chem. *5*, 30 (1951)
8. Venuto, P. B.; Habib, E. T.: Catal. Rev. *18* (1), 1 (1978)
9. Oblad, A. G.: Oil and Gas J. *61* (13), 124 (1963)
10. Blazek, J. J.; Grey, F. E., Jr.; Baker, R. W.: Proc. API Div. Refining *42* (3), 277 (1962)
11. Emmett, P. H.: Catalysis (Vols. 1—7). New York: Reinhold Publ., 1954—60
12. Thomas, C. L.: Catalytic processes and proven catalysts. New York: Academic Press 1970. pp. 87ff
13. Gunness, R. C.: Adv. Chem. *5*, 109 (1951)
14. Rabo, J. A.; Pickert, P. E.; Stamires, D. N.; Boyle, J. E.: Proc. 2nd. Intern. Congr. Catalysis, Edition Tech., Paris, 1960, p. 2055
15. Eastwood, S. C., Plank, C. J., Weisz, P. B.: Proc. 8th. World Petroleum Congr., Edinburgh: Neil & Company 1971
16. Plank, C. J.; Rosinsky, E. J.; Hawthorne, W. P.: I & EC Product Res. and Devel. *3* (165) (1964)

17. Burke, D. P.: Chem. Week *124* (13), 42 (1979)
18. Oblad, A. G.: Oil Gas J. *70* (13), 84 (1972)
19. Weisz, P. B.; Frilette, V. J.; Maatman, R. W.; Mower, E. B.: J. Catal. *1*, 307 (1962)
Weisz, P. B., Frilette, V. J.: J. Phys. Chem. *64*, 382 (1960)
20. Chen, N. Y.; Maziuk, J.; Schwartz, A. B.; Weisz, P. B.: Oil and Gas J. *66* (47), 147 (1968)
21. Meisel, S. L.; McCullough, J. P.; Lechthaler, C. H.; Weisz, P. B.: Recent advances in the production of fuels and chemicals over zeolite catalysts. Leo Friend Symposium, ACS, Aug. 1977;
Chem. Tech. *6*, 73 (1976)
22. Heinemann, H.: Catal. Rev. *15* (1), 53 (1977)
23. Chen, N. Y.; Gorring, R. L.; Ireland, H. R.; Stein, T. R.: Oil and Gas J. *15* (23), 165, 170 (1977)
24. Chang, C. D.; Silvestri, A. J.: J. Catal. *47* 249 (1977)
25. Ciapetta, F. G.; Dobres, R. M.; Baker, R. W.: In P. H. Emmet (Ed.), Catalysis (Vol. 6). New York: Reinhold Publ., 1958. Pp. 495ff
26. Mills, G. A.; Heinemann, H.; Milliken, T. H.; Oblad, A. G.: Ind. Eng. Chem. *45*, 134 (1953)
27. Weisz, P. B.; Prater, C. D.: Advan. Catal. *9*, 575 (1957) and Science *126*, 31 (1957)
28. Ciapetta, F. G.; Wallace, D. N.: Catal. Rev. *5*, 67 (1972)
29. Klucksdahl, H. E.: U.S. Pat. 3, 617, 520; 3, 558, 477
30. Sinfelt, H. H.: J. Catal. *29*, 308 (1973)
31. Allison, E. G.; Bond, G. C.: Catal. Rev. *1*, 233 (1973)
32. Flank, W. H.; Beachell, H. C.: J. Catal. *8*, 316 (1967)
33. Thomas, C. L.: Catalytic processes and proven catalysts. New York: Academic Press 1970. pp. 87ff
34. Hatch, L. F.: Hydrocarbon proc. *48* (2), 77 (1969)
35. EPRI Report AF-190, 1976
36. Rylander, P. N.: Catalytic hydrogenation over platinum metals. New York: Academic Press, 1967
37. Van Hook, J. P.: Catal. Rev. *20* (2), (1980)
38. McMahon, J. F.: U.S. Pat., 3, 119, 667 (1961)
39. Oblad, A. G.; Heinemann, H.; Friend, L.; Gamero, A.: Proc. 7th World Petroleum Congr. *5*, 197 (1967) Galliard, Ltd.
40. Moe, J. M.: Chem. Eng. Prog. *58* (3), 33 (1962)
41. Nielsen, A.: Catal. Rev. *4*, 1 (1971)
42. Royal, M. J.; Nimmo, N. M.: Hydrocarbon Proc. *48* (3), 147 (1969)
43. Kearby, K. K.: In P. H. Emmett (Ed.); Catalysis, *3*, 453 (1955) New York: Reinhold Publ.
44. Boundy, R. H.; Boyer, R. F.: Styrene. New York: Reinhold Publ. 1952
45. Lee, E. H.: Catal. Rev. *8*, 285 (1973)
46. Pier, M.: Z. Electrochem. *53*, 291 (1949)
47. Anderson, R. B.: In P. H. Emmett (Ed.), Catalysis, *4*, 29 (1956), New York: Reinhold Publ.
48. Storch, H. H.: Advan. Catal. *1*, 115 (1948)
49. Kölbel, H.; Ackermann, P.; Engelhardt, F.: Proc. 4th World Petroleum Congr. Sect. 4-C, Paragraph 9, 1955
50. Kölbel, H.; Ralek, M.: In J. Falle (Ed.), Chemische Rohstoffe aus Kohle, Stuttgart: Thieme 1977
51. Office of Coal Res. Departm. Interior, Res. and Development Rep. No. 38, 1968
52. Mills, G. A.; Steffgen, F. W.: Catal. Rev. *8*, 159 (1973)
53. Miller, S. A.: Ethylene and its industrial derivatives. London: E. Bern 1969
54. Hancock, E. G.: Propylene and its industrial derivatives. London: E. Bern 1973
55. Hearne, G. W.; Adams, M. L.: U.S. Pat. 2, 451, 485 (1948)
56. Veatch, F.; Callahan, J. L., Idol; J. D., Jr.; Milberger, E. C.: Chem. Eng. Prog. *56* (10), 65 (1960)
57. Anonymous: Hydrocarbon Proc. *46* (11), 141 (1967)
58. Anonymous: Hydrocarbon Proc. *46* (11), 239 (1967)

59. Bedell, K. R.; Rainbird, H. A.: Hydrocarbon Proc. *50* (11), 141 (1971)
60. Spector, M. L.; Heinemann, H.; Miller, K. D.: Ind. Eng. Chem., Process Des. Develop. *6*, 327 (1967)
61. Logan, R. S.; Banks, R. L.: Oil and Gas Journal *66* (*21*), 131 (1968)
62. Logan, R. S.; Banks, R. L.: Hydrocarbon Proc. *47* (*6*), 135 (1968)
63. Bailey, G. C.: Catal. Rev. *3*, 37 (1970)
64. Parshall, G. W.: J. Molecular Catalysis *4*, 243 (1978)
65. Wender, I.; Sternberg, H. W.; Orchin, M.: In P. H. Emmett (Ed.), Catalysis *5*, 73 (1958)
66. Anonymous.: Chem. Eng. News *39* (16), 52 (1961)
67. Lowery, R. P.; Aguilo, A.: Hydrocarbon Proc. *53* (11), 103 (1974)
68. Roth, J. F.; Craddock, J. H.; Hershman, A.; Gulick, F. E.: Chem. Tech. *1*, 600 (1971)
69. Stobaugh, R. B.; Calarco, V. A.; Morris, R. A.; Stroud, W. W.: Hydrocarbon Proc. *52* (1), 99 (1973)
70. Oblad, A. G.: Ind. Eng. Chem. *61* (7), 23 (1969)
71. Oblad, A. G.; Mills, G. A.; Heinemann, H.: In P. H. Emmett (Ed.), Catalysis *6*, 341 (1958)
72. Lenz, R. W.: Organic Chemistry of Synthetic High Polymers New York: Interscience 1967
73. Thomas, C. L.: Catalytic processes and proven catalysts. New York: Academic Press 1970 pp. 72—73
74. Peters, E. F.; Zletz, A.; Evering, B. W.: Ind. Eng. Chem. *49*, 1879 (1957)
75. Hogan, J. P.; Banks, R. W.: U.S. Pat. 2, 825, 721 (1958)
76. Lenz, R. W.: Organic Chemistry of Synthetic High Polymers, New York: Interscience 1967, p. 631
77. Farkas, A.; Mills, G. A.; Erner, W. E.; Maerker, J. B.: Ind. Eng. Chem. *51*, 1299 (1959)
78. Wei, J.: Advan. Catal. *24*, 57 (1975)
79. Fickett, A.: EPRI Journal, April 1976, 17

Chapter 2

An Introduction to the Theory of Catalytic Reactors

J. C. R. Turner

Department of Chemical Engineering,
University of Exeter,
North Park Road,
Exeter EX4 4QF, UK

This chapter is meant to introduce to the catalytic chemist those aspects of chemical reaction engineering which will be involved in any industrial application of a catalytic chemical reaction. The design and operation of full-scale chemical reactors involves scale-up from laboratory results, and introduces new factors of mixing, and of mass-transfer and heat-transfer. The procedures to deal with these are not self-evident, and the chemist should be aware of the problems, particularly since they may occur even at the bench scale, where supposedly "chemical" rate data may be seriously affected by "physical" transfer processes.

After a review of elementary chemical reactor theory, as applied to *non*-catalytic reactions, the chemical reaction engineering of catalytic reactions is considered, with particular emphasis on the use of porous solid catalysts. The chapter ends with a brief discussion of the formulation and testing of catalysts, and also introduces the subject of catalyst deactivation.

Contents

1. Introduction

This chapter will treat the subject of catalytic reactors from the point of view of a chemical engineer, who is less concerned with the physical and chemical behavior of reactant molecules than with the consequences of such behavior when dealing with large quantities of material in an industrial context.

It is, of course, true that much of scientific interest in the field of catalysis can and should have interest for the chemical engineer designing a plant to produce a substance via a catalytic route, but frequently the details of the molecular mechanics are not of primary concern. This may be because such detail may have consequences of no significant financial weight, or perhaps because other factors than those being considered become dominant as the scale of operation is increased from bench to plant size. It may be that the precision of measurement, or control, of quantities such as temperature or composition may be inadequate, on the large scale, to demonstrate, or turn to advantage, a property of the chosen system which may be of great interest theoretically, or of use on the bench scale. In such cases, expenditure on further research can be a waste of money. It is not always easy for a research chemist in industry to comprehend what is needed for effective design of a plant. Equally, the chemical engineer should realise that he is unlikely to originate a new product route, even if he may be skilled at developing that route.

In the consideration of industrial catalytic reactors, it is helpful to outline first some of the properties of continuously-operating large scale reactors. These, even in the idealisation of the simple models usually chosen, present characteristics which are not self-evident, and the neglect of which can lead to grievous errors of conception and design. While it is true that the non-steady-state performance of reactors is of great importance, it is not appropriate for a chapter in this volume to consider such behavior, save perhaps to note the simplest consequences of the fact that such plants have to be started and stopped!

In these first considerations of reactors, the catalytic nature of the reaction may not be of concern; the conclusions may hold as well for homogeneous reactions as for heterogeneous, for catalysed as for non-catalysed. The consequences of catalysis as it affects reactor design will be discussed after outlining these first parts of what is known as chemical reaction engineering.

Many of the matters treated here will be considered in much greater detail by other authors in later chapters of this series, where fuller references to original papers will be found. There are many texts on chemical reaction engineering, of which, for present purposes, a selection of four can be suggested for general reading. All contain numerous examples, either worked out in the text, or for the reader to test his progress. Denbigh and Turner [1] is a short text, primarily for senior undergraduates, with a lead in to graduate course work. Of several texts by Levenspiel, his Chemical Reaction Engineering [2] must be mentioned. This is a much longer text, and contains valuable chapters on fluidized-bed reactors, and deactivating catalysts. A more austers text is that by Aris [3], in which concentration is laid on the mathematical analysis of reactor behavior. Finally, the longest text of these four is by Carberry [4], who has chosen as his principal target catalytic reactor behavior. There is a wealth of information in this treatment, which should be regarded as a post-graduate text. In the present author's opinion a reading of this work should follow a proper digestion of such texts as the first two mentioned.

2. Reactor Theory

A. The Balance Equations and Rate Equation

1. General Considerations

To describe the performance of a reactor it is necessary to consider the application of the following three basic equations to the particular reactor:

i) The material balance equation, which describes the conservation of matter.
ii) The energy balance equation, which describes the conservation of energy.
iii) The kinetic rate equation, which describes the conversion of one species into others.

Because conditions may vary with time and position in a reactor, the above equations should first be applied to an elementary volume, but before laying them out in such a fashion, some comments can be made on them. Firstly, if only one reaction occurs in a system of known initial composition and known reaction stoichiometry, then a material balance on a single species may suffice, in that a knowledge of the concentration of that species will enable that of all others to be calculable. Secondly, it is useful, and usual, to separate thermal energy from chemical energy in considering the energy. From one point of view there is no change in the total enthalpy of the stream flowing through an adiabatic reactor in the steady state. However, there is often a change of temperature, and it is more profitable to consider the thermal energy separately, allowing an input to the 'energy' balance equation from any chemical reaction, via the 'heat of reaction'. Lastly, a correct appreciation of what is meant by the 'rate of reaction' is vital. As an example, the rate of a first-order chemical reactions is *not* in general given by an expression of the type $dc/dt = -kc$. There is indeed a case where this formulation is correct, but students (and even others who should know better, to judge from the author's experience) persist in applying it to situations where it is incorrect. Sometimes such a malefactor can get by for a while by defining what he 'means' by 'time', but this device will usually collapse when placed under the strain of considering a reactor in the non-steady state, where performance is a function of 'real' time.

The above prompts one to consider the time scales which might be of interest to catalytic chemical reaction engineers.

i) The time scale of molecular motions, such as vibration. On this scale, a microsecond is a very long time, and however interesting such events may be to catalytic scientists, they are far too quick to be of direct concern to reaction engineers.

ii) The chemical kinetic time scale — e.g. the half-life for a first-order batch reaction. This is, of course, very variable but reactor design becomes difficult, or cumbersome, if it does not lie between a millisecond and an hour. Clearly a reactor producing substantial conversions will have to be sized to give a holding time of at least the order of a half-life.

iii) The operational, or control time scale. This will be important in determining the speed of start-up or shut-down of a process, and in determining the response-time of a reactor to changes, whether deliberately imposed or not. This time-scale is usually about the same as ii) above, or larger, but an important proviso should be made. If a change leads to an increase in temperature this can cause 'runaway' either because the increased temperature causes a great reduction in the time scale of the desired reaction, or because a new (invariably undesired) exothermic reaction becomes dominant. In both cases the effect of iii) is actually to reduce ii) to a value far smaller than design.

iv) The time-scale of decay of catalyst activity. It is much easier to design and operate a process when this time-scale is much larger than ii) or iii) above. When a catalyst lasts for a year or more, giving adequate, though reduced, performance at the end of such a period, it is usually possible slowly to change operating conditions so as to compensate for this effect while keeping the costs within acceptable limits.

   If the life-time of the catalyst falls below a month, say, it may become imperative to have standby, or duplicate reactors. In this case the process is switched from a dying reactor to a regenerated one after an appropriate time.

   If the catalyst life-time is shorter than an hour, say, it may be that the dead catalyst cannot be revived fast enough for two reactors to suffice for 'continuous' operation. Alternatively, the process may not take kindly to switching of streams so frequently. In that case some method of continuously adding fresh catalyst, and removing used catalyst, should be found. Such processes as the 'moving-burden bed' achieve this, but the best-known case is surely the 'cat-cracker', in which fluidized-bed technology is used to cope with the rapid coking reactions which occur on cracking catalysts.

The above considerations may convince the reader that a correct treatment of 'time' in reactor designs is important — and with that in mind, we revert to considering the balance equations.

2. The Balance Equations Applied to an Elementary Volume

The mass balance equation for a given species can be stated as follows, for a volume element dV, over a small increment of time, dt:

Moles entering element =
1

Moles leaving element	+	*Moles reacting in element*	+	*Change of moles within element*
2		**3**		**4**

A similar enthalpy balance is customarily also written (see the immediately previous section)

Heat entering element =
5 (2)

Heat leaving element	+	*Heat absorbed by reaction*	+	*Change of heat within element*
6		**7**		**8**

Let us consider the above eight terms. Numbers **1**, **2**, **5** and **6** are transport terms. Usually convection is the dominant component, but we shall frequently have to consider diffusion (in the case of **1** and **2**) and conduction, or even radiation (in cases **5** and **6**).

Terms **3** and **7** are simply related, via the heat of reaction. It is essential to express the rate term, **3**, correctly. It will involve a velocity constant and concentration (or pressure) terms appropriate to the kinetics of the reaction. The definition of the rate expression should lead to the same numerical value of the velocity constant whatever species is chosen for equation (1) (by using, for example, the concept of extent of reaction-see references 1–4). This has by no means always been the case in the past. For instance, in the classic reaction of hydrogen and iodine to give hydrogen iodide there may be an ambiguity of a factor of 2 depending on whether the rate of formation of HI, or of destruction of H_2, is used to define the rate constant.

The terms **4** and **8** give the changes in hold-up within the element, and it is they, not the rate terms **3** and **7** which lead to time appearing in the final expressions.

For reactor design we wish to be able to integrate these elementary balance equations to calculate the performance of a reactor (integrating over volume) at any time (integrating up to that time). The ideal reactor models which form the start of chemical reaction engineering possess the advantage of eliminating certain of the terms in the balance equations (1) and (2).

3. *Special Cases of the Balance Equations*

a) The constant-volume batch reactor

If the reactor is well-mixed, then integration over the whole volume of the reactor is simple. The degree of stirring required for the reactor to approximate to well-mixed behavior will depend upon the nature of the reaction, or reactions, occurring therein. Great care is required with heterogeneous reactions involving mass transfer between solid (catalyst) particles, or gas bubbles, and a liquid phase. A degree of mixing adequate to give apparent uniformity of behaviour throughout the reactor volume may still be insufficient to reduce undesired effects in the 'boundary layers' close to such interfaces. Where there are successive reactions involved, a desired intermediate product may undergo decomposition to an extent dependent upon the 'micro-mixing' within these boundary layers. This behavior is quite likely to occur in reactors continuously fed with a gaseous reactant, which will be discussed in the next section on 'semi-batch' reactors.

However, for homogeneous batch reactions term **3**, which can be written as $r_c \mathrm{d}V$, can be integrated to $r_c V$. Here r_c is the number of moles of the reference species destroyed by reaction per unit volume and time. The element is now the reactor volume, and for a batch reactor terms **1** and **2** are both zero. The change of moles within the reactor can be written as $\mathrm{d}(cV)/\mathrm{d}t$, and we arrive at

$$r_c V = -\frac{\mathrm{d}(cV)}{\mathrm{d}t} \tag{3}$$

where c is the concentration of the reference species. To produce equation (3), it should be noted, we have divided by the time increment $\mathrm{d}t$. If the reaction volume is constant, which for liquid reactions can almost invariably be taken to be the case, we end up with

$$r_c = -\frac{\mathrm{d}c}{\mathrm{d}t}. \tag{4}$$

For a first order decomposition r_c can be written as kc, where k is a velocity constant. Hence it is correct to say that for a first-order reaction $\mathrm{d}c/\mathrm{d}t = -kc$, but *only* for a constant-volume batch reactor. This type of reactor, though common, or even usual, in chemistry laboratories, is uncommon on the large scale.

Before leaving the constant-volume batch reactor we must consider the heat balance equation (2). If the reaction is isothermal, this equation is irrelevant. If the reactor is well-mixed, we can again integrate over the whole volume. Term **7** will require the heat of reaction, conventionally written as $-\Delta H$. If the reactor is adiabatic, terms **5** and **6** are zero for the whole reactor, and we arrive at

$$r_c \Delta H V = -\frac{\mathrm{d}(\varrho C V T)}{\mathrm{d}t} \tag{5}$$

in which T is the temperature, while ϱ is the density, and C the specific heat, of the reactor contents. If the reactor is of constant volume this reduces to

$$r_c \Delta H = -\varrho C_v \frac{\mathrm{d}T}{\mathrm{d}t} \tag{6}$$

in which ΔH is positive for an endothermic reaction, and C_v is the specific heat at constant volume.

The temperature will thus change with time, affecting the velocity constant k, in r_c, and hence giving a further implicit time-dependence in equation (4). If heat transfer to the reactor is provided (e.g. by a steam jacket, or by cooling coils) terms **5** and **6** will be non-zero. Very effective heat transfer can lead to isothermal behavior, even for reactions with a large ΔH, but such heat transfer requires design calculation.

To sum up, a constant volume batch reactor will, for a single reaction, obey an equation of the form of (4), and show variation of concentration (and temperature) with *time* but not with *position* within the reactor. But we note the following provisos:

i) where the reactor contents are a heterogeneous mixture, and more than one reaction is involved, selectivity can be a sensitive function of the intensity of mixing.
ii) where the reactions have a significant heat effect (which usually means exothermic) the details of heat transfer can be of great importance in determining the yield, or selectivity.

b) The semi-batch reactor

Where one of the reagents for a 'batch' liquid-phase reaction is supplied as a gas (e.g. air) it may be necessary to feed it continuously to provide the required reagent (e.g. oxygen) in sufficient quantity for the desired extent of reaction. Such a gas may be 'sparged' into the liquid through a submerged distributor, which procedure both stirs the liquid and produces a cloud of bubbles with a large surface area for rapid dissolution of the gaseous reagent. It may often prove necessary to provide extra stirring by means of an impeller, and if this is so the gas is usually sparged into the liquid just below the impeller. This means that close to the impeller is a region of high turbulence, much bubble break-up, and greatly increased mass transfer.

The simplest design procedure would be to assume that the gaseous reagent is at its saturation concentration in the liquid, i.e. at equilibrium with the incoming gas. In that case, the concentration of this component can be taken as constant in the r_c term in equations (4) and (6). The reactor thereafter can be considered as a batch reactor for design purposes.

This procedure will often be inadequate. The reason for high agitation (which would be avoided for cost reasons, if no other) is to provide a mass-transfer rate capable of satisfying, at least partly, the demands of comparatively rapid chemical reaction. In such cases design becomes very difficult, for performance becomes sensitive to the details of such mass transfer, which details are extremely difficult to 'scale-up' from laboratory to plant size. Three examples may be discussed to illustrate this point.

i) Industrial fermentation — e.g. to produce penicillin. Here a broth is made up in a large batch (maybe 100 m^3 at a time), innoculated with a live culture, and thereafter fed continuously with sterile air. The mycelial growth increases in extent, and the broth increases in penicillin content until the process is stopped, perhaps after a week or so. Although the aerated mixture is 'well-stirrid', the oxygen content of an element of (fluid + mycelium) may fall from a high value near the impeller to zero if the element subsequently spends a long time before reappearing at the impeller. The details of the mass-transfer and circulation within the tank have to be known for accurate prediction of fermenter performance.

ii) The oxidation of cyclohexane by air — a first step in the production of nylon. Though carried out continuously for production, the laboratory experimentation could well be semi-batch; the following considerations apply to both. The conversion is kept low to prevent the undesired reaction of intermediate products with further oxygen close to an air bubble. The selectivity improves, at the same conversion, if mass transfer away from the bubble is increased, since such mixing lowers the maximum concentration of reactive intermediate close to the bubble. This may well require a greater degree of agitation than is necessary to produce uniformity of rate throughout the reactor. This reaction is in practice homogeneously catalysed.

iii) The reaction of a gaseous reagent with, or on the surface of, a solid suspended in a liquid. Here we have the problems of i) and ii) compounded by the presence of three phases in the reactor. As examples, we might have the carbonylation reaction of an olefin with ($CO + H_2$) in the presence of a finely divided metal catalyst, or the hydrogenation of a suspension of coal. Here the range of products, or selectivity, may depend markedly on the mass transfer and mixing phenomena.

The above examples should persuade the chemist working on the laboratory scale that his kinetic results would require much skill for interpretation and use on the plant scale. Furthermore, even on the laboratory scale, it may be quite wrong to believe that the results obtained are "kinetic", or "chemical", and not affected, or even dominated, by "mass-transfer", or "physical" factors.

c) The continuous stirred-tank reactor

The term 'continuous stirred-tank reactor' (CSTR) is used by reaction engineers to imply a reactor in which all elements of volume contain fluid identical in properties, and in which stirring, or circulation, leads to the fluid being 'completely mixed'. Homogeneous reaction in a liquid contained in a stirred tank may well approximate to this ideal model, and hence the choice of name. Of course, the reactor must be continuously fed with reagents, and products continuously removed.

How 'complete' does mixing have to be for the tank to be 'completely mixed'? As a rough rule, if the circulation time of a liquid element is, say, one hundredth of the average time spent within the tank then mixing may well be 'complete' for present purposes. For a flow rate through the system of v m^3 s^{-1}, and a reactor volume of V, this average residence time is V/v. For a given rate of production V will be large if the reaction is slow, and small if the reaction is fast, as we shall shortly see. But one would not have a larger vessel is necessary, nor stir it more vigorously than necessary, so it is to be expected that such reactors in practice may not approximate to 'completely mixed' behavior all that closely.

If we consider the mass balance equation (1) we can — as for batch reactors — integrate over the whole reactor volume, and we can thus say, for any given reaction species,

$$v_f c_f = v_{out} c_{out} + r_c V + V\,\mathrm{d}c_{out}/\mathrm{d}t\,. \tag{7}$$

Here we note that v_{out}, for a gaseous reaction with change in number of molecules, may be different from v_f, the volume rate of feed. We also note that the concentration, which equals c_f in the feed, falls on entry to the well-mixed tank to the value of concentration inside the tank. This latter value is that in the fluid coming out of the tank.

If for the present we take v_f to equal v_{out}, and if we consider the reaction to have attained the steady state — so that $\mathrm{d}c_{out}/\mathrm{d}t$ is zero — and if we further consider a first-order decomposition, we obtain

$$vc_f = vc_{out} + kc_{out}V \tag{8}$$

in which we again note that the c in the r_c term is c_{out}, since that is the value of c inside. Thus we have

$$\frac{c_{out}}{c_f} = \frac{1}{(1 + kV/v)}\,. \tag{9}$$

A common student error is to take the batch reactor equation (4) and integrate it from $t = 0$ to the residence-time V/v obtaining

$$\frac{c_{out}}{c_f} = \mathrm{e}^{-kV/v}\,. \tag{10}$$

This equation is wrong, of course, and does not even have the attraction of being simpler than equation (9). It is wrong because the mass balance equation has not been correctly used (in fact two terms have been put equal to zero which are not, and a third, which is zero, has been used instead). Put another way, in a CSTR all elements of fluid do not have the same residence-time V/v, which is only an average value. For reactions of other orders the appropriate form of r_c must be used in equation (7).

Equation (9) gives the output of a CSTR as a function of (kV/v). This is shown in Figure 1, which also shows the performance of a similar batch reactor as a function of (kt), where t is the time of a batch reaction. For this latter, of course, the correct equation is

$$\frac{c_{out}}{c_f} = e^{-kt} \tag{11}$$

for which equation (10) is the (incorrect) analogue. For a second-order decomposition the appropriate reaction parameters are (kVc_f/v) for a CSTR, and (kc_ft) for a batch reaction, and the CSTR performance curve is also shown on Figure 1. It can be seen that for high conversions (low c_{out}/c_f) a CSTR may be inconveniently large. Before discussing a way of reducing this inconvenience, let us consider our CSTR as it approaches the steady-state — say at start-up. For that situation we will have, for a first-order reaction,

$$vc_f = vc_{out} + kc_{out}V + V\frac{dc_{out}}{dt} \tag{12}$$

If we take v, c_f, k, and V to be constant with time, equation (12) is a linear first-order differential equation in c_{out} and t. Taking, for an example, an initial concentration for c_{out} equal to zero (perhaps the reactor was started up after filling with inert solvent), the solution is

$$\frac{c_{out}}{c_f} = \frac{1}{(1 + kV/v)}\left[1 - e^{-\left(\frac{v}{V} + k\right)t}\right]. \tag{13}$$

Equation (13) does involve an exponential term, but this arises from the hold-up term, **4**, in the mass-balance equation, not from the reaction term **3**. The steady-state value given by equation (9) is seen to be contained within equation (13).

It was seen earlier that for high conversions the volume V becomes inconveniently large, for fixed v, k, and c_f. This is actually due to 'bypassing', by which is meant the chance of some unreacted molecules being able to

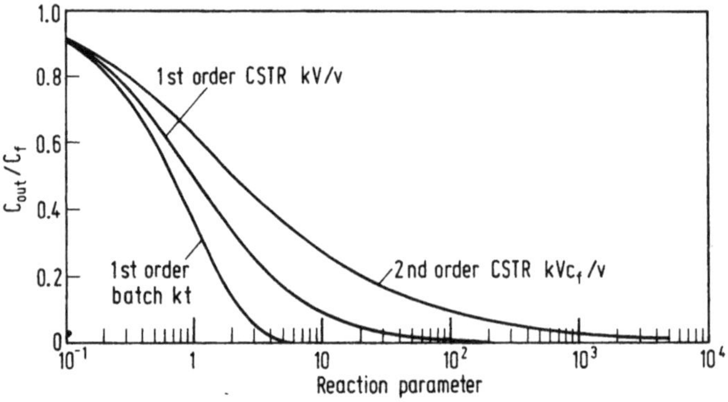

Figure 1. Product composition vs. reaction parameter.

find their way to the exit in very short times. This is another way of saying that there is a wide spread of residence-times (which spread is actually another exponential function of V/v). A further explanation for the large value of V is that in a CSTR the whole of the reacting to be done has to be done at the exit, or lowest, concentration of reagent. For this reason, as higher conversions are required so does the necessary volume, V, increase, proportionally, much more for a higher-order reaction than in the first-order case. The second-order example in Figure 1 shows this clearly.

These large reactor volumes for high conversion can be reduced by using two CSTR's in series (and further reduced by using yet more). By 'reduced' is meant a reduction in *total* volume required. The CSTR's need not be all of the same size, though this is actually optimal for a first-order reaction, but the gain by choosing unequal sizes for other orders of reaction is small and would not outweigh the advantage of using similarly sized equipment.

As an illustration, Figure 2 shows the advantage, in terms of total volume, of using two tanks in series instead of one single CSTR, for first- and second-order reactions, as a function of desired conversion. Note that the volume is markedly more reduced for a second-order reaction than for a first-order one.

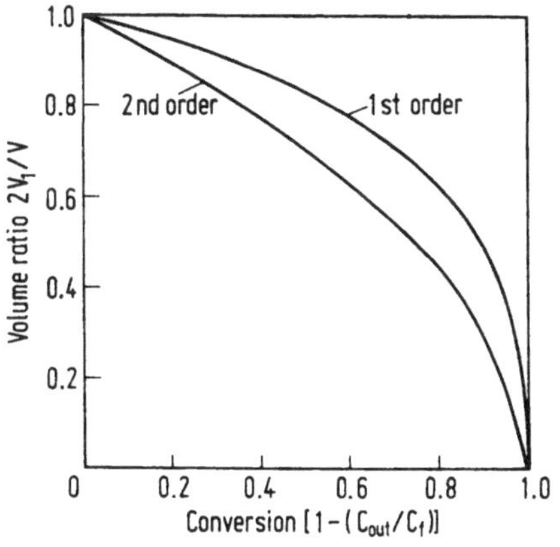

Figure 2. Ratio of volume of 2 CSTR in series to volume of single CSTR vs. conversion

d) The plug-flow, or tubular reactor

The plug flow reactor, PFR, (or tubular reactor — both terms are used) can be regarded as an opposite ideal to the CSTR. In a PFR all elements of fluid experience an identical sequence of pressure, temperature and concentration changes during their passage through the reactor, which passage takes the same time for all elements. If the reactor is indeed a straight tube, which is frequently the case, then all properties of the flow are uniform across a cross-section, and there is no mixing backwards of forwards in the direction of flow. (One could imagine perfect radial mixing and zero axial mixing, however impractical such a concept might be.)

Turbulent flow through an empty tube (as in an ethylene cracker) might approximate to a PFR, but a much commoner practical situation is that when the tube is packed with a solid, which is frequently a heterogeneous catalyst for the desired reaction. The fluid flows between the solid particles, and although this is bound to introduce some axial mixing the effects of such mixing can often be neglected.

If we consider an element of reactor volume and apply the balance equations to it, equation (1) will give us

$$0 = \mathrm{d}(\textit{molar flow}) + r_c\,\mathrm{d}V \tag{14}$$

in which we have assumed a steady-state in the reactor (this will be reconsidered later). For the present, let us assume isothermal behavior, so the heat balance, equation (4), is inapplicable.

In equation (14) the r_c term will, as before, include concentration terms (or pressure terms). The molar flow at any cross-section of the PFR can be expressed in a variety of ways; an obvious choice is the volumetric flow v times the concentration c. Thus equation (14) becomes

$$\mathrm{d}(vc) = -r_c\,\mathrm{d}V\,. \tag{15}$$

For a liquid-phase reaction, it is almost always acceptable to assume that v is a constant, but for a gas-phase reaction it frequently is not. Although there is a drop in pressure across the reactor (to produce the flow through it) such a pressure drop is usually small in comparison with the mean pressure, particularly for high pressure reactors. Thus it is usual to model a PFR as a constant-pressure reactor, and for a gas-phase reaction which results in a change in the number of moles the volume flow rate will change.

As a simple example of equation (15) let us assume, for the present, that v is constant, and that we have a first-order reaction, then equation (15) is readily seen to integrate to

$$c_{out} = c_f\,\mathrm{e}^{-kV/v} \tag{16}$$

which is equation (10), seen to be wrong for a CSTR. For this special case of a PFR it is correct, for all elements can be regarded as small batch reactors passing through the PFR in identical times V/v, and discharging their product at exit. Thus the insidious temptation to use the "time" of a batch reaction is not punished by error here — because all elements have the same residence time *and* there is no volume change. If a volume change does arise, equation (16) is incorrect, and the residence-time is not immediately obvious. This is because both v and c in equation (15) vary with position in the reactor.

In such a case we have various options open to us. We could base all our compositions on some constant flow rate (e.g. the mass flow rate, or the entering molar flow rate), in which case the concentrations in r_c must be expressed, via the reaction stoichiometry, in the same composition variable. Or we might relate the flowrate v, via the reaction stoichiometry, to the concentration in r_c. After integration the final relation between reactor volume V, and outlet concentration c_{out}, will naturally be the same.

As an example, let us take an n'th-order irreversible gas-phase reaction of a single component, in which each mole of reagent produces $(1 + \varepsilon)$ moles of product. We shall choose to work in terms of the conversion x of our reagent. It is then readily seen that the molar flow rate of reagent, vc, is given by $m_f(1 - x)$, where m_f is the molar flow rate of feed. The concentration of our reagent, c, at any point where the conversion is x, is given by the mole fraction times the total molar concentration. Hence

$$c = \frac{1 - x}{1 + \varepsilon x} \cdot \frac{p}{RT} \tag{17}$$

provided we can assume ideal gas behavior; here p is the pressure, R the gas constant, and T the temperature. Substituting into equation (15) and integrating gives

$$V = \frac{m_f}{k}\left(\frac{RT}{p}\right)^n \int_0^{x_{out}} \left(\frac{1 + \varepsilon x}{1 - x}\right)^n \mathrm{d}x \tag{18}$$

from which the conversion, x_{out}, of the product leaving a PFR of volume V can be calculated.

To illustrate the effect of ε, Figure 3 shows dimensionless volume as a function of x for second-order reaction, $n = 2$. The examples chosen are

a) $\varepsilon = -{}^1/_2$ e.g. a dimerisation
b) $\varepsilon = 0$ e.g. an isomerisation
c) $\varepsilon = 1$ e.g. a decomposition into two molecules.

For a conversion of 90 percent, the dimensionless volume for $\varepsilon = 0$ is 9. This is reduced to $3 \cdot 63$ if $\varepsilon = -{}^1/_2$, and increased to $27 \cdot 7$ if $\varepsilon = 1$. Clearly for PFR design one cannot in general use procedures appropriate for constant

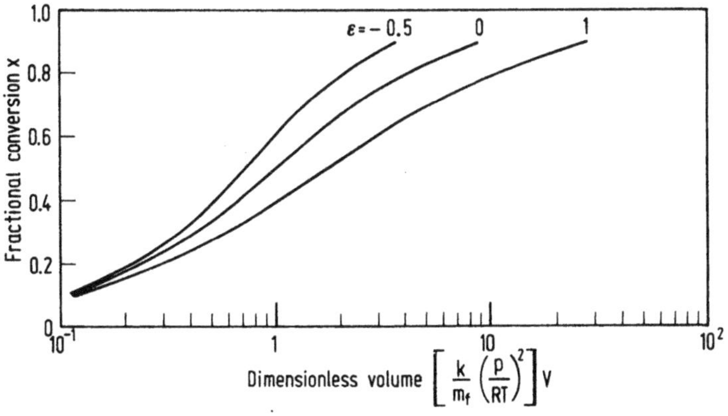

Figure 3. Conversion vs. PFR volume. 2nd-Order reaction with change in number of molecules

volume batch reactions. Such procedures, for $\varepsilon \neq 0$, are wrong for two reasons. Firstly the gas (at approximately constant pressure, be it remembered) speeds up or slows down as it passes through the PFR. This alone would rule out a simple relation between volume and residence time. Furthermore the kinetics are changed by such an expansion or contraction, so a constant volume batch reactor is an inappropriate analogue on that score as well.

To be able to use equation (18) requires that ε and n be known. The stoichiometry will give us ε, but if the kinetics are not known, the problem may be how to use a PFR to determine them. A series of experiments with different m_f (or V) would be carried out, and x_{out} measured in each case. Let us assume, for an example, that m_f is varied (T, p, V being kept constant) and x_{out} is obtained as an acceptably smooth function of m_f. Define a variable τ by

$$\tau = \frac{V}{m_f}\left(\frac{p}{RT}\right) = \frac{1}{k}\left(\frac{RT}{p}\right)^{n-1}\int_0^{x_{out}}\left(\frac{1+\varepsilon x}{1-x}\right)^n \mathrm{d}x. \tag{19}$$

We shall be able to draw a curve of τ against x_{out}, the slope of which can be obtained:

$$slope = \frac{\partial \tau}{\partial x_{out}} = \frac{1}{k}\left(\frac{RT}{p}\right)^{n-1}\left(\frac{1+\varepsilon x_{out}}{1-x_{out}}\right)^n. \tag{20}$$

Finally, taking logarithms,

$$\ln(slope) = constant + n\ln\left(\frac{1+\varepsilon x_{out}}{1-x_{out}}\right), \tag{21}$$

so n and k are both obtainable from the data. In view of experimental errors it would be best to use equation (21) to give a first estimate of n, and then use equation (18) to calculate k.

The variable τ is actually a time, and equals the residence-time the feed would spend in the PFR if no expansion effects occurred. It is called the *space-time*. Because it is always known to an experimenter, it is a useful variable to employ in the consideration of experimental results. Its inverse is known as the *space velocity*, a most unfortunate choice of terminology since it is not a velocity, but has units of $time^{-1}$. The space velocity is the number of times per unit time the total reactor volume would seem to be swept out by the given feed rate. Since it is usually based upon the total reactor volume it is similar to a superficial velocity in that respect.

The space-time is not equal to the residence-time unless i) it is based on the reactor void volume, ii) $\varepsilon = 0$, and iii) the reaction is isothermal. When condition ii) is not satisfied, the residence-time in a PFR can be obtained by changing the variable in equation (15) from $\mathrm{d}V$ to $\mathrm{d}t$ and performing the slightly different resultant integration. As an exercise for students this has some point, but it need not concern us here, since the residence time in fact tells us nothing we need to know that the volume V and feed flow rate v do not already tell us.

Let us suppose now that the PFR is not isothermal. If that is the case, equation (18) becomes more complicated:

$$V = m_f \left(\frac{R}{p}\right)^n \int_0^{x_{out}} \left(\frac{1 + \varepsilon x}{1 - x}\right)^n \frac{T^n}{k} \, dx \,. \tag{22}$$

It can be seen that T and k occur in the integral as variables; they have to be known as functions of x for the integration to be performed, numerically. The general case, with heat transfer to or from the reactor, is complicated indeed. The special case of the adiabatic reactor is more tractable, if still messy, but it illustrates some valuable points.

We have to relate T, and hence k, to the conversion, x. For an adiabatic reactor that can be done by a thermal balance from the entry to the reactor to the point in question. Thus the procedure is to choose a value of x, and use the heat balance to calculate T. The relation between k and T must be known (an Arrhenius-type expression is usually assumed). Thus the integrand in equation (22) can be plotted as a function of x and the graph integrated, numerically or by quadrature.

The variation of k with T is usually by far the major component of the variation of the integrand with x. The resultant graph will be sharply 'spiked' at the low temperature end of the reactor (entry for an exothermic reaction, exit for an endothermic reactor). Thus the volume of reactor necessary for a desired conversion is rather inaccurately known, and this is reflected in a very real control problem with adiabatic reactors. Sometimes it is necessary, with exothermic reactions, to restrict the conversion so as to obtain satisfactory selectivity. If the feed to an adiabatic reactor in such a case is preheated by only a degree or two too much, the conversion is far too high, whereas conversely if it is a little too cool, almost no conversion will occur. This is not strictly a case of instability, the term *parametric sensitivity* has been coined to describe it. There is analogous behavior with endothermic reactions, but they are far less important in practice.

A packed bed is very frequently chosen as the configuration for an industrial reactor involving heterogeneous reaction. In such a case, reactor performance many well depend to a great extent on mass- and heat-transfer effects to be discussed later. In any such considerations, it is essential that the fundamentals of an idealised PFR should be thoroughly understood, and such an understanding has been the aim of this section.

B. Further Developments in Reactor Theory

1. More Complicated Reactor Types

We have previously considered some idealised types of reactor, to which the balance equations can be fairly easily applied to obtain concise expressions for the performance of such reactors. In this section we shall briefly consider a few examples of reactors in which the material flows are more complex, and hence the application of the balance equations presents some difficulty. Such reactors frequently involve catalysis, or heat-transfer problems, and may indeed have been conceived to get round such difficulties which would arise with simpler designs. We shall not consider here these catalytic, or heat-transfer aspects, which we shall defer until later.

A simple extension of the earlier treatments arises with auto-catalytic reactions. Here it is necessary to have some product present in the reactor close to the entry of the reagents — or the reaction never 'gets going'. This can be done either by internal mixing (leading in the limit to a CSTR) or by recycling some product, adding it to the feed stream entering, e.g., a PFR. Such design exercises are simple, if the reaction kinetics are known. If high conversion is desired, it may be desirable to have a CSTR operating at maximum reaction rate, followed by a PFR to produce the final high conversion. A single CSTR operating at high conversion might be inconveniently large.

Where the reaction involves reagents in different phases, contact between the phases is essential for reaction to proceed, and the different phases may have very different holding times. This has been touched on already, when dealing with semi-batch reactors. Examples are legion, and many in fact involve catalysis. They all involve mass-transfer, and often heat-transfer as well.

'Bubble reactors' bring about a reaction between a cloud of gas bubbles and the liquid through which they rise. The bubbles may cause "gulf-streaming", by which is meant the tendency for bubbles distributed uniformly across the vessel at the bottom to emerge from the liquid at the centre of the cross-section. The liquid passes up the centre and down the sides of the vessel. The bubbles may coalesce, and if they do they will rise faster, and spend a shorter time in the reactor. The process of mass-transfer may increase, or decrease, the tendency to coalesce. These phenomena are not well understood, and have led to serious deficiencies in performance of reactors scaled up from laboratory-sized equipment. If the fraction of the reactor volume occupied by liquid is only a half of that predicted in design, the reactor's performance may well be very inadequate. Surface tension effects, and superimposed mass-transfer phenomena (leading, for example, to foaming) would perhaps be unexpected by chemists working at bench scale.

It can readily be seen that reactions involving two liquid phases can present very similar problems. Extraction columns, and the manufacture of fats and soaps provide examples.

If a solid is involved, either as a reagent or as catalyst, then a new range of factors requires consideration. The PFR is a simple case, of wide application; ammonia production, SO_2 conversion, and xylene isomerisation are three examples. There may be three phases involved, as in a 'trickle-bed reactor'; the solid-catalysed hydration of propylene to give isopropanol is an example. Here the two fluid phases are passed together through a catalyst bed, and one question quickly comes to mind — should these flows be co-current or counter-current?

If the solid particles are small, they may be kept suspended in the liquid by the flow of the fluid, or two fluid phases. Such a case is the 'slurry reactor', used for carbonylation reactions.

A final example of a complex flow system is the 'fluidized-bed reactor'. The subject of fluidization has evolved a prolific literature, and a substantial band of practitioners within chemical engineering. The physics of the flow is an interesting study in itself. For particles fluidized by a liquid, increasing the liquid flowrate causes the whole bed to expand and the solid particles to move further away from each other on average. There are unusual exceptions, which behave like gas fluidized beds. Here increasing the gas flow rate causes the appearance of 'bubbles', which rise through the bed, bursting through the liquid-like top surface of the bed. However, the gas in a bubble at the top of the bed may not be the same as that present in the 'same' bubble earlier and lower down. The mass-transfer is complex. It can be seen that the addition of chemical reaction further complicates the problem. In practice not all solids fluidize easily; size, size-distribution, shape, 'stickiness', all can have gross effects on the fluidizing behavior.

The effective design of large-scale reactors of any of the above types requires careful study. A design based solely on laboratory-scale experiments to determine the 'chemistry' can lead to catastrophic failure.

2. The Effects of a Spread of Residence-Times

As shown earlier, the performance of an ideal PFR with a single fluid phase may be simply related to that of a batch reactor because in such a PFR all elements of fluid experience the same environments for the same periods of time, and no mixing in the direction of flow occurs. All elements have the same residence-time. In every other case considered there is a spread of residence-times, a CSTR providing an extremely wide spread. Experiments with nonreacting tracer inputs can be carried out fairly easily [1, 2] to determine the residence-time distribution or RTD. If we know this, what can it tell us about the performance of the system when a reaction is taking place?

Let us define the residence-time distribution function $E(t)$ by saying that $E(t)\,\mathrm{d}t$ is the fraction of the outlet flow from a reactor which has been in the reactor for times between t and $t + \mathrm{d}t$. Its measurement, other residence-time functions, and relationships between them are described in chemical engineering texts — e.g. [1–4], and need not be gone into. The concepts of the spread, and the function $E(t)$ will here suffice. As an example, for a CSTR it can be shown that

$$E(t)\,\mathrm{d}t = \frac{v}{V}\,\mathrm{e}^{-vt/V}\,\mathrm{d}t \tag{23}$$

in which v is actually v_{out}, and both sides of equation (23) are dimensionless. This latter point is mentioned to advise caution if changes in the scale of time are proposed (e.g. to dimensionless time $T \equiv vt/V$). For a PFR $E(t)$ is actually a sharp spike, or δ-function, since all material coming out has the same residence-time, equal to V/v, and none has residence-times greater or less than this value.

In practice the RTD found may be very different from the above. Figure 4 shows some examples. What might we deduce from them? Case A is fairly close to the (ideal) CSTR, case B not far from a PFR. Case C is like neither, and in fact shows that a small, but significant amount of material is spending far longer in the reactor than average. That in itself can be diagnostically very valuable in showing how 'off-spec' product may arise. The author has personally come across examples involving the treatment of heat-sensitive materials, and in glass manufacture, where such a 'peak' gave the clue to improvement. In the case of the glass plant, the peak actually occured at a time much shorter than the average, and showed that most of the material was only in the process for about half the design time, a smaller fraction of the material spending very much longer, to no useful effect.

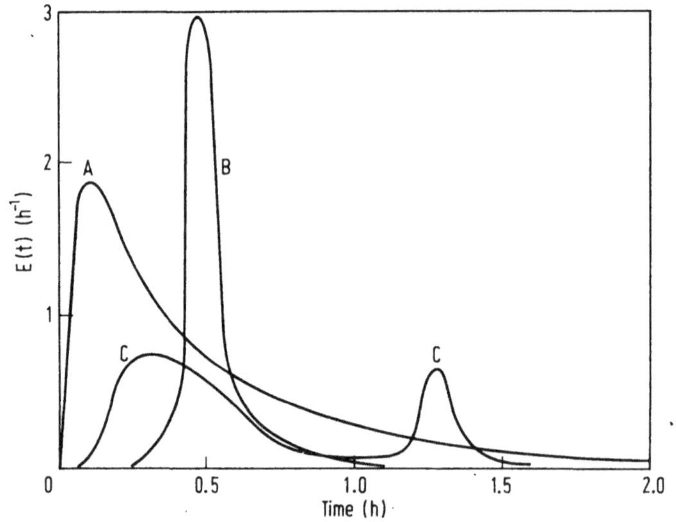

Figure 4. Three typical residence-time distributions.
A, approximates to a CSTR
B, shows plug-flow with dispersion
C, is binodal
All three have the same mean residence time, V/v = 0.5 hour

A procedure for estimating the effect of the RTD on reactor performance suggests itself as follows: If we regard each element of fluid in the reactor outflow as being a separate batch reactor, then perhaps the mean concentration of a species in this effluent would be given by

$$\bar{c} = \int_0^\infty c(t) \cdot E(t) \cdot \mathrm{d}t \tag{24}$$

which follows the usual definition of a mean. Here $c(t)$ is the concentration to be expected in a batch reactor after time t. We know from earlier that this is only likely to be correct for the isothermal, $\varepsilon = 0$ case.

Even for the above restrictions this can readily be shown to be inadequate. Consider the ideal CSTR, for which equation (24) becomes

$$\bar{c} = \int_0^\infty c(t) \cdot \frac{v}{V}\, e^{-vt/V}\, dt\,. \tag{25}$$

Let us first take a simple irreversible first-order decomposition. Equation (25) leads straight to the conclusion

$$\frac{\bar{c}}{c_f} = \frac{1}{1 + kV/v}$$

which is exactly the same as the answer found by applying the simple mass balance, equation (9).

So far, so good, but consider now the second-order case. Equation (25) becomes

$$\frac{\bar{c}}{c_f} = \frac{v}{V} \int_0^\infty \frac{e^{-vt/V}}{1 + kc_f t}\, dt \tag{26a}$$

which can be computed numerically, or from tables of exponential integrals. If we apply the balance equation (3) to our CSTR, the answer obtained is

$$\frac{\bar{c}}{c_f} = 1 - \left(\frac{kVc_f}{v}\right)\left(\frac{\bar{c}}{c_f}\right)^2 \tag{26b}$$

which is clearly not the same as equation (26a).

This difference requires explanation, and it is not that either is wrong. They actually refer to different processes. The RTD by itself is insufficient to determine the extent of 'micromixing', it describes only the distributions of times spent by reagent molecules in the reactor, and does not say much about their immediate neighbors while in the reactor. Equation (24) describes the 'completely segregated' case; the reagent molecules see the same near neighbors for the whole of their time in the reactor. In the first order case the distribution of times suffices, since the chance of any given reagent molecule reacting in any element of time is independent of the molecules around it (which immediately suggests caution here for the pseudo first-order case). For orders other than one, equation (24) is insufficient.

However, equation (24) provides a bound to performance — the completely segregated case. The simple balance equation provides another bound — for the CSTR. If we have a known RTD (which is not that of one of our idealised models) what can we say about this other bound? A paper by Zwietering [5] discusses this point, and shows how to use the RTD to put limits between which the performance as a reactor must lie. For a PFR the bounds close on the completely segregated case, equation (24). For a CSTR, equation (24) is one bound; the balance equations give the other. For a general RTD, equation (24) remains one bound; the other is obtainable by Zwietering's treatment. The central point is whether that mixing which is allowed by the RTD occurs as early, or as late, as possible in the reactor. The difference between these bounds is generally not large, and attempts experimentally to verify them have not been very succesful (the close temperature control necessary for this purpose is difficult to achieve in steady-flow reactors). Equation (24) is a useful guide for design, however.

Two last examples of RTD may be shortly mentioned. In the processing of a viscous fluid, it is possible that the flow in a tubular reactor may be laminar, and thus very far from plug-flow. For fully-developed laminar flow in a tubular reactor we have

$$E(t)\,\mathrm{d}t = \frac{2\,t_o^2}{t^3}\,\mathrm{d}t \tag{27}$$

for $t > t_o$, being zero for $t < t_o$. Here t_o is the time for the center-line fluid to pass through the tube; it is half the average time spent by the fluid in the reactor. Three-quarters of the process fluid spends less than the average time in the reactor; the fluid near the walls spends very much longer than average. To neglect this spread in, e.g. a polymeriser, could lead to serious consequence.

If the flow in an unpacked tubular reactor is highly turbulent, the RTD may be close to that of a PFR, but may show a spread of the PFR's 'spike' which can be described by a diffusion-type or dispersion model. The same kind of spread can also be shown by a packed-bed reactor, for entirely different physical reasons. For catalytic reactors the packed-bed case is more important, and here it can be shown [1, 2, 4] that such a dispersive axial spread can be neglected for all practical purposes if the bed is more than 100 catalyst particles deep. If it is not, the effect would in principle be worth considering, but in beds which are shallow (either because the reaction is very fast, or because a small laboratory reactor is being used) the dispersion model is likely to be inappropriate. This model assumes that the fluid flow is uniformly distributed across the bed, and with shallow beds maldistribution, or bypassing, can rather easily occur. A relevant examples is the catalytic conversion by the methanation reaction of $CO + CO_2$ in ammonia synthesis gas. It is necessary to reduce the concentration of carbon oxides to below one part per million, to prevent poisoning of the catalyst in the following ammonia synthesis. The reaction is not difficult to bring about, but bypassing must be avoided if the virtually-complete conversion to methane is to be achieved.

If, as is frequently the case, more than one reagent is necessary, and they are mixed in the reactor, the RTD's of the different feeds may be different. This complication is increased if there are two fluid phases. In such cases equation (24) may be clearly inapplicable, but for single fluid streams, well mixed before entry, it is a useful guide.

3. Non-Steady Behavior of Reactors

So far we have mainly considered reactors operating in the steady state, though batch, and semi-batch reactors clearly are not of that description. We must now consider the behavior of 'continuous' reactors when they are not operating steadily. By 'steady-state' we mean that conditions are invariant with time, though they may well be different with position in the reactor.

Continuous reactors have to be started up (and shut down), and the time required to get to 'steady' production of acceptable product is clearly of important economic concern. The sensitivity of reactor performance to change of feed or of process conditions, and the speed of response of the reactor to such changes, are of vital concern to the control engineer. It may be necessary to be able to shut down the plant very rapidly ('crash shut down') in the event of some equipment failure, if disaster is to be avoided. These considerations are the province of the instrumentation and control engineers. We shall only consider briefly 'normal' start-up and shut-down procedures.

To get to the steady operating state both the flows and the chemical reactions must have become steady. The two may well interact, particularly if the reaction leads to substantial changes of temperature of the flow. For single-phase reactors, steadiness of the physical flows may be quite rapidly attained, though the substances flowing may take longer to settle down either because of chemical reaction or, for example, because an inert stream is slowly replaced by a reacting one. It is clearly more difficult to set up steady flows when more than one phase is involved.

Even if the flow is steady the chemistry will take time to become steady. For a CSTR the calculation can be simple — and equation (13) is such an example, in which the mass balance equation has been readily solved. It was there seen that a CSTR approaches the steady-state asymptotically. It might in fact be better to adjust the flow (a rapid process) to decrease the time producing off-specification product. As an extreme example, one could fill the reactor, run it as a batch reactor until the concentration reached the desired steady-state value, and then start running as a CSTR. With more complicated stoichiometry and kinetics this might not be quite so simply done, but some procedure of this sort could be helpful.

With a PFR, in principle it only takes one charge of reagent going through the reactor for it to reach steady state. However if the change to reacting flow is a sharp and large one (as it may well be in respect of temperature for a strongly exothermic reaction) the system may take some time to adjust (e.g. thermal lags), or it may be unsafe to start that way (e.g. thermal shock). In that case, a more gradual start-up procedure is required.

Similar considerations apply for planned shut-down, for example for annual maintenance checks (provision for safe crash shut-down must also be allowed for in design). The planned shut-down may also be as the result of a gradual fall in catalyst performance, which can often be counteracted for a while by increased 'severity' of operation — perhaps by raising the temperature of the feed.

These points have only been touched on. The purpose has been to show that continuous large-scale reactors not only have different characteristics from batch laboratory-scale reactors, but that their starting-up and shutting-down also require thought.

3. Catalysis in Practice

A. Homogeneous Catalysis

So far we have concentrated attention on chemical reactor theory, and we have mainly used as examples non-catalytic chemical reactions. If the reaction involves *homogeneous* catalysis then what further considerations apply?

Firstly, if the reaction occurs in a single fluid phase then the concentration of the catalyst may, in most cases, be combined with the velocity constant to give an apparent velocity constant, and a pseudo-reaction-order less than the true order of reaction. If the reaction occurs in one phase, but involves mass-transfer from another phase, things are not so simple. If coupled reactions are involved, matters can become very complicated indeed. As an example consider the oxidation of ethylene to acetaldehyde using air as the oxidant and an aqueous medium containing a mixture of catalysts dissolved therein. This, the Hoechst-Wacker process, uses chlorides of copper and palladium as 'catalysts', though it might be more helpful to regard them as reagents which are regenerated and chemically recycled. Of course a catalyst takes part in a reaction by forming some intermediate species which later breaks down to release the catalyst. But in the Hoechst-Wacker process the palladium oxidises the ethylene, being itself reduced, and this would rapidly remove all the palladium from the scene of action. Air will not readily re-oxidise the palladium, so this is done by means of cupric ion, which in turn is reduced to cuprous ion. Air will re-oxidise cuprous to cupric ion, and so the cycle can repeat itself.

The processes can be brought about in two separate stages with liquid transferred to and fro. It can also be done in a single stage, which is simpler but has the disadvantage of air and ethylene being together in a single reactor.

With such a coupled scheme of chemical reactions it may not be a fruitful concept to consider the "rate-determining step". The kinetics are complicated and it is obvious that mass transfer from the gas phase to the liquid phase has a role to play which may be dominant. In the single-stage process both gaseous reactants (oxygen and ethylene) diffuse into the liquid, where all four catalyst species (cupric, cuprous, palladium in oxidised and reduced states) are to be found. Acetaldehyde will diffuse into the gas stream and thereby be carried out of the reactor.

This reaction happens to be one with a high selectivity for the desired product. If a seriously-competing reaction were involved, the selectivity could be sensitive to variations in the mixing phenomena. It is for this reason that scale-up of such reactions from small-scale to production scale is fraught with danger.

So in summary, homogeneous catalysis may, for single-phase reactions, be simple to engineer on the basis of batch kinetic results. If there is mass-transfer between phases involved, this may have effects on the catalyst, as well as all the effects already mentioned for non-catalytic reactions. Where more than one phase is involved in a reaction, one can of course describe the reaction as heterogeneous, but we shall reserve the description "heterogeneous catalysis" to the case where the catalyst is not dissolved in the phase where reaction is taking place. Such a heterogeneous catalyst is invariably a solid, or a coating on a solid, and it is that situation that we now shall proceed to consider.

B. Heterogeneous Catalysis

1. General Considerations

The reaction of a substance (or substances) in a fluid phase may be catalysed by a solid surface in contact with the phase. In the absence of that solid surface, the reaction proceeds negligibly slowly. Clearly the surface is involved in the reaction, and the study of how the surface is so involved is a large part of the subject of catalytic chemistry. Equally clearly the reactants in the fluid phase must find themselves "on", or at the very least within atomic distances of, the surface. Hence the five steps usually proposed as necessary for such a reaction:

1) Diffusion of reactant(s) "to" the surface.
2) Adsorption of reactant(s) "onto" the surface.
3) Reaction "on" the surface.
4) Desorption of product(s) "off" the surface.
5) Diffusion of product(s) "from" the surface.

Adsorption and desorption "on" the surface are involved because the surface must affect the substances "chemically" if the reaction path is to be so radically eased by the presence of the solid.

Chemists are understandably interested in the chemical physics of the surface and its interaction with the reagents. Experiments using a selected crystal face of the catalyst material aim to obtain such information unambiguously. Vigorous stirring of the fluid phase may reduce the effects of steps 1) and 5) to negligible proportions.

However, practical catalyst materials will present a range of crystal faces, dislocations and all, to the fluid phase. To obtain the necessary surface area to produce an acceptable reaction rate per unit of overall reactor volume may require porous catalysts. There are some reactions which go so fast, per unit area of catalyst surface, that non-porous catalysts, e.g. iron chips or silver wire gauze, can be used. But it is far commoner that porous catalysts, with surface areas measured in $m^2\ g^{-1}$ are required. Such areas could be obtained by using very small particles, but if used in a packed-bed these would give far too high a pressure drop. So the small particles are pelletised into porous pellets, commonly cylinders of 5 mm size, or thereabouts.

Stirring between the pellets will arise from the flow through the bed, and will often not be adequate to remove consideration of steps 1) and 5), while within the pellets there will be no stirring at all.

If the reaction involves a significant temperature change, due to the heat of reaction, then it may be that a chemist using a single-crystal metal can both know, and control, the reaction temperature. With a packed-bed of porous catalyst, it may be difficult to calculate what the temperature is at any point, and impossible to control it to be the same everywhere.

Catalytic chemical reaction engineering is deeply concerned with these transport properties of the system, and perhaps rather less concerned with the molecular chemistry at the surface. This is very properly the province of the catalytic chemist; the chemical engineer has to take the chemist's active catalyst surface and supervise the production of it in a physically-robust and long-lived form which will be useable in an industrial-scale reactor. He has also to estimate the performance of this reactor, using the chemist's laboratory results. For this he will have to consider

i) the diffusion of reactants to the surface of the pellets from the bulk fluid flowing between the pellets.
ii) the diffusion of reactants within the porous pellets, for it is there that the vast majority of the active surface will be found.
iii) the transport of heat between the pellets and bulk fluid flow.
iv) the transport of heat within the pellet.

Since reaction rate is in general so dependent on temperature, iii) and iv) may be more important than i) and ii). The sight of a glowing coal particle surrounded by much cooler gas and ash particles in a fluidised-bed combustor is convincing evidence that large temperature differences can exist in a heterogeneous reactor.

We shall now consider these transport phenomena in more detail. If, as is usually the case, it will be convenient, for illustration, to choose chemical kinetics of great simplicity, it is done to focus attention on these transport phenomena. Of course, matters can be more complicated, and cases will doubtless be discussed in other chapters of this work. For present purposes, let us ask catalytic chemists to bear with this simplification of the chemistry.

2. Transfer of Reactants to the Pellet Surface

As a result of reaction at the catalyst surface, the concentrations of reagents will tend to be lower there than in the bulk fluid, and this lowering of concentration will be counteracted by transport of further reagents to the surface. Vigorous stirring, or turbulent flow, will facilitate this transport, but the effectiveness of such procedures needs to be examined. Let us consider a non-porous catalyst pellet, with reaction at its surface.

A convenient model, which is oversimplified, but still useful for illustrative purposes, is the 'stagnant layer' model. Close to the surface a stagnant (or laminar) layer is said to exist, and reagent has to diffuse through this layer to the reaction surface. The fluid mechanics (stirring, flow regime) are said to determine the thickness of the layer, the diffusion properties of the materials determine their transport through the layer.

It is clear that if the reaction rate per unit area is small, then the necessary transport of material will be readily obtained, but that if the catalyst is more effective it will become more difficult to provide the material for the increased rate of reaction. The concentration at the surface will drop, which will both reduce the reaction rate and increase the transport rate, bringing the two into balance. If we assume a steady state, so that at a particular location no changes in concentration with time are occurring, then the reaction rate and transport rate are equal. We can then write

$$k_g(c_b - c_s) = kc_s^n \tag{28}$$

in which we have assumed, for simplicity, a single reagent. Here k_g is a mass-transfer coefficient, c_b and c_s are the concentrations of the reagent in the bulk fluid and at the surface, respectively. The reaction rate kc_s^n is per unit area of catalyst surface, and we assume that the reaction rate depends upon the concentration "at" the catalyst surface to the n'th order. We are not here concerned with steps 2), 3) and 4), when we write the rate of reactions as in equation (28). Step 5) is also assumed not to be a problem; in fact diffusion of product from the pellet surface can be said to be involved (e.g. by equal counter-molar diffusion) in the defining of the diffusive flow of reagent *to* the surface.

It is clear from equation (28) that if k_g is large and k small, c_s will tend towards c_b, and transport limitation will not be noticed. But an effective catalyst has high k, while k_g is determined by the fluid mechanics. In the limit of improved catalyst $c_s \to 0$, and the rate cannot exceed $k_g c_b$. Such a diffusion-limited reaction appears to be first order, whether n is unity or not. Further effort to improve the catalyst may be pointless; one should concentrate on increasing k_g.

Two pieces of evidence can be obtained which may point convincingly to such 'mass-transfer limitation'. The first is to examine the temperature dependence of the rate. The Arrhenius expression can be used to provide an energy of activation of the apparent rate constant. The energy of activation is lower for transport processes than characterisitically found for chemical reaction. Secondly, it is possible to estimate the mass-transfer rate from correlations in the literature: as an example see Mixon and Carberry [6]. The physical process of mass transfer does not depend on the chemical kinetics at the surface. Clearly the observed rate should not be greater than $k_g c_b$. If it is considerably less than this quantity, then 'kinetic' limitation is indicated (and in such a case the observed order, n, may well be different from unity, too).

We may define an effectiveness factor η by the relation

$$\eta = \frac{\textit{observed rate}}{\textit{rate obtainable if } c_s \textit{ were equal to } c_b} \tag{29}$$

This can be obtained from equation (28) very readily for the case $n = 1$. We have

$$\eta = \frac{kc_s}{kc_b} = \frac{k_g}{k + k_g}. \tag{30}$$

We may define a Damköhler number, D_a, by

$$D_a = kc_b^{n-1}/k_g \tag{31}$$

and hence equation (30) becomes

$$\eta = \frac{1}{1 + D_a} \tag{32}$$

for a first-order reaction. The reader may care to show for himself that if $n = 2$, then

$$\eta = \left[\frac{1}{2D_a}((1 + 4D_a)^{1/2} - 1)\right]^2 \tag{33}$$

The Damköhler number can be regarded as the ratio of the maximum rate of chemical reaction to the maximum rate of mass transfer. If D_a is very large, the effectiveness factor, η, is small. If D_a is small, then $\eta \to 1$. One should note that a higher rate may well be associated with a lower effectiveness factor. What that means is that one may not obtain the full expected gain from an improvement of catalyst.

In all of the above it has been assumed that the temperature of the surface is constant. If the heat of reaction is large, and in practice exothermic reaction is most likely, then the temperature of the catalyst surface may be noticeably higher than that of the bulk fluid. Since k is in general markedly temperature-dependent, the value of η may seem to be much greater than unity. The lowering of the reagent concentration at the surface is more than compensated by the increased velocity constant at the higher surface temperature.

If the reaction is "slow", so that to obtain a worthwhile rate a porous catalyst is necessary, with an active surface perhaps several orders of magnitude greater than the pellet external surface area, then diffusional phenomena within the pellet will be of much greater importance than those external to the pellet.

For both the above reasons, the simple case discussed here is unlikely to be of much practical relevance, but it serves as an introduction to those cases which we shall now move on to consider.

3. *Diffusion of Reactants Within a Pellet*

If the reaction rate per unit area of catalyst surface is 'slow', in the sense of the previous section, then a useful reaction rate can be obtained if a porous catalyst material is used. As explained earlier, such a pelleted material will have a catalytically-active surface vastly greater than its external area, but to get to this surface reactants have to diffuse into and through the pellet. Such transport is diffusive, since external flow or stirring of the fluid have no effect within the pellet.

Transfer to the external surface of the pellet from the fluid stream will thus be only insignificantly affected by mass-transfer resistance, since the choice of porous material was indicated by the 'slow' chemical reaction. It is certain that, in practice, if diffusion to the pellet causes the reaction rate to be lower than expected, then the catalyst area inside the particle will be almost totally ineffective.

Within a porous pellet reagent will diffuse towards the center, reacting on the walls of the "pores", the product meanwhile diffusing out. If this diffusion is effective the concentration of reagent will be almost the same at the center of the pellet as in the fluid outside the pellet. If the reaction is somewhat faster, the concentration of reagent will be lower towards the particle center, and the reaction rate lower too (unless the order of reaction is negative with respect to this species). To calculate this lowering of rate, we have to model the diffusion and reaction within the particle. For the present, we will assume isothermal conditions.

Two models suggest themselves. The first is that of diffusion down a cylindrical pore, of radius r_0, with concurrent reaction on the pore walls. It is assumed that diffusion across the pore is rapid, so that the concentration of reagent, c, is a function of axial distance, x, from the pore mouth. The appropriate equation, for the steady state, is then

$$D\frac{\partial^2 c}{\partial x^2}\cdot \pi r_0^2 = kc^n \cdot 2\pi r_0 \tag{34}$$

to be solved for boundary conditions $c = c_s$ at the pore mouth ($x = 0$) and $\partial c/\partial x = 0$ at the end of the pore, of length $x = l$. This is easily solved (for $n = 1$), but we shall not do this, for the following reasons:

i) All pores will not be of the same length.
ii) To represent the spaces between particles in a compressed pellet by straight-sided cylindrical tubes, is somewhat unrealistic.
iii) The results are of similar form to those obtained from the second model, which follows.

For this, we assume that diffusion occurs through the porous volume, with concurrent reaction. Both diffusion and reaction are formulated as though the catalyst were a homogeneous phase. For illustration let us choose a spherical catalyst particle, for which we can then write

$$\frac{\partial}{\partial r}\left(-D'\frac{\partial c}{\partial r}\cdot 4\pi r^2\right) = -k'c^n \cdot 4\pi r^2 \tag{35}$$

in which r is the radial position within the pellet. D' and k' are the apparent (volume-based) diffusion coefficient and velocity constant, respectively. If R_p is the pellet radius, equation (35) can be written as

$$\frac{\partial^2 c}{\partial r^2} + \frac{2}{r}\frac{\partial c}{\partial r} = \frac{\Phi^2}{R_p^2} c \tag{36}$$

where $\Phi^2 = R_p^2 k' c^{n-1}/D'$. Φ is called the Thiele modulus [7], and is constant *for the case n* = 1. For such first-order kinetics, equation (36) is well-known. Its solution, subject to the boundary conditions:

$$c = c_s \text{ at the pellet surface, } r = R_p$$

$$\frac{\partial c}{\partial r} = 0 \text{ at the pellet center, } r = 0$$

is

$$c = \frac{c_s R_p \sinh(\Phi r/R_p)}{r \sinh \Phi}. \tag{37}$$

The effectiveness factor, η, is the reaction rate observed, divided by that which would be observed if $c = c_s$ everythere. This is most simply obtained by noting that the total reaction within the sphere equals the diffusion into the sphere from outside, and that hence

$$\eta = \frac{4\pi R_p^2 D' \left(\frac{\partial c}{\partial r}\right)_{r=R_p}}{\frac{4}{3}\pi R_p^3 k' c_s}. \tag{38}$$

Using equation (37), we obtain

$$\eta = \frac{3}{\Phi}\left(\coth \Phi - \frac{1}{\Phi}\right). \tag{39}$$

The same result can, of course, be obtained by integrating the rate of reaction over the whole volume of the pellet, using equation (37).

For small values of Φ, equation (39) becomes

$$\eta = 1 - \Phi^2/15 \tag{40}$$

while for large values of Φ

$$\eta \rightarrow 3/\Phi \tag{41}$$

Before drawing further conclusions from these results, it is useful to make the following comments:

i) If the heat of reaction gives rise to significant temperature differences, these can seriously affect the results. Such phenomena are discussed in the next section.

ii) If the pellet is not spherical, the above treatment is not applicable in detail. Certain simple geometries — such as the straight pore, the planar slab, or the cylinder with sealed ends — can be analytically treated. For these cases, if we choose to define the Thiele modulus, Φ, in terms of a characteristic length L given by

$$L = \frac{\textit{volume of particle}}{\textit{external area for diffusion of particle}}$$

then for all the above geometries (and one suggests for any regular-shaped particles) the plots of η against Φ are very closely the same.

Figure 5 shows this "common" plot. If we choose L as defined above, then for a slab L = half the thickness, for a cylinder with sealed ends L = half the radius, and for a sphere L = one third of the radius. Using L instead of R_p in the definition of Φ, in all three cases

$$\eta \to 1/\Phi \qquad \text{for large } \Phi \tag{42}$$

which is to be compared with equation (41) for the sphere. Equation (39), again for the sphere, becomes

$$\eta = \frac{1}{\Phi}\left(\coth 3\Phi - \frac{1}{3\Phi}\right) \tag{43}$$

when L is used in the definition of Φ.

For values of Φ between about 0.4 and 2, the plots of η for different geometries are different, but only by a few percent. For values of Φ below 0.4 or greater than 2, Figure 5 gives an accurate picture for all geometries.

iii) If the reaction is not first-other, Φ depends on c, and equation (36) is not so readily solved. For simple orders and geometries, solution is possible, but Aris further shows (Ref. 3 p. 136–138) that if the Thiele modulus Φ is defined somewhat differently, then, whatever the kinetics, the asymptotic

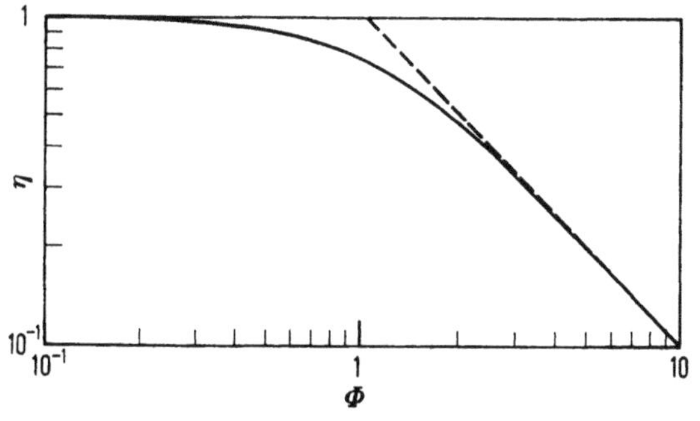

Figure 5. "Common" plot of Effectiveness Factor, η, vs. Thiele Modulus, Φ, for particles of different shapes.

form of η becomes equation (42). Thus we can say that Figure 5 is a useful plot for any kinetics or geometry, provided that Φ is appropriately defined with respect to the kinetics and characteristic length.

We can now consider the uses to which Figure 5 can be put. If Φ is known, the effectiveness factor is given by Figure 5, and this will aid the designer of a catalytic reactor, by indicating how much more catalyst will be required for a desired extent of reaction. To calculate Φ, we need to know L, k', and D'. The size and shape of the pellets will give L; the other two quantities require comment.

The velocity constant k' in equations (35) and (36) is a volume-based one. It is more informative on physical grounds to use a k based on the reaction rate per unit area of catalytically active surface. Firstly, the value of k would then be directly comparable with that used in the definition of the Damköhler number, equation (31), and a correct appreciation of the relative importance of internal and external mass-transfer resistance will then follow. Secondly, the relation between equations (34) and (35) becomes clearer. Thirdly, comparison between catalyst pellets with different active areas per unit volume becomes more meaningful. The relationship between k' and k is

$$k' = a'k \tag{44}$$

where a' is the active area per unit volume of pellet. It has units of m^{-1}, and is analogous to the quantity a in mass-transfer terminology, the interfacial area per unit volume. It is equal to $\varrho_b S_g$ in Aris [3].

If we do not know k, and are using a catalytic reactor to try and determine it, the arguments can become circular unless care is used. The product $\eta\Phi^2$ is calculable from quantities which can be measured, or estimated a priori. As an example, for the first-order case,

$$\Phi^2 = L^2k'/D' \tag{45}$$

and the observed reaction rate per unit volume of pellet, $\dot{N}$, is given by

$$\dot{N} = \eta k' c_s \tag{46}$$

so that

$$\eta\Phi^2 = L^2\dot{N}/D'c_s\,. \tag{47}$$

Thus if $\dot{N}$ is measured, and L and c_s are known, we can obtain $\eta\Phi^2$ (provided D' can be estimated — see later). From equation (43) we obtain, for first-order reaction in a spherical pellet,

$$\eta\Phi^2 = \left(\Phi \coth 3\Phi - \frac{1}{3}\right). \tag{48}$$

Thus a value of $\eta\Phi^2$ will give Φ, and hence η from equation (43). A plot of η against $\eta\Phi^2$ is given in Figure 6. It is somewhat similar in shape to Figure 5.

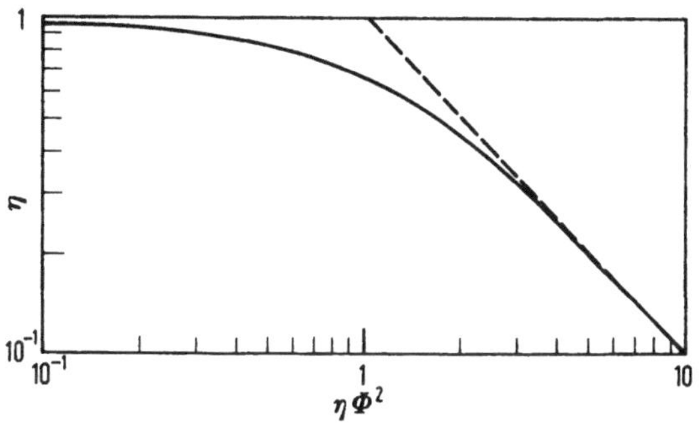

Figure 6. Effectiveness Factor η, vs. the calculable product, $\eta\Phi^2$

In the limit of large Φ, equation (48) shows that $\eta\Phi^2 \rightarrow \Phi$, and since $\eta \rightarrow 1/\Phi$, we have $\eta \rightarrow 1/\eta\Phi^2$. As was the case with Figure 5, Figure 6 should be adequate for any kinetics or geometry (for isothermal conditions).

If mass-transfer limitations are severe, and are not taken into account, then a fallacious order of reaction and velocity constant will be derived from a 'kinetics' experiment. Equation (47) shows that measurement of $\dot{N}$ gives $\eta\Phi^2 c_s D'/L^2$. For severe mass-transfer limitation, η tends to $1/\Phi$, and hence $\dot{N}$ is proportional to Φ. Substituting for Φ from equation (36) we obtain

$$\dot{N} = \frac{\sqrt{k'D'}}{L} \cdot c_s^{(n+1)/2} . \qquad (49)$$

Thus apart from the magnitude of any derived velocity constant being affected by the values of D' and L, even the order of the reaction will seem to be different from its true value — unless n happens to be unity, in which case a truly first-order reaction will also seem to be first-order. However, for all reactions $\dot{N}$ will be proportional to $\sqrt{k'}$, which means that an Arrhenius plot will give an energy of activation equal to half the true value.

If experiments are carried out with particles of similar shape but different size, measurement of $\dot{N}$ can be useful in clarifying the position. It is a large assumption that making pellets of different size does not change any other property, but let us make that assumption. If the size ratio of the particles is denoted by S_R, then we can deduce from the above results the following:

i) If $\dot{N}$, the measured rate per unit volume of catalyst material, is independent of S_R, then the values of Φ in both cases are small, mass transfer limitation is negligible and the results are truly kinetic in nature.

ii) If $\dot{N}$ is proportional to S_R^{-1} (i.e. the larger particles give a lower value of $\dot{N}$), then mass transfer limitation is severe in both cases. The values of Φ are both large, in the ratio S_R, but we cannot say what they are.

iii) If $\dot{N}$ varies with S_R in a manner between i) and ii) then Φ for one size of particle (or both) is such that $\eta\Phi^2$ lies in the transition part of Figure 6. Then use of equations (45), (47) and (48) enables both values of Φ to be obtained, as well as the desired value of k'. A similar procedure can be adopted if the reaction is not first-order.

Lastly, in this section, we must consider the quantity D', which was defined after equation (35) as the apparent, or volume-based, diffusion coefficient of reagent in the porous medium. Estimation of this quantity is not simple; one can refer to Satterfield's book [8], written in 1969, and to papers in the Chemical Reaction Engineering Symposium, in 1980 [9], which show that doubts remain in areas which one might hope to be comparatively simple. The problems arise from a combination of physics and geometry. The three types of diffusion generally considered are

i) Surface diffusion. Transfer occurs by diffusion within a 2-dimensional adsorbed phase on the surface of the pellet material. This is not clearly understood.

ii) Knudsen diffusion. At low pressures, or in small-diameter pores, the diffusion of any species becomes unaffected by the pressure, or diffusion, of any other species.

iii) At higher pressures, or in larger pores — which one might expect to be encountered in practice — diffusion is similar to that in the bulk fluid phase subject to two geometrical complications. These are

a) If we draw a plane normal to the gradient of concentration of the diffusing species, only a fraction of this plane (equal to the void fraction, or volume-fraction of fluid phase, ε_f) is available for diffusion. Note that the concentration of reagent is a fluid-phase concentration and not a concentration based on total pellet volume.

b) The path for diffusion is tortuous, since the reagent has to diffuse around particles within the pellet. Thus to get to the center will require a longer path than the radius, longer by a "tortuosity factor" τ.

As a result of these two factors, the diffusion coefficient D' will be given by

$$D' = D\varepsilon_f/\tau\,, \tag{50}$$

and will be considerably less than D. The voidage, ε_f is measurable, but τ is only calculable for simple geometries, and real porous media are not that simple.

If the reaction velocity constant is known, from experiments which are not diffusion-limited, then measurement of effectiveness factors from experiments which are diffusion-limited may enable D' to be calculated. But it is a great assumption that other factors assumed to be the same in the two sets of experiments really are the same.

It may be possible to estimate, or measure, D' from data derived from non-reacting systems, such as steady-state diffusion experiments, e.g. measuring the flux through a single pellet. However, such measurements will not be affected by dead-end, closed pores, but if these pores are catalytically active in a reacting system, they will contribute to the reaction rate, and the diffusion within them may be less rapid than that obtained in the larger channels passing right "through" the particle.

Thus the present position seems unsatisfactory, in that detailed predictions, or even post-hoc explanation, cannot be made with confidence. It should, however, be clear from the earlier part of this section that mass-transfer can be of dominant importance in the behavior of catalytic reactors. The procedure for describing the importance of mass-transfer is clear, even if the parameters cannot be precisely estimated beforehand.

The whole of the above section required isothermal conditions. We have now to consider temperature effects.

4. Heat Transfer Within and From a Pellet

In the previous sections we have considered mass-transfer within and to a pellet, and it was shown how the concentration gradients involved led to a lowering of the effectiveness factor. If the reaction rate is sufficiently large for mass-transfer to find it 'difficult to keep up', and the reaction also evolves heat, heat transfer to remove the heat from the pellet may similarly require significant temperature gradients. It is true that *endo*thermic reactions occur, but catalytic reactions of practical interest are usually *exo*thermic, and it is therefore heat transfer *from* a pellet which will concern us most.

In the discussions of chemical reactor design, particularly in part 3d of Section IIA, the effect of temperature changes was pointed out; the primary consideration is the effect on the velocity constant, k, which usually increases very rapidly with temperature.

It is clear from the above that the center of a catalyst pellet may be significantly hotter than its surface, and this latter temperature may be noticeably higher than that of the bulk fluid passing the pellet. We have to consider firstly how much higher these temperatures may be, and secondly whether such temperatures affect the rate significantly.

If we consider diffusion and heat transfer in a catalyst slab (this geometry simplifies the treatment but in no way affects the conclusion), the balance equations applied to an element give, in the steady state

$$D' \frac{\partial^2 c}{\partial x^2} = r_c \tag{51}$$

and

$$\lambda' \frac{\partial^2 T}{\partial x^2} = -r_c(-\Delta H)\,. \tag{52}$$

Here r_c is, as before, the rate of reaction, λ' is the thermal conductivity, and $(-\Delta H)$ is the heat of reaction, being positive for an exothermic reaction, for which ΔH is negative.

The boundary conditions for these equations are similar. By symmetry, at the center-plane of the slab,

$$\frac{\partial c}{\partial x} = \frac{\partial T}{\partial x} = 0 \quad \text{at} \quad x = 0 \tag{53}$$

while at the surface (the slab is of thickness $2L$) we have

$$c = c_s\,, \qquad T = T_s \quad \text{at} \quad x = L\,. \tag{54}$$

Equations (51) and (52) can be added to give

$$D' \frac{\partial^2 c}{\partial x^2} + \frac{\lambda'}{(-\Delta H)} \frac{\partial^2 T}{\partial x^2} = 0\,. \tag{55}$$

This can be integrated twice, using equations (53) and (54) to give

$$T = T_s + \frac{D'(-\Delta H)}{\lambda'}(c_s - c)\,. \tag{56}$$

This remarkably simple relationship, valid for any kinetics or geometry, was first established by Prater [10], and the temperature, or temperature rise, is in consequence called the Prater temperature. The maximum value the Prater temperature can have is given by $c = 0$, which will occur towards the center of a catalyst particle severely limited by diffusion.

Equation (56) is a relation between temperature and concentration, and can be used in equation (51) to eliminate the temperature-dependence of k' in the reaction rate term r_c. This results in a highly non-linear equation between c and x, which has to be solved numerically. This was done by Weisz and Hicks [11] for the case of first-order kinetics in a spherical pellet. Their paper remains an excellent account of the factors involved, though many other treatments of more complicated problems have since appeared.

Equation (51) can be cast into dimensionless form, and the results are then seen to depend upon the following dimensionless groups. Firstly, there is

$$\beta = \frac{D'(-\Delta H)\,c_s}{\lambda' T_s} \tag{57}$$

which is readily from equation (56) to be the dimensionless Prater temperature rise.

The second group is

$$\gamma = E_A/RT_s \tag{58}$$

which is the dimensionless Arrhenius activation energy. If γ is small, the effect of temperature change is small, and if β is small, the temperature change is itself small. It is not surprising to find, as Aris shows [3], that the product $\beta\gamma$ can be used, as a single parameter, to describe the results with fair accuracy.

The third group is, as before,

$$\Phi = L\sqrt{k' c_s^{n-1}/D'} \tag{59}$$

about which three things can be said. Firstly, we choose $L = R/3$, for reasons given earlier; this means that values of Φ in Weisz and Hicks' paper [11] should be divided by three for comparison here. Secondly, $n = 1$ is the case dealt with by Weisz and Hicks; and thirdly, Φ is based on surface conditions, c_s as well as T_s, which gives k'.

Weisz and Hicks' results can be illustrated by the example shown in Figure 7. Three main conclusions can be drawn:

(i) If β is negative, i.e. the reaction is endothermic, η is lower than the isothermal value, at the same Φ.

(ii) If β is positive, i.e. the reaction is exothermic, η is higher than the isothermal value, and may be higher than one. This means that the overall rate of reaction is greater than would have been expected at T_s and c_s. The decrease in c as we move towards the pellet center is more than made up for by the increase in k' resulting from the increase of temperature.

(iii) If β is high enough, there may be more than one value of η at a given value of Φ, though the range of Φ showing this effect may be small. We shall not discuss this further here, though it will be mentioned in the concluding section.

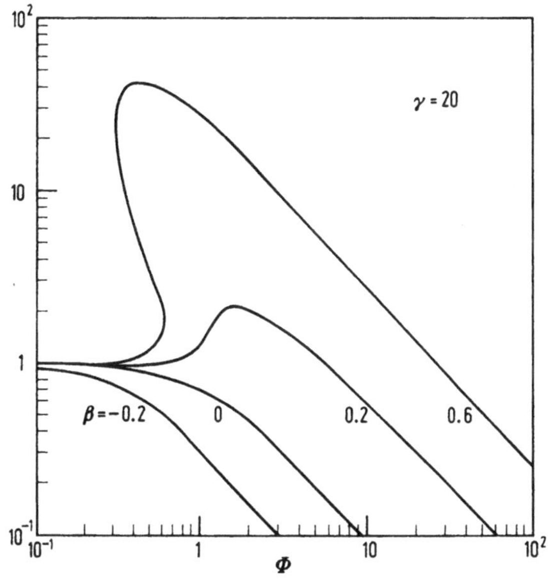

Figure 7. Effectiveness Factor, η, vs. Thiele Modulus, Φ, for Various Values of β (adapted from Weisz and Hicks [11]).

Weisz and Hicks give criteria for the absence of significant thermal, or mass-transfer effects. For the non-isothermal situation these are

$$\Phi \leqq \frac{1}{3} \exp\left(-\beta\gamma/2(1 + \beta)\right) \tag{60}$$

or, in terms of the observable $\eta\Phi^2$,

$$\eta\Phi^2 \leqq \frac{1}{10} \exp\left[-\beta\gamma/(1 + \beta)\right]. \tag{61}$$

We note that if $\beta = 0$, equations (60) and (61) give the isothermal criteria for insignificant mass transfer effects. We also note that since β is usually <1, these criteria are primarily affected by the value of $\beta\gamma$, which is a consequence of the rough dependence of the results on the product $\beta\gamma$.

These considerations lead us to examine the values of β and γ, and to see what their maximum values are likely to be.

For gas-solid reactions $D'c_s$ is roughly constant with pressure, except at low pressures, when this product will fall. D' will be somewhat less than D, by virtue of the voidage and tortuosity factors described earlier. The thermal conductivity λ' is that of the porous solid, and though reagent cannot diffuse through the solid matrix, heat *can* be conducted through it. Putting an upper limit of 300 kJ mol^{-1} for $(-\Delta H)$, and substituting values for λ' and for $D'c_s$, Weisz and Hicks arrive at the conclusion that β_{max} will be about unity, and β will usually be somewhat smaller. For liquid-solid reactions λ' is likely to be somewhat larger and D' some orders of magnitude smaller, so that β will be too small to give significant thermal effects.

The value of γ might well lie between 10 and 100, and the value of $\beta\gamma$ might be as high as 10 for a gas-solid reaction, being much less for a liquid-solid reaction.

If the kinetics are not simple but, for example, follow Langmuir-Hinshelwood form with coverages which are not small, then the effectiveness factor is not so easily estimated. Hutchings and Carberry [12] treat such a case, which also involves competing adsorption by the product.

Finally, in this section, we must consider the possibility of heat-transfer resistance between the pellet surface and the bulk fluid flowing past it. This is discussed by Carberry [4], and by Hutchings and Carberry [12]. The effectiveness factor η is related to conditions, c_s and T_s, at the pellet surface. If heat transfer from the pellet to bulk fluid is poor, the pellet, though perhaps at acceptably constant temperature, equal to T_s, throughout its volume, may be, as a whole, at a significantly higher temperature than the bulk fluid.

At a given point in a packed-bed reactor the bulk fluid values of c_b and T_b may be known, or even calculable from the total reaction up to that point. The surface concentration c_s will be lower than c_b, so η based on c_b will be lower than η based on c_s. But T_s can well be higher than T_b, and sufficiently higher for η, given by the ratio of reaction obtained to that at T_b, to be yet higher than η based on reaction at T_s. In fact, if heat transfer exterior to the pellet is sufficiently poor then this could be a very large effect. The existence of a glowing particle of coal in a much cooler fluidised bed indicates that the 'effectiveness factor', based on T_b, is much higher than unity, even though the particle might internally be virtually isothermal.

For reasons similar to those applying to internal heat transfer, external heat-transfer effects are unlikely to be serious for liquid-solid reactions.

This is an introduction to the field, and it is thus appropriate to stop at this point. However, it should be clear that the behavior of exothermic catalytic chemical reactors can well be grossly affected by heat-transfer effects. We are forced to conclude, yet again, that a knowledge of the catalytic chemistry is not enough to predict reactor behavior. It is of course important, but it will not suffice.

4. Some Consequences for the Design of Catalytic Reactors

A. Catalyst Formulation

When a particular solid surface is found to act as a catalyst for a desired reaction, it becomes necessary to consider how this catalytic activity can be effectively harnessed for the required purpose at practical production rates and acceptable cost.

The rate per unit area of catalytically-active material may be small. Thus it becomes necessary to produce high surface areas of catalyst per unit weight of the material. Very small particle sizes are thus indicated, but consideration of pressure drop (pumping costs) rules out the use of beds of such particles.

Thus pellets of material which are porous are generally used. The pellets are large enough (say 5 mm in dimension) for flow through a bed of them to be possible at acceptable pressure drop. They are porous, to provide the large surface area per unit volume.

Even so, the catalytically-active material may be so expensive that porous pellets of this material alone are economically unacceptable. In such a case (e.g. platinum metal) the catalyst may be finely dispersed on the surface of the pores of pellets of a porous support material, such as silica, or alumina, or a blend. There are reactions for which silica/alumina is itself a catalyst, but in other cases the support is chemically inert.

A pelleted catalyst must satisfy two physical requirements. It must be strong enough for the pellets at the bottom of a bed not to collapse under the bed's weight; and the physical strength must not be weakened by the leaching effect of the fluid. Some years ago the author was involved with a tungstic-oxide catalyst used in the manufacture of isopropanol. Addition of poly-tetra-fluorethylene as a binder in the pelleting process was found to improve the physical life of the catalyst, which otherwise crumbled in service to fine dust.

The large surface area of an active catalyst must not be rendered ineffective by the mass-and-heat-transfer effects described earlier. Thus close control of the size of the support particles, and of their pelleting, is demanded.

Promoters may be added. These are substances which, while not catalytically active themselves, enhance the activity of the catalyst. They may act by retaining the original activity of the catalyst, for example by reducing the rate of sintering of small finely-divided metal crystals into larger crystals, with a consequent reduction in catalytically-active area.

The catalyst material may be in an inert from when charged to the reactor, and 'activation' will then be required in situ. This may be done by, e.g., steam treatment, or reduction by hydrogen may be required. In some cases the reagent stream may itself activate the catalyst by changing its chemical nature. This is the case with some metal catalysts which are charged as oxides and are reduced to active metal by the feed stream. Extreme care on shutdown may then be necessary, since highly-activated metal catalysts can be pyrophoric in air.

In practice, cost factors are paramount. One looks for the minimum usage of expensive catalyst, and a maximum period of its effective utilization. Catalytic activity is frequently a matter of comparative rates of desired and undesired reactions (rather than between desired reaction and none at all). Thus catalyst development may lead to higher selectivity, and more effective use of feed material, even at the cost of lower rate. It is often not too serious, economically, to overdesign a reactor in terms of its size. If the reactor makes the wrong product, however, that can destroy the viability of the whole process.

Thus it can be seen that the establishment of the chemical activity of a catalyst is only the first step in bringing about its use in a manufacturing process.

B. Catalyst Testing in the Laboratory

In the previous section we saw that the useful application of a catalytically-active material ordinarily requires that this material be formed into porous pellets. For design purposes there *must* be experimental data on the performance of the catalyst in the form in which it is actually going to be used. These data have to be obtained on the laboratory scale, for use in the design of full-scale plant. Even on the laboratory scale such experimentation is expensive, and time-consuming, but it is important that reduction in the scale of operation, to reduce cost, should not lead to results which cannot be relied upon for full scale design. It is particularly useful to set up a proper experimental facility for catalyst evaluation when it is envisaged that many variants on the formulation (e.g. promoters) or activation (e.g. pretreatment by steam) of the catalyst will be examined.

It is essential that the catalyst particles tested should be those actually to be used. It is indeed undesirable to use full-size catalyst pellets in small-bore laboratory tubular reactors from the point of view of deviations from plug flow behavior, but these defects are minor compared with the misleading results which can arise if experiments are done with catalyst particles of a smaller size than, or formulated differently from, those that will be used in practice. This is because the mass-and-heat transfer effects can be very different for particles of different size or porosity.

In a plug-flow, packed-bed reactor the changes of composition and of temperature from entry to exit can make the understanding of the *course* of a reaction difficult to obtain from measurements on the *product* only. The effectiveness factor can vary widely throughout the reactor, so the procedure leading to equation (21) may be inapplicable.

If the catalyst deactivates, or ages, at a significant rate, this further complicates the position. The obtaining of chemical-kinetic information from reactor measurements is discussed by Weekman [13], who categorises the advantages and disadvantages of many different laboratory reactors. We shall describe here only one type of reactor which has been found to be useful. We shall assume that catalyst ageing is not rapid, though in fact this type of reactor is better suited than most to the investigation of rapidly-deactivating catalysts.

A useful characteristic of a CSTR reactor is that conditions are uniform throughout the reactor, and thus its overall performance relates to a particular set of reaction conditions (concentration, temperature). By contrast, the product from a PFR is the result of reaction occurring over a range of conditions from entry to exit. The problem for a laboratory reactor is how to obtain uniform conditions for all the catalyst particles in the reactor. A possibility, for small catalyst particles, might be a fluidised bed, but for pellets of a practical size there are obvious difficulties.

A packed-bed reactor with a high recycle rate approximates to a CSTR. The amount of reaction per pass may be so slight that reaction conditions are effectively constant throughout the bed. The flow rate is high, so the amount of reaction per unit time may be considerable, and can be accurately determined from the known, small, flowrates of fresh feed, and of product, and the difference of concentration between them. This difference can be much larger than the difference in concentration across the reactor.

Two means of bringing this about are the spinning-basket reactor described by Carberry [14], also used in liquid-solid ion-exchange systems by Turner and co-workers [15], and the fixed-bed pumped-recycle reactor described by Bennett, Cutlip and Yang [16]. In the former variants, the pellets are contained in a basket stirrer which is used to stir the fluid violently. It is essential that the "blades" of the stirrer should allow easy passage of the fluid through them, and that the pellets should not mask each other from the fluid. Because the fluid flow through the bed of particles comprising the stirrer may be less adequate than at first sight (and thus there may be significant differences in the fluid properties seen by different particles), the latter type of reactor would seem preferable, if perhaps more complicated to construct. In this design [16], the catalyst pellets are contained in a fixed bed through which the fluid is pumped at a high (recycle) rate, the catalyst bed and circulating pump being contained together within a vessel in which the desired conditions of temperature and pressure can be maintained.

From reactors such as the above the "kinetics" of the catalysed reaction can be obtained for point values of reactor conditions. These "kinetics" will be affected, of course, by the mass and heat-transfer effects appropriate for these conditions. The behavior of a full size fixed-bed reactor can then be predicted by an "integration" procedure, allowing for the changes in conditions as the fluid passes through the reactor. Kinetics and transfer effects *within* the particles will have been properly obtained from the laboratory reactor, but transfer effects *to* the particles from the flowing stream will not have been accurately matched, since the circulation within the laboratory reactor is probably at a much higher Reynolds number than would be the case in the full scale reactor. However, such transfer effects external to the particles are often of small effect, as seen in Section 3 above, though with extremely exothermic reactions care has to be exercised in using such laboratory reactor data for fullscale design.

C. Multiple States and Limit Cycles

It is shown in text-books on chemical reaction engineering [1–4] that the product stream from an adiabatic chemical reactor supplied with a defined feed may not be uniquely determined by that feed. For example, with an exothermic reaction two stable aternatives may be possible, a low reaction case giving a low temperature product, and a high reaction case giving a high temperature product. Transfer from the former to the latter can be regarded as an 'ignition' phenomenon, and can occur as the result of quite small changes in the conditions. Obviously systems of this kind pose problems in design, and also in control, particularly if one of the states gives rise to hazards.

For exothermic catalytic reactions, we saw in Figure 7, section 3B4, that the effectiveness factor, η, may, for a limited range of the Thiele modulus, Φ, have more than one value. In the case shown the middle value is unstable, and that value of the other two which is obtained will depend upon whether the reactor conditions have been approached from the left or right direction on the diagram.

The above are cases of alternative, but steady, states in which the reactor may stably operate. It is also possible, though rather unusual, for a reactor to oscillate in what could be called a 'steady' fashion. In such cases the composition and temperature may vary together round a 'limit cycle' which can be approached from start-up conditions lying either inside or outside the limit cycle.

There is a considerable literature on this topic, which has an attraction for the more mathematically-minded chemical engineers. Reports of experimental work are fewer by comparison, but increasing in number. As examples, Bush [17] described temperature and compositions oscillations in the gas-phase reaction of chlorine and methyl chloride, and more recently Heemskerk et al. [18] have obtained oscillations in a liquid-phase acid-catalysed hydrolysis. Cutlip and Kenney [19] observed multiple steady states of the carbon monoxide-oxygen reaction over a supported platinum catalyst, and the addition of 1-butene to the reaction mixture produced oscillatory behavior. These limit cycles had periods from about one minute to thirty minutes depending on conditions.

Oscillatory behavior is not only interesting in itself, but depends upon the details of the coupled reaction scheme, and thus can be used to investigate these details. This can be valuable in the elucidation of the kinetics, even if the designer wishing to make use of the reaction will naturally take care to choose operating conditions leading to a controllable steady state of his plant.

D. Catalyst Deactivation

The activity of a catalyst changes with time. Contact with the process fluid may initially cause the catalyst to become more active, though it would be more useful to say that this process of activation of the catalyst is part of its preparation. Whether or not this activation takes place, the catalyst activity will show a decline thereafter, and this deactivation may take place over times measured in seconds or in years. It is appropriate to consider here some of the causes of such deactivation (or 'ageing'), as well as its effect on the performance of the reactor. Finally a brief description will be given of how the operation of the reactor may be altered so as best to take account of this deactivation.

A common cause of deactivation is called "poisoning". The presence in the feed stream of small quantities of certain substances can cause the activity to fall markedly. This poisoning is due to strong chemisorption of these substances on the active catalyst sites, thereby blocking them for reaction. If the poisoning substance is removed from the feed, activity may recover as the poison is removed from the surface by reaction with the feed, or by slow desorption. Such poisoning is called reversible, and may be tolerated in operation if it occurs only infrequently after maloperation of feed purification. If the activity does not recover in this way, the poisoning is said to be irreversible. Here again it may be possible to restore activity by treatment for a while with another feed (e.g. hydrogen to reduce an oxide poison, or steam reactivation of alumina). If this fails, and the catalyst stays 'dead', it will have to be replaced. The author recalls the case of a platinum reforming catalyst which after operating satisfactorily for about two years lost most of its activity over a period of a week. Restorative measures of increasingly severe types were tried, to no avail, and the catalyst had to be replaced at great cost, even after allowing credit for the recoverable platinum. The cause of this failure was never precisely determined; it should be borne in mind that the number of poison molecules required to block the active sites of the catalyst is not great, and that 1 part per million of strongly absorbed poison in a process stream can rapidly kill a catalyst. The catalyst in question had in fact performed satisfactorily for over twice its design life, so its eventual demise could not be classified as a disaster.

Fouling is rather different from poisoning, in that comparatively large amounts of material are laid down on and in the catalyst, blocking pores, and physically covering up active sites. Fouling may arise from material deposited physically (e.g. water impurities carried over in process steam), but a much more frequent cause is undesired chemical reaction producing the foulant. A common example is coking, the laying down of carbonaceous material formed by undesired cracking-type reactions. This can occur very rapidly with cracking catalysts, not surprisingly, with activity declining in times of the order of a second. It was this rapid fall-off in activity which led to the development of fluidised-bed catalytic crackers, in which the coke formed in the reactor is burnt off the catalyst in an adjacent regenerator.

A less severe example of coking occurs in the isomerisation of xylenes (in the manufacture of p-xylene). As well as the isomerisation, disproportionation reactions (e.g. to toluene and trimethylbenzene) occur. These lead to lower-value products, and are undesirable, but they do not lead to fouling. However a much slower cracking-type reaction leads to coking, with activity declining over a period of weeks. Here the fixed-bed catalyst is regenerated in situ, with a second reactor switched into the process.

If the catalyst is finely-divided on a support material the conditions of the desired reaction may cause sintering of these fine particles, with a loss of surface area and a fall in activity. This does not occur as the result of impurity in the feed, and such physical transformations are irreversible — the catalyst cannot be regenerated. Remedy must be sought in the catalyst formulation. The addition of "promoters" can prevent migration of catalyst molecules, thus stopping the sintering process.

The decline in catalyst activity can itself be modelled as a chemical reaction, and an introduction to this is given by Levenspiel [2]. Such 'reactions' may be autocatalytic; coking often occurs more rapidly on laid-down coke than on clean catalyst. Whatever mechanistic interpretation may be put on such models, they can provide a quantitative description of the behavior of a catalyst during its life time.

If a catalyst's activity declines with time, this may be compensated for by an increase in reaction temperature — the lower active area being balanced by a higher velocity constant. However it is usually the case that undesired reactions, and also the deactivation reaction itself, are preferentially favoured by increased temperature (or else the higher temperature might have been chosen to start with). So the optimal program of reactor conditions over a life-cycle of catalyst may be complicated to work out — it is often preferable to start with a temperature rather lower than might seem best, so as to prolong the life of the catalyst. The subject is one made for mathematical modelling, but one suspects that plant managers have a volume of experience to inform their practice, and that for them modelling comes *after* the practice.

Symbols

a'	Active area per unit volume of catalyst	m^{-1}
c	Concentration	$mol\ m^{-3}$
C	Specific heat	$J\ kg^{-1}\ K^{-1}$
D	Diffusion coefficient	$m^2\ s^{-1}$
$E(t)$	Residence-time distribution function — see II.B.2	
E_A	Energy of activation	$J\ mol^{-1}$
k	Velocity constant (depends on reaction order)	
k_g	Gas-film mass transfer coefficient	$m\ s^{-1}$
l	Length of pore	m
L	Characteristic particle size	m
m_f	Molar flowrate of feed	$mol\ s^{-1}$
$\dot{N}$	Reaction rate per unit volume of pellet	$mol\ m^{-3}\ s^{-1}$

p	Pressure	Pa
r	Radial coordinate	m
r_c	Rate of reaction	$mol\ m^{-3}\ s^{-1}$
R_p	Pellet radius	m
R	Gas constant	$J\ mol^{-1}\ K^{-1}$
S_R	Size ratio of particles	
t	Time	s
T	Temperature	K
v	Volume flowrate	$m^3\ s^{-1}$
V	Volume of reactor	m^3
x	Fractional conversion of reagent (section II)	
x	Distance coordinate (section III)	m
β	Dimensionless Prater temperature — see equation (57)	
γ	Dimensionless activation energy — see equation (58)	
$-\Delta H$	Heat of reaction	$J\ mol^{-1}$
ε	Fractional increase in moles due to reaction	
ε_f	Void fraction of porous catalyst	
η	Effectiveness factor	
λ'	Thermal conductivity	$J\ m^{-1}\ s^{-1}\ K^{-1}$
ϱ	Density	$kg\ m^{-3}$
τ	Space time	s
Φ	Thiele modulus — see equation (36)	

References

1. Denbigh, K. G., Turner, J. C. R.: Chemical reactor theory. Cambridge University Press 1971.
2. Levenspiel, O.: Chemical reaction engineering. New York–London–Sydney–Toronto: J. Wiley 1972.
3. Aris, R.: Elementary chemical reactor analysis. Englewood Cliffs, New Jersey: Prentice Hall, Inc. 1969.
4. Carberry, J. J.: Chemical and catalytic reaction engineering. New York: McGraw-Hill 1976.
5. Zwietering, T. N.: Chem. Eng. Sci. *11*, 1 (1959).
6. Mixon, F. O., Carberry, J. J.: ibid. *13*, 30 (1960).
7. Thiele, E. W.: Ind. Eng. Chem. *31*, 916 (1939).
8. Satterfield, C. N.: Mass transfer in heterogeneous catalysis. Cambridge, Mass.: M.I.T. Press 1969.
9. Chemical Reaction Engineering, Nice (Villermaux, J., Trambouze, P., ed.), Chem. Eng. Sci. *35* (1980).
10. Prater, C. D.: Chem. Eng. Sci. *8*, 284 (1958).
11. Weisz, P. B., Hicks, J. S.: ibid. *17*, 265 (1962).
12. Hutchings, J., Carberry, J. J.: A.I.Ch.E.J. *12*, 20 (1966).
13. Weekman, V. W.: ibid. *20*, 833 (1974).
14. Carberry, J. J.: Ind. Eng. Chem. *56*, 39 (1964).
15. Patell, S., Turner, J. C. R.: J. Sep. Proc. Tech., *1* (2), 31 (1980).

16. Bennett, C. O., Cutlip, M. B., Yang, C. C.: Chem. Eng. Sci. *27*, 2255 (1972).
17. Bush, S. F.: Proc. Roy. Soc. A *309*, 1 (1969).
18. Heemskerk, A. H., Dammers, W. R., Fortuin, J. M. H.: Chem. Eng. Sci. *35*, 439 (1980).
19. Cutlip, M. B., Kenney, C. N.: Amer. Chem. Soc. Symp. Ser. No. 65, 1978, pp. 475—486.

Chapter 3

Catalytic Activation of Dinitrogen

A. Ozaki and K. Aika

Research Laboratory of Resources Utilization
Tokyo Institute of Technology
Nagatsuta 4259, Midori-ku, Yokohama, Japan

The fixation of nitrogen is an indispensable process for survival of human being. A large number of studies have been devoted to the catalytic synthesis of ammonia since the discovery of Haber process. Many new thoughts and tools which can be applied to other field of catalysis have been developed through those studies. To learn this reaction is almost equal to learning basic issue on metal catalysis.

This chapter is a comprehensive and critical review of studies on the catalytic activation of dinitrogen, including chemisorption and coordination of dinitrogen, kinetics and mechanism of ammonia synthesis, and chemical and instrumental characterization of active catalysts. The nature of dinitrogen chemisorption is elucidated by reference to nitride formation as well as to instrumental evidences. A full account of well-known iron catalyst, including structural feature and promoter action, is given in the second section in addition to an elemental survey of ammonia synthesis catalysts. The mechanism of ammonia synthesis is demonstrated in terms of kinetics and isotope effect, giving reference to the elementary steps involved in the reaction. The present status of studies is mostly directed to elucidation of the nature of active surface as summarized in the fifth section with respect to surface structure as well as to electronic feature. The last section gives a concise account of homogenous activation including metal-dinitrogen complexes, which will be correlated with or fed back to the heterogeneous activation in future work.

Contents

1. Introduction

The fixation of nitrogen is a very important process for the living things on the earth. The discovery of the catalytic synthesis of ammonia has made it possible to increase enormously world cereal grain production. The extensive use of nitrogenous fertilizer has given rise to a large scale production of ammonia, which has been effective in reducing the ammonia price. Although there exists a large potential demand, the ammonia price by the conventional process is largely controlled by the price of energy, requiring a further improvement in the nitrogen fixation process.

As is well known, the global fixation of nitrogen is largely made by nature, mainly by bacteria. The global consumption of nitrogenous fertilizer was estimated to be 16 percent of the global fixation with 69 percent being biological fixation in the year 1974 [1]. Thus much attention has been directed to the mechanism of fixation by nitrogenase as well as to nitrogen fixation with metal complexes. Athough no catalytic fixation process has been found besides the Haber process, information obtained in the homogeneous activation of dinitrogen should be taken into account in designing a heterogeneous catalysis. One unique finding in the homogeneous system is the formation of hydrazine, although it is not catalytic. The catalytic synthesis of hydrazine is an important target in nitrogen fixation.

This chapter is mainly devoted to the well established heterogeneous catalysis of dinitrogen, while a brief introduction will be made for the homogeneous activation.

2. Adsorption of Dinitrogen

A. Physico-Chemical Properties of Dinitrogen

The high stability of dinitrogen is well represented by the bond energy of 941 kJ mol^{-1} which is the largest among the diatomic molecules. The ionization potential for N_2, 15.58 e.v. (or 1503 kJ mol^{-1}), is close to that for Ar, 15.76 e.v., and larger than that for O_2, the more electronegative molecule [2], demonstrating the high stability of N_2 against oxidation. The electron affinity of N_2 is estimated to be —352 kJ mol^{-1} [3] as compared with that for O_2, 43 kJ mol^{-1} [2]. Although O_2^- has been detected, N_2^- from N_2 has not been detected. The high negative value of the electron affinity of N_2 suggests that the reduction of N_2 needs a strong reducing agent.

In general the bond energy is reduced by converting the triple bond to a double bond. The extent of the decrease is 1/2.25 for N≡N as compared with 1/1.32 for C≡C or 1/1.43 for C≡N [4, 5], suggesting that the most important step to activate dinitrogen is the first reduction of triple bond.

B. Dinitrogen Adsorption as Referred to Nitride Formation

Every metallic element forms a nitride except for noble metals. Elements of Groups IA and IIA form ionic nitrides. Those of Li and alkali earth elements are stable and readily formed by reaction with N_2, whereas the heavier alkali metals do not react with N_2. Nitrides of Groups IB and IIB metals are mostly unstable, while Groups IIIA and IIIB metals form stable covalent nitrides. Nitrides of Group IVB to VIIB, and of Fe, Co and Ni are interstitial compounds [6, 7]. Those of Fe, Co, and Ni are only formed from NH_3 and not from N_2.

Chemisorption of dinitrogen takes place on evaporated metal films at room temperature. Those metals that can chemisorb N_2 at room temperature are; IIA (Ca, Sr and Ba) IVB (Ti, Zr and Hf), VB (V, Nb and Ta), VIB (Cr, Mo and W), VIIB (Re) and VIII (Fe) [8, 9]. Note that they are mostly found in Group IVB through VIII, which are known to form interstitial nitrides.

The much lower ability of other metals to chemisorb dinitrogen seems to come firstly from the difficulty in activating the dinitrogen molecule. Even a copper surface can chemisorb dinitrogen when the copper surface is activated by ion bombardment [10], although copper nitride, Cu_3N, is unstable. Recently chemisorption of dinitrogen was found on reduced cobalt oxide promoted by potassium oxide at room temperature [11] and even on noble metals (Ru, Rh, Os and Ir) promoted with alkali metals (cf. 5. B. 2).

The second reason for an inability for dinitrogen chemisorption may come from an insufficient metal-nitrogen bond energy. The heat of formation of nitride seems to be endothermic with the noble metals as suggested by no nitride formation [12].

The heat of chemisorption of N_2 has been measured over metals. The differential heat of chemisorption generally decreases with an increase in the surface coverage as illustrated in Figure 1 for Fe surfaces [13, 14, 15, 16]. Although the initial heat of chemisorption is much larger on evaporated film than on crystal or supported iron, the value on the evaporated film decreases more steeply, presumably because of a larger degree of disorder in crystallinity.

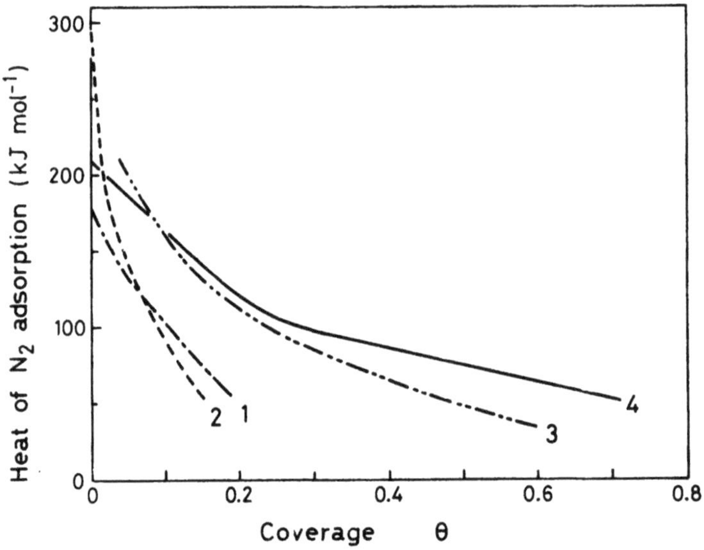

Figure 1. Heat of chemisorption of dinitrogen on iron as a function of surface coverage (1) [13] and (2) [14] on evaporated film, (3) [15] on $Fe-Al_2O_3-K_2O$ and (4) [16] on $Fe-Al_2O_3$

Table 1. The initial heat of chemisorption of dinitrogen over metals

Metal	State	Temperature K	Initial heat of chemisorption kJ mol^{-1}	Method	Ref.
Ta	Evap. film	Room temp.	590	cal[a]	[17]
W	Evap. film	Room temp.	397	cal	[13]
	Powder	673–1023	314	iso[b]	[24]
	Ribbon	1150–1670	485	iso	[25]
	Ribbon	1200–1400	370	iso	[26]
Fe	Evap. film	Room temp.	167	cal	[13]
	Evap. film	Room temp.	293	cal	[14]
	Foil	973–1173	203	kin[c]	[18]
	Single crystal (100)	373– 773	222	kin	[19]
	(111)	373– 773	213	kin	[19]
	(110)	373– 773	205	kin	[20]
	$-Al_2O_3$	300– 773	209	kin	[16]
	$-Al_2O_3-K_2O$	573– 673	213–251	iso	[23]
	$-Al_2O_3-K_2O$	573– 773	209	iso	[15]

[a] calorimetric; [b] isosteric; [c] kinetic

The initial heat of chemisorption of N_2 over metals so far reported are summarized in Table 1 [17]. Most of the values on Fe surfaces are in accord with each other within 220 ± 20 kJ mol^{-1}, irrespective of the methods of measurement, calorimetric [13, 14], isosteric [15, 23], or kinetic [16, 18, 19, 20], while the values on W are rather scattered in a range between 310 and 490 kJ mol^{-1} [13, 24, 25, 26].

Two different approaches have been made to evaluate the heat of chemisorption of gases (A_2) on metals (M). The first one is based on Pauling's equation for single bond energy $E_{M-A,S}$ between a metal atom M and an atom A:

$$E_{M-A,S} = (E_{M-M} + E_{A-A,S})/2 + 23\,(X_M - X_A)^2 \tag{1}$$

where E_{M-M} and $E_{A-A,S}$ are the single bond energies between metal atoms and between A atoms respectively. The heat of dissociative adsorption of an A_2 molecule, Q_{A_2}, may be given by

$$Q_{A_2} = 2\,E_{M-A,S} - E_{A-A,S} \tag{2}$$

as shown by Eley [27] for chemisorption of hydrogen, estimating the electronegativity term $(X_M - X_H)$ from the contact potential of adsorbed hydrogen on the metal and the E_{M-M} from the atomization energy of the metal. The work on the chemisorption of dinitrogen has been less extensive along this line [9, 27, 28].

Recently Miyazaki and Yasumori made a more empirical approach by introducing a parameter δ to give the single bond energy for the surface metal atom M_S [29]:

$$E_{M_s-A,S} = \delta E_{M-A,S} \tag{3}$$

and taking the multiplicity of chemisorption bond into consideration. The heat of dissociative adsorption is given by

$$Q_{A_2} = 2nE_{M_s-A,S} - D_{A_2} \tag{4}$$

where n is the bond order of $M_s - A$ and D_{A_2} is the dissociation energy of the A_2 molecule. From the above equations (1), (3), and (4) it follows:

$$(Q_{A_2} + D_{A_2})/2n = \delta[(E_{M-M} + E_{A-A,S})/2 + 23(X_M - X_A)^2]\,. \tag{5}$$

The plot of the quantity $(Q_{A_2} + D_{A_2})/2n$ against the right side of equation (5) without δ is shown in Figure 2 for the dissociative adsorption of dioxygen and dinitrogen on metals. The values of n were taken as 2 for oxygen and 3 for nitrogen. The term E_{M-M} was evaluated by dividing the sublimation energy of the metals by 4 for bcc metals and 6 for the others. The $E_{A-A,S}$ values were taken as 142 kJ mol^{-1} for oxygen and 159 kJ mol^{-1} for nitrogen. The $(X_M - X_A)$ values were taken from the Pauling's assignment.

The linearity as expected by equation (5) is almost established in Figure 2, giving a value of δ close to 0.7 for both gases, (0.66 for oxygen and 0.71 for nitrogen). Thus the heat of dissociative adsorption of dinitrogen on metals can be evaluated using the value of δ, as shown in Table 2 which lists the calculated and the observed values of initial heat of dinitrogen chemisorption.

The second approach is based on correlations between the initial heat of chemisorption and the heat of formation of corresponding compound as demonstrated for hydrides, nitrides and oxides by Sachtler and Van Reijen [30]. Tanaka and Tamaru showed that the initial heat of chemisorption of gases (H_2, NH_3, N_2, C_2H_4, and O_2) on metals are linearly correlated with the heat of formation of metal oxides per gram atom of metal, $-\Delta H^0_O$, referred to

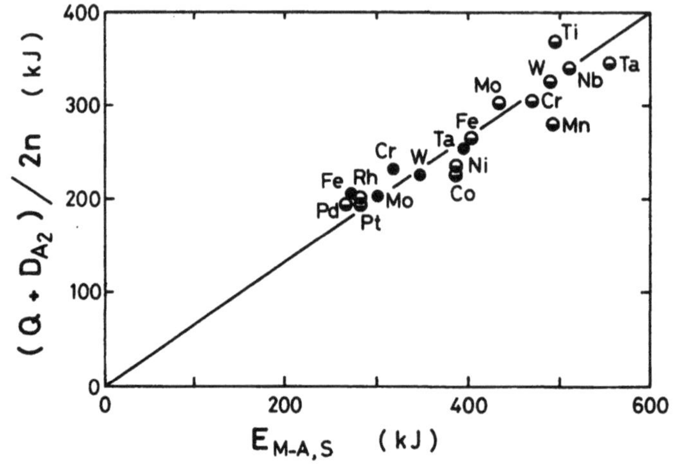

Figure 2.
Plots according to equation 5 for O_2 (◒) and N_2 (●).

Table 2. Observed and calculated heats of dissociative adsorption of dinitrogen[29]

Metals	Heat of adsorption/kJ mol^{-1}		Ref.
	Calculated	Observed	
Ti	481		
V	469		
Cr	410	439	[13]
Mn	465		
Fe	205	293	[14]
		ca. 220	Table 1
Co	134		
Ni	138		
Zr	657		
Nb	582		
Mo	335	264	[21]
Tc	126		
Ru	−117		
Rh	−146		
Pd	−209		
Hf	816		
Ta	732	590	[17]
W	536	397	[13]
Re	167		
Os	−67		
Ir	−109		
Pt	−142		

the highest oxidation state [31]. The initial heat of chemisorption of nitrogen is also correlated with the heat of formation of nitrides per gram atom of nitrogen. Both correlations are shown in Figure 3 from which it is possible to predict unknown values of the heat of dinitrogen chemisorption on metals, although the rather scattered plot, as well as the small number of observed values, gives rise to uncertainty.

It is notable that the heats of dinitrogen chemisorption on noble metals are negative according to equation (5), while they are positive according to the correlation shown in Figure 3. The negative values are refered to the dissociative adsorption of dinitrogen, and are caused by the large value of D_{N_2} as compared with the metal-nitrogen bond energy, whereas the available observed values of initial heat of dinitrogen chemisorption is not necessarily referred to the dissociative adsorption.

If the heat of chemisorption is really negative, the adsorbed state of nitrogen atoms on noble metal should be unstable. Mimeault and Hansen showed that nitrogen atoms (desorbed from a hot tungsten filament (2000 K)) can be adsorbed on iridium or rhodium filment at 300 K, as demonstrated by the desorption of dinitrogen upon flashing [32]. Since contamination of the filament with tungsten vapor was carefully avoided, the nitrogen atoms must be held by the iridium or rhodium surface. However the result does not exclude a possibility that the nitrogen atoms stay on the surface because of a very

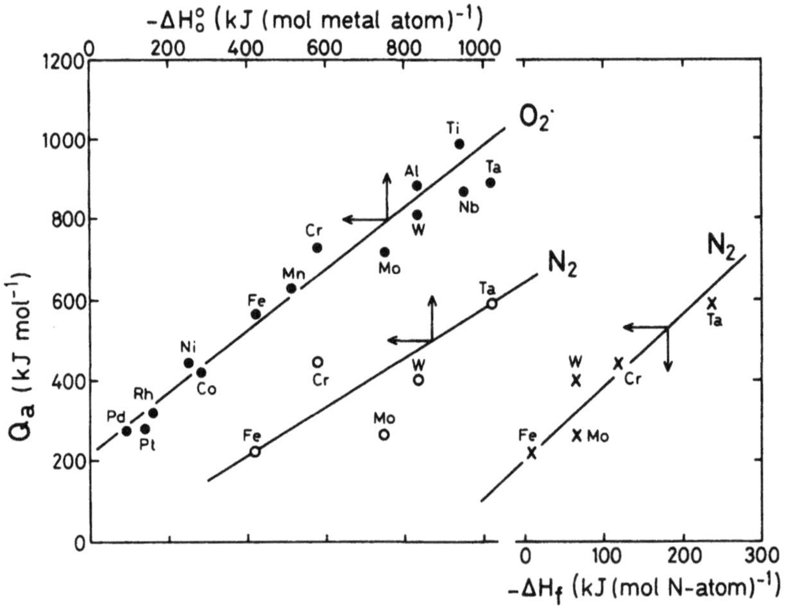

Figure 3. Correlations between the initial heat of chemisorption (Q_a) and the corresponding heat of oxide formation (for O_2 (●) and N_2 (○)) or nitride formation (for N_2 (×))

slow rate of the second order desorption at lower temperatures, or in the form of N_2. The nature of dinitrogen adsorption on noble metals awaits further clarification.

C. Kinetics and Isotherm of N_2 Chemisorption

The chemisorption of N_2 was first observed by Emmett and Brunauer in 1934 on promoted iron catalysts [22]. The chemisorption is in agreement with the Freundlich isotherm

$$\theta_N = K p_{N_2}^{1/n} \tag{6}$$

where θ_N is the coverage by nitrogen, K a constant and $n = 1/6$ on Fe—Al_2O_3—K_2O and Fe—Al_2O_3 at around 673 K. Later a theory of N_2 chemisorption was formulated by Temkin and Pyzhev in 1940 [23, 33], in which they introduced the Frumkin-Slygin isotherm [34]:

$$\theta_N = \left(\frac{1}{f}\right) \ln\,(a_O p_{N_2}) \tag{7}$$

where f and a_0 are constants, and are explained it in terms of the Zeldowitch equation (35):

$$V_a = k_a p_{N_2} \exp(-g\theta_N) \tag{8}$$

for the rate of adsorption V_a and the Langmuir equation (36):

$$V_d = k_d \exp(h\theta_N) \tag{9}$$

for the rate of desorption V_d, where g and h are constants. At the adsorption equilibrium, where $V_a = V_d$, equation is obtained from equations (8) and (9) with $f = g + h$ and $a_0 = k_a/k_d$.

The Temkin-Pyzhev theory was reinterpreted by Brunauer, Love and Keenan [37] in terms of nonuniform surface, assuming a linear dependence of heat of desorption on θ_N:

$$Q = Q_0 - fRT\,\theta_N \,. \tag{10}$$

Even on a nonuniform surface, the coverage θ_s on surface elements ds may be given by the Langmuir isotherm:

$$\theta_s = ap_{N_2}/(1 + ap_{N_2}) \,. \tag{11}$$

Based on the above assumption of linear dependence, equation (11) can be integrated to give θ_N for the total surface:

$$\theta_N = (1/f) \ln [(1 + a_0 p)/(1 + a_0 p \exp(-f))] \,. \tag{12}$$

In the range of adsorption where

$$a_0 p \gg 1 \gg a_0 p \exp(-f)$$

equation (12) reduces to equation (7), the Frumkin-Slygin isotherm. It is obvious that equation (7) is only applicable to intermediate degree of coverages, while equation (12) can cover whole range of coverages.

The above assumption of a linear dependence of heat of adsorption on θ_N corresponds to the assumptions for the activation energies:

$$\text{for adsorption} \quad E_a = E_{a_0} + gRT\,\theta_N \tag{13}$$

$$\text{for desorption} \quad E_d = E_{d_0} - hRT\,\theta_N \tag{14}$$

from which the equations (8) and (9) can be derived respectively for the intermediate range of θ_N.

It is to be noted that the Freundlich isotherm can be derived on the assumption of an exponential decrease in heat of adsorption with coverage [38]. The heat of chemisorption of N_2 given in Figure 1 discloses an exponential decrease with coverage on evaporated iron films, and a rather linear decrease on Fe—Al_2O_3, suggesting a larger degree of surface heterogeneity on evaporated films. This is reasonable considering the lower treatment temperature for the films [13, 14] than for the supported iron which was reduced at around 823 K [15, 16].

Scholten, et al. [16] performed gravimetric measurements of adsorption and desorption rates of N_2 on Fe—Al_2O_3 to examine the validity of the Temkin-Pyzhev theory. From the effect of temperature on the rates, they determined the activation energies, E_a and E_d, for adsorption and desorption respectively as functions of coverage. The results are given in Figure 4 together with those from other sources in which measurements were made manometrically [16, 19, 24, 39]. The heat of adsorption of N_2 on Fe—Al_2O_3 shown in Figure 1 is the value estimated by Scholten et al. from the relation:

$$Q = E_d - E_a \,. \tag{15}$$

Figure 4 shows that both E_a and E_d change linearly with an increase in coverage as expected by the Temkin theory, while E_a tends to reach a plateau value above $\theta = 0.2$, suggesting that the iron surface is nearly homogeneous with respect to the adsorption of N_2 in the high coverage region. For such a region the rates of adsorption and desorption can be expressed in terms of Langmuir model i.e.

$$V_a = k_a p_{N_2}(1 - \theta_N)^2 \,, \tag{16}$$

$$V_d = k_d \theta_N^2 \,. \tag{17}$$

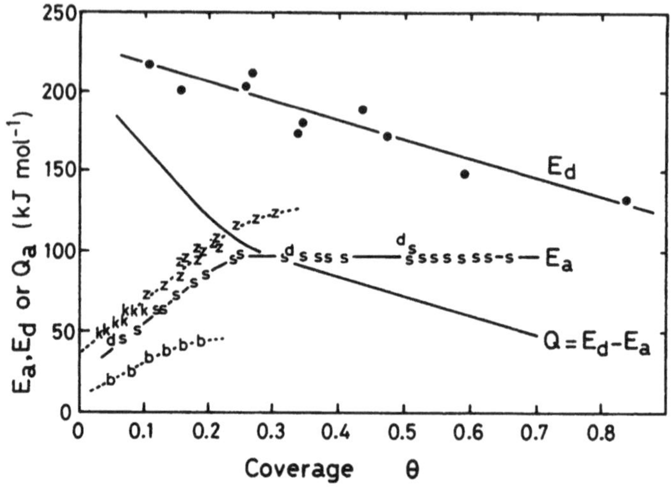

Figure 4. Activation energies of adsorption (E_a) and desorption (E_d) of N_2 on iron catalysts as function of surface coverage.
Fe powder: d [24], Fe—Al_2O_3: ● [16], s [16] and z [39], Fe—Al_2O_3—K_2O: k [15], Fe (100): b [19]

In fact the steady rate of ammonia synthesis or isotopic equilibration can be expressed by equation (16) (cf. 4.E.2).

Another feature to be noted is the large difference in E_a between clean single crystal surfaces and promoted iron catalysts. The values on Fe (100) [19] are smaller than on promoted iron by about 21 kJ mol^{-1} or more, and those on Fe (111) [19] are very small (varies between 0.4 and 4.2 kJ mol^{-1}), while about 29 kJ mol^{-1} at low coverages on Fe (110) [20]. This lower activation energy made it possible to observe the adsorption at room temperature or lower [19]. The higher value on promoted iron is presumably due to an oxygen contamination as shown with polycrystalline iron [96].

D. Absorption of N_2 by Metals

As described earlier iron is transformed into nitride by treating with NH_3 at above 573 K [41, 42], while the nitride readily decomposes on removal of NH_3 to leave a small amount of dissolved nitrogen, which can be detected by electric resistance of the iron sample [42]. The solubility of N in α-Fe reaches a maximum value of 0.1 wt. percent at 863 K, while the solubility at equilibrium with 101 kPa N_2 is much less, of the order of 10^{-3} wt. percent [43]. Grabke has investigated the kinetics of dissolution of N in α-Fe from NH_3 [42] or from N_2 [18] and obtained a rate equation for the dissolution from N_2 as follows:

$$R = k\frac{p_{N_2}}{1 + K(\mathrm{N})} - k'\frac{K(\mathrm{N})^2}{1 + K(\mathrm{N})} \tag{18}$$

where (N) denotes the concentration of dissolved nitrogen and k, k' and K are constants at a given temperature. The second term which represents the desorption rate was separately examined, as illustrated in Figure 5 [43]. With a lower value of (N) at higher temperatures the desorption rate appears to be second order in (N), while with high concentrations or at lower temperatures the rate is consistent with first order kinetics as expected from the second term of equation (18). Thus Grabke interpreted the kinetics in terms of two different adsorbed atoms N_α and N_β. N_β is a strongly adsorbed species that blocks the surface for the chemisorption of N_2, while it is in equilibrium with the dissolved nitrogen, N_s, so that the surface coverage of N_β is given by

$$\theta_\beta = K(\mathrm{N})/(1 + K(\mathrm{N})) \tag{19}$$

N_α is a weakly adsorbed species that has no effect on the rate of chemisorption, while the surface coverage of N_α, θ_α, is proportional to (N) or the activity of dissolved nitrogen, so that the rate of dissolution is given by

$$R = kp_{N_2}(1 - \theta_\beta) - k'\theta_\alpha\theta_\beta \tag{20}$$

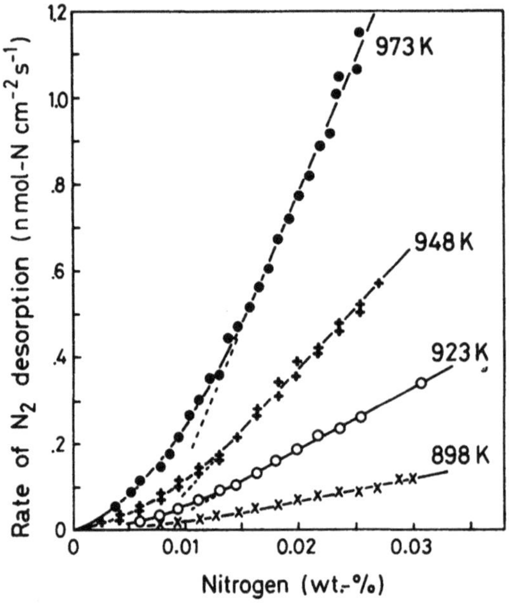

Figure 5. Rate of N_2 desorption as a function of nitrogen concentration in iron at temperatures between 898 and 973 K [43]. (mol denotes mole).

which implies that N_2 is adsorbed on a site such that N_β can be dissociated into N_α and N_β before dissolution. Assuming that $\theta_\alpha \ll 1$, equation (18) is obtained from equations (19) and (20).

In the above model the constant K represents an equilibrium constant for the segregation reaction:

$$N_s \rightleftharpoons N_\beta + Q_s \quad (21)$$

The heat of segregation Q_s as determined from the temperature dependence of K is 109 kJ (mol N)$^{-1}$ (exothermic). Using a known value for the heat of dissolution (—43 kJ (mol N_2)$^{-1}$), the heat of chemisorption to give 2 N_β is estimated to be $2 \times 109 - 43 = 175$ kJ (mol N_2)$^{-1}$, which is a reasonable value on Fe (cf. Table 1). The large value of the heat of chemisorption led Grabke to suggest that the more labile N_α instead of N_β is the intermediate for ammonia synthesis.

E. Adsorbed State

1. Types of Adsorbed State

A number of studies have been made on the adsorbed state of nitrogen on metals, mostly on W and Fe and some on Mo or Re. It is generally observed that there are three types of adsorbed nitrogen γ, α and β for tungsten. The nomenclature is traced back to the thermal desorption study by Ehrlich [44, 45] in which dinitrogen was adsorbed on a tungsten filament at 115 K and desorbed at increasing temperatures, monitoring the desorption in terms of pressure. The three types thus observed were named after the three types of adsorbed hydrogen on an iron catalyst, C, A and B as found by Emmett and Harkness in the adsorption isobar [46]. Thus the desorption temperature increases in the order γ, α and β.

The typical features of the three types of nitrogen on a W surface are summarized in Table 3. The first order kinetics with respect to the nitrogen coverage, as well as the lower energy of desorption observed with γ and α types, suggest a molecular or dinitrogen nature of the adsorbed state, which was confirmed by the absence of isotopic mixing in dinitrogen desorbed from the surface on which a mixture of $^{28}N_2$ and $^{30}N_2$ was preadsorbed [47, 48]. On the other hand, the type β was accepted as an atomic state on the basis of the second order desorption kinetics, the higher energy of desorption, and the occurrence of isotopic mixing [26, 47, 49, 50, 51, 52].

Analogous types of adsorbed nitrogen have been found or suggested on Mo [53–57], Re [58] and Fe [59–61] in addition to W by thermal desorption studies.

Table 3. Adsorbed states of nitrogen on tungsten surface

		γ-state	α-state	β-state
Desorption temperature/K		150	300	1,000
Desorption energy/kJ mol^{-1}		8–40	80	300–320
Desorption kinetics		1st order	1st order	2nd order
Isotopic mixing of N_2		no	no	yes
Suggested form		molec.	molec.	atomic
Plane selectivity	(110)	formed	no	a little
	(111)	formed	formed	a little
	(100)	formed	no	formed

2. *Plane Selectivity of Nitrogen Adsorption*

As expected from the weak interaction, the γ state is found on any of three planes, W (100) [50], (110) [62, 63], and (111) [64], while the α state is only found on W (111) as confirmed by FEM [50] or probe-hole emmission [64], and the β state is predominant on (100) as demonstrated by the large differences in sticking probability of nitrogen between W (100) (0.3–0.6) [65, 66, 67] and W (110) (0.01–0.005) [50, 67, 68] or W (111) (0.08) [66, 67]. Adams and Germer demonstrated by LEED techniques that the β nitrogen atoms on W (100) are located in an interstitial position surrounded by 4 W atoms as illustrated in Figure 6 [65, 69].

Analogous LEED study of adsorbed nitrogen on single crystal Fe planes was first made by Ertl and his coworkers [19, 20], monitoring the contamination level of the surface by AES. The sticking coefficient of N_2 on Fe (100) and (111) is of the order of 10^{-6}–10^{-7} as compared with 0.6 on the W (100) plane. The relative rates of N_2 adsorption on (111), (100) and (110) planes are 60:3:1 respectively above room temperature, demonstrating Fe (111)

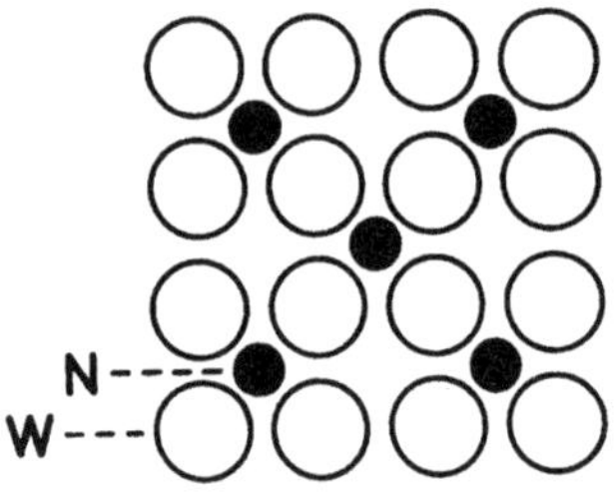

Figure 6. Surface structure of W (100) with nitrogen atoms

to be the most active plane. The adsorbed states on both (100) and (111) are identified to be atomic or the β state, on the basis of following results. i) A single desorption peak at 860 K when pretreated with N_2 at 700 K for 12 hr. ii) Large desorption energies: 243 kJ mol^{-1} on (100) and 213 kJ mol^{-1} on (111). iii) Isotopic mixing in desorbed N_2.

The LEED pattern observed with the β state on Fe (100) was analogous to that on W (100), giving evidence for a surface nitride Fe_4N. Indeed the location of the N atom in Fe (100) is very close to that in the (002) plane of Fe_4N (fcc), thus resulting in incorporation of N atoms into a subsurface layer of Fe (111) or Fe (110).

3. Work Function Change on Nitrogen Adsorption

The work function change on adsorption of nitrogen is another important source of information. The atomic adsorption (β state) on Fe (111) and Fe (100) above room temperature gives rise to an increases in work function by 0.25 and 0.33 ev respectively [19], which is reasonable in view of the electronegative nature of nitrogen. The situation is more complex with the β state on tungsten. The β state on W (100) results in decreases in work function as found by Adams and Germer [69] and later confirmed by Singh-Bopari et al. [67, 70]. Adams and Germer also found that the same adsorption gives rise to different changes on different planes as illustrated in Figure 7

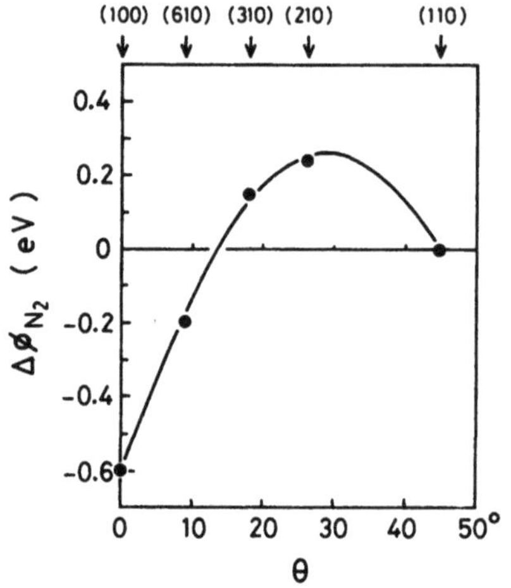

Figure 7. Work function changes ($\Delta\Phi_{N_2}$) by chemisorbed nitrogen on different planes of tungsten. θ is the angle between (hk0) and (100) [69].

as a function of the angle θ between ($hk0$) and (100): namely a large decrease on (100), with increases on (310) and (210).

The different changes on various planes are reasonably accounted for by a model as illustrated in Figure 8. Since the adsorbed N atom lies d pm above the (100) plane but x pm below the ($hk0$) plane, the sign of the work function change can be different between (100) and ($hk0$). An equation was derived to give the work function change ($\Delta\Phi_{hk0}$) on ($hk0$) as a function of the change on (100), ($\Delta\Phi_{100}$).

$$\Delta\Phi_{hk0} = \Delta\Phi_{100}(h - k)\,[a_0 k/2d - h]/(h^2 + k^2) \tag{22}$$

where a_0 is the lattice constant of tungsten. By assuming $d = 38$ pm, a good fit was obtained as shown in Figure 7 by the solid line.

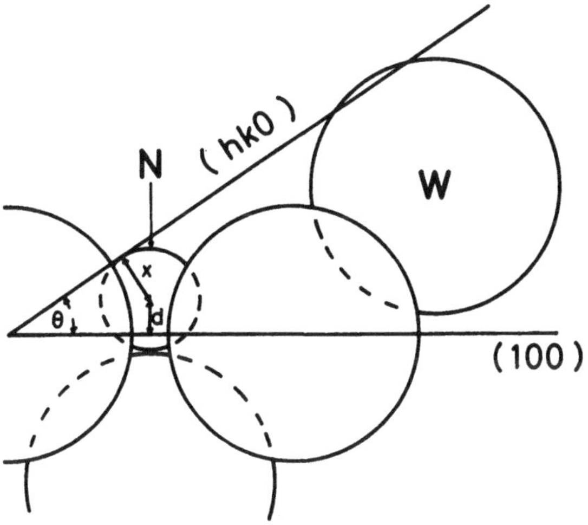

Figure 8. Projection of superficial tungsten atoms on the (100) plane, showing location of an N atom on a (100) site [69].

In this way the atomic radius of adsorbed nitrogen was found to be 59 pm on the basis of $d = 38$ pm. This values of the atomic radius is smaller than 62 pm for the double bond covalent radius, while larger than 55 pm for the triple bond radius.

4. Infrared Absorption of Adsorbed Nitrogen

Although nickel had been known to be inactive in the chemisorption of dinitrogen, Eischens and Jacknow first found an IR active adsorbed dinitrogen on $Ni-SiO_2$ which disappeared on evacuation at 303 K. The adsorbed state was confirmed to be molecular as evidenced by the isotope shift (2195, 2160, and 2123 cm^{-1} for $^{28}N_2$, $^{29}N_2$ and $^{30}N_2$ respectively) and was assigned to $Ni-N{\equiv}N^+$ [71]. It was later found by Hardeveld and Montfoort that the IR active species is formed only when the size of the nickel particles on silica is in a range 1.5 to 7.0 nm. The effect of particle size was interpreted in terms of the "B5" site (inverted square pyramid) for the dinitrogen adsorption [72].

This discovery on Ni—SiO_2 has induced successive findings of analogous IR active species on other metals as summarized in Table 4. These bands were also shown to be molecular by the isotope shift [71, 73, 74] as well as the first order kinetics of desorption [75]. The large extinction coefficient associated with the absorption bands compard to the value for chemisorbed CO has given evidence for an end-on type of coordination rather than a side-on type [76].

The adsorption was interpreted formerly by an interaction with electric field on the metal surface [72]. However later findings disclosed a chemical nature of adsorption as suggested by; 1) a fairly high initial heat of adsorption on Ni, 50 kJ mol^{-1}, 2) the fact that the shift in IR band frequency was large, 3) the effect of hydrogen which causes a shift toward higher wave numbers [71, 73, 74]. In fact the effect of strong electric field is opposite on ZnO [77] or zeolite [78] on which adsorbed dinitrogen gives an absorption band at higher wave number than the normal dinitrogen molecule.

Although iron is well known as a dinitrogen-activating metal, analogous IR active species observed on iron are formed by treatment with ammonia [79] or a N_2—H_2 mixture [80] at elevated temperatures (cf. 4. E. 4). A very stable and IR active dinitrogen species which can be converted to ammonia by hydrogen treatment has been found on Ru—K—Al_2O_3 at 2020 cm^{-1} by N_2 treatment above 473 K [81]. Since $^{15}N_2$ gives rise to an isotope shift to 1997 cm^{-1}, and replacement of K with Na results in a shift which is in-

Table 4. IR data of adsorbed dinitrogen on metals

Sample	Support	Wave No/ cm^{-1}	Deviation from gaseous N_2/cm^{-1}	Heat of adsorption/ kJ mol^{-1}	Ref.
Fe	Al_2O_3	2200	131		[79]
	MgO	2200	131		[79]
	MgO	2050	281		[79]
Ni	SiO_2	2202	129	50	[71]
	SiO_2	2202	129		[72]
	SiO_2	2195	136		[73]
	SiO_2	2202–2220	129–111		[86]
	(evaporated)	2202	129		[84]
Ru	K-Al_2O_3	2020	311		[81, 82]
		1935	396		[40]
Rh	SiO_2	2236	95		[74]
Pd	Al_2O_3	2260	71		[72]
Ir		2185	146	17	[83]
Pt	SiO_2	2230	101		[72]
	SiO_2	2230	101	15	[75]
	(as Pt black)	—	—	18	[85]
ZnO		2337	−6	19	[77]
Na-A-Zeolite		2339	−8		[78]

compatible with azide KN_3 or NaN_3, the band is ascribed to N≡N coordinated to Ru. The large extinction coefficient associated with the band suggests an end-on type coordination. In fact another band at 520 cm^{-1} which can be ascribed to Ru—N bond is observed an Ru—K/KBr after N_2 treatment at 623 K [82]. More recently the band at 2020 cm^{-1} is ascribed to a bulk compound $(KN_2Ru)_n$ (cf. 5. B. 3). A new band at 1935 cm^{-1} which is formed by treatment of the 2020 cm^{-1} species with H_2 at above 423 K is assigned to adsorbed species on the surface of $(KN_2Ru)_n$ [40].

5. *Adsorbed States under Higher Pressure and Temperature*

The foregoing discussions are mostly referred to the adsorbed species formed at low temperature and low pressures. The situation can be different at higher temperature and pressures where real catalysis of dinitrogen is carried out. Some techniques have been worked out to study the adsorbed state of nitrogen at higher temperatures and pressures.

Scholten et al. calculated the differential entropy of dinitrogen adsorption from their experimental values of the heat of adsorption as a function of surface coverage on a singly promoted iron (Fe—Al_2O_3) at 573 K and compared this with theoretical values based on different models; atomic or molecular, and mobile or immobile [16]. In this way it was suggested that the adsorbed state is immobile and probably atomic at low coverage (below 10 percent) whereas a mobile, and probably molecular, state predominates at higher coverage (above 30 percent).

It was previously shown that dinitrogen preadsorbed on iron catalysts does not inhibit successive chemisorption of CO at 78 K. This result was taken as evidence for the atomic state of adsorbed nitrogen, because if the adsorbed state is molecular, there would be no room for successive adsorption of CO [87]. More recently the effect of adsorption temperature on this behaviour of preadsorbed dinitrogen has been examined by Emmett and his coworkers on promoted [88] as well as unpromoted [89] iron catalysts. The result on a single promoted iron catalyst is illustrated in Figure 9 where the extent of the inhibitive effect, that is, a decrease in chemisorption of CO per amount of preadsorbed dinitrogen (p_{N_2} = 50.7 kPa), is plotted against the temperature of preadsorption. It is clear that the inhibitive effect increases with a decrease in the adsorption temperature, suggesting a change in the adsorbed state from an atomic state at 723 K to a molecular state at lower temperatures [88]. An analogous result was obtained on unpromoted iron, giving a greater inhibitive effect on pure iron than on promoted iron [89].

Another source of information is found in the displacement of adsorbed dinitrogen with isotopically labelled gaseous dinitrogen. If the adsorbed state is molecular, the displacement does not result in isotopic mixing. It was found on unpromoted iron that the rate of displacement is much faster than the rate of isotopic exchange, indicating that more than 80 percent of preadsorbed dinitrogen at 623 K is undissociated [90] (cf. 4. E. 2). On the other hand the rate of displacement on a doubly promoted iron is in agree-

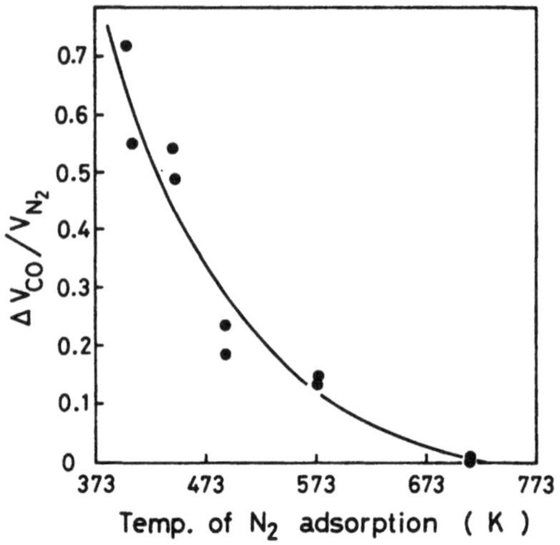

Figure 9. Effect of prechemisorbed N_2 on chemisorption of CO at 78 K as a function of temperature of prechemisorption on a promoted iron catalyst [88].

ment with the rate of isotopic exchange, giving no indication of dinitrogen species [91].

The above results are qualitatively in accord with the findings of the inhibitive effect of preadsorbed dinitrogen in that the promoted iron holds a larger fraction of the dissociated species. The displacement method should give a larger fraction of molecular species because the experimental procedure does not require any change in gas phase pressure which is necessarily made in the inhibition experiment.

Although the rate of displacement was in agreement with the rate of isotopic exchange on the promoted iron, this does not necessarily mean that no molecular state exists on the surface, as discussed by Temkin [92]. Thus Amenomiya and Pleizier examined desorption spectra of dinitrogen preadsorbed at 573 K or 470 K on a promoted iron catalyst [93]. They obtained, however, a single peak at around 770–820 K as observed with β state of nitrogen on Fe (111) [19]: this result was obtained irrespective of the temperature of preadsorption 573 K or 470 K. It would be possible that the molecularly adsorbed species were already desorbed by evacuation before the desorption run.

In this way, molecularly adsorbed species are suggested on pure iron, while it is likely, that they are less abundant on promoted iron.

6. Electron Spectroscopy of Adsorbed Nitrogen

Electron spectroscopy of adsorbed species has been developed in recent years. Both AES (Auger electron spectroscopy) and XPS (X-ray photoemission spectroscopy) can give information about the electronic state of atoms in the outermost layer of a solid surface, thus being adequate to study

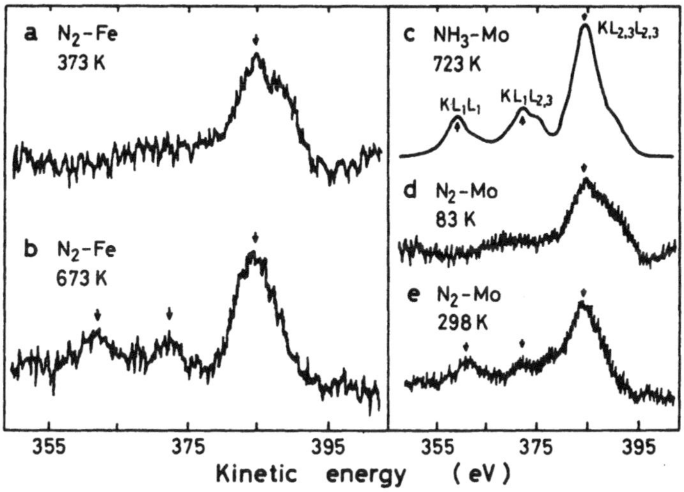

Figure 10. Nitrogen KLL Auger spectra of adsorbed N_2 on Fe (a and b) and on Mo (d and e) and of Mo nitride (c) [57]

adsorbed species. Figure 10 shows Auger spectra of adsorbed nitrogen on Mo and Fe [57]. When dinitrogen is adsorbed at 83 K on Mo or 373 K on Fe, at which temperatures molecular adsorption is expected, a single peak is observed at around 385 ev while, when it is adsorbed at 273 K on Mo or 673 K on Fe, three peaks are observed. Since Mo_2N gives three similar peaks, as shown in Figure 10c, the adsorbed species obtained at higher temperatures are assigned to atomic nitrogen [57].

Analogous Auger spectra were observed on W (110) for N_2 adsorbed at 100 K [63]. The amount of adsorbed nitrogen can be monitored by the peak height of the Auger spectra as was done on W by 390 ev peak [26] and on Fe by 380 ev peak [19].

In AES, the solid surface is excited by electron bombardment so that the adsorbed species are apt to be damaged, whereas less possibility of damage is expected for XPS. Moreover, the energy distribution is narrower with XPS (0.7 ev) than with AES (3–4 ev) so that XPS is more relevant to study chemical shift which is usually within a range of 10 to 20 ev at most.

Dinitrogen adsorbed on W, Ni or Fe at 100 K or lower gives two peaks in XPS at around 400 and 406 ev, which were interpreted in different ways: 1) N_2(a) for 406.6 ev and N(a) for 397.6 ev on W because the latter remained while the former disappeared at 300 K [94]. 2) Two types of N_2(a) [95]; end-on and bridge [96]. 3) Two nitrogen atoms of an end-on type dinitrogen: W—N≡N [63].

There has been agreement that the atomically adsorbed nitrogen gives rise to a peak at around 400 ev, that is, 397.2 and 398.0 ev for N(a) on W at 300 K [94]; 397.2 ev for N(a) on Fe at 290 K [96]; and 397.5 ev for N(a) on promoted iron at 673 K [97].

3. Catalysts for Synthesis of Ammonia

A. Elemental Survey of Ammonia Catalysts

The catalytic synthesis of ammonia was first found on iron at 1073 K and atmospheric pressure by Perman [98] and was confirmed by Haber and Oordt at 1273 K [99] and also under elevated pressures (5–7 MPa) by Nernst [100]. The discovery was followed by exploration of catalysts for high pressure operation [101]. Haber found osmium and uranium to have high activity, whereas osmium was too expensive for commercial use and uranium was too sensitive to poisons such as O_2 and H_2O [102]. Further extensive work at BASF by Bosch and Mittasch led to the discover of the well-known doubly promoted iron, Fe—Al_2O_3—K_2O in 1910 [103].

The development at BASF remained unpublished for many years, except in the patent literature, while exploratory work in the US was reported by Larson et al. since 1922 [104]. In a review article by Mittash [103], the relative efficiencies of pure metals as catalysts for ammonia synthesis under 10 MPa at 823 K were given. They are shown in Figure 11 b as plots of percent NH_3 in the exit gas (60 litres h^{-1} over 2 g of metals) against the parameter ($-\Delta H_0^0$) for the heat of chemisorption (see Figure 3). It is obvious that osmium and iron are the most effective metals under these conditions. On the other hand, the relative efficiencies of metals for ammonia decomposition are known

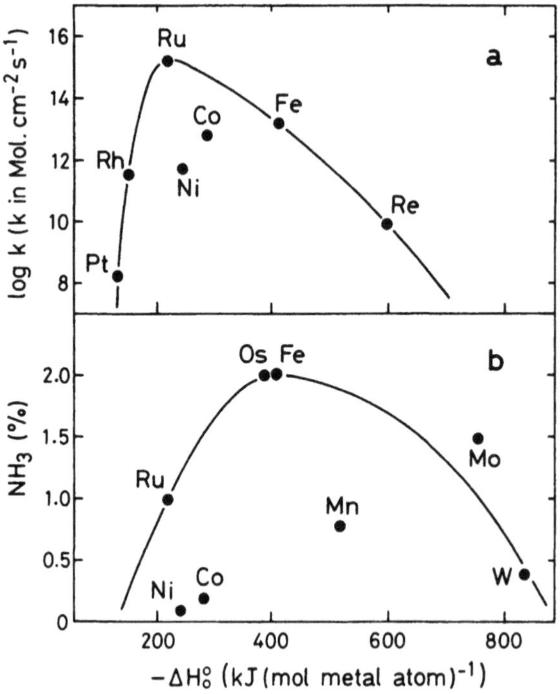

Figure 11. The rate constants of ammonia decomposition (a) [172] on and the ammonia synthesis capacities (b) [107] of metals as a function of $-\Delta H_0^0$. (Mol. denotes molecule, mol denotes mole)

and are shown in Figure 11a as an analogous plot of rate constants [105], where ruthenium is the most active metal. Since the two reactions are forward and backward reactions of the same reaction, the most active metals should be the same, if the reaction condition including the surface state of catalysts is identical. The difference disclosed above is probably caused by some discrepancy in the reaction conditions. It is obvious that the H_2/NH_3 ratio is much larger in the synthesis reaction than in the decomposition reaction. The effect of such variables will be discussed later in 5. A.

More recently the authors have developed a catalyst system promoted by metallic potassium [106, 139]. The ammonia synthesis rates at 79.8 kPa and 523 K over transition metals supported on active carbon and promoted by metallic potassium, are given in Figure 12 as analogous plots [107]. In this case, ruthenium is the most active metal. There is a common trend that those metals of too low or too high values of $-\Delta H_O^0$ are less active.

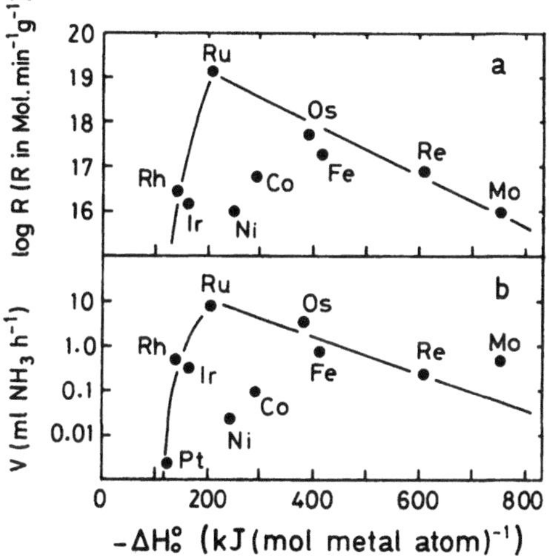

Figure 12. Catalytic activity of carbon-supported metals promoted by metallic potassium as a function of $-\Delta H_O^\circ$.
(a) The rate of isotipic equilibration of N_2 at 623 K, 20 kPa of p_{N_2} [243]. (Mol. denotes molecule, mol denotes mole)
(b) The rate of ammonia synthesis at 523 K, 80 kPa of total pressure [107].

B. Promoted Iron Catalysts (Unreduced Form)

The promoted iron catalyst for ammonia synthesis contains 0.6–2.0 percent Al_2O_3 and 0.3–1.5 percent K_2O and is called "doubly promoted iron". "Singly promoted iron" contains only Al_2O_3. Industrial catalysts sometimes contain CaO, MgO, or SiO_2 in addition to Al_2O_3 and K_2O.

These catalysts are generally prepared by fusion of premixed powders of iron oxide, potassium nitrate and alumina at above 1873 K. Alumina is dissolved into magnetite in the fusion process. Thus the lattice constants of a series of fused iron oxide samples with increasing amounts of alumina decrease linearly with the $Al^{3+}/(Fe^{3+} + Al^{3+})$ ratio in conformity with the Vegard rule [108, 109, 110]. When the straight line is extrapolated to $Al^{3+}/(Al^{3+} + Fe^{3+}) = 1$, the composition corresponding to $FeAl_2O_4$, the lattice constant thus obtained, 814 pm, is very close to the observed value for $FeAl_2O_4$ [109].

Since a solid solution is usually formed substitutionally, the solubility of metal oxides into magnetite depends on the size and valence of substituting ion. Goldschmidt radii (pm) for the relevant ions are 133 (K^+), 106 (Ca^{2+}), 83 (Fe^{2+}), 78 (Mg^{2+}), 67 (Fe^{3+}), 57 (Al^{3+}), and 39 (Si^{4+}). It is unlikely that K^+ is incorporated in the magnetite matrix, although it was reported that potassium ferrite ($KFeO_2$) is soluble to some extent in magnetite [111]. In fact electron probe micro analysis shows that potassium is concentrated at grain boundaries, while aluminum is uniformly spread through the whole [112, 113]. The location of potassium can be found by techniques based on grinding, separation according to particle size or ferromagnetic properties, and chemical analysis. More potassium is found in smaller particle (5 μm) [114] or in less ferromagnetic particles [115]. Interestingly, the content of K_2O is 0.7 percent in larger particles (50–100 μm) [114] and at most 0.5 percent in more magnetic particles [115], while it is much larger (about 4 percent) in smaller [114] or less ferromagnetic particles [115]. It is likely that the smaller particles come from grain boundary so that they have more potassium and less ferromagnetism, with the larger particles being magnetite phase in which smaller amounts of K_2O are incorporated.

According to Kobayashi and Nishijima [116], some part of the K_2O in the fused iron catalyst is water-insoluble, and an increase in the content of the water-insoluble potash up to 0.5 percent results in an appreciable decrease in the activation energy of ammonia synthesis. This is reasonable if the allowable concentration of K_2O in the magnetite phase is about 0.5 percent and excess K_2O separates out of the magnetite grain.

Recent XPS observations of unreduced iron catalysts discloses that the concentration of K is 6–20 percent in the surface region, a much larger

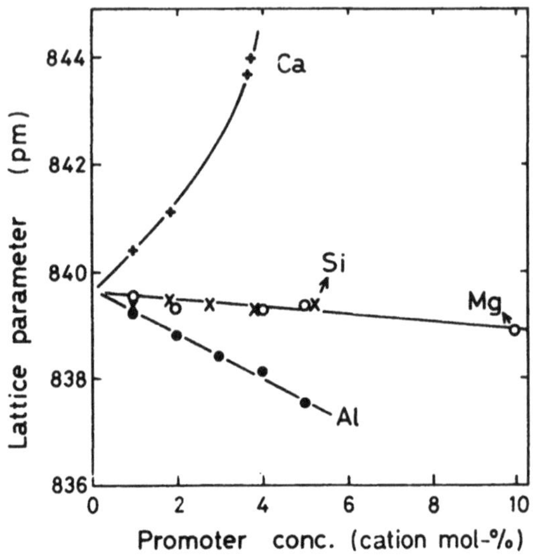

Figure 13. Variation of the lattice parameter of magnetite with the promoter concentration [110].

value than that in the bulk 0.35 percent [97]. This is in agreement with the classical view of potash accumulation in the grain boundary.

Effects of other metal oxides on the lattice parameter of magnetite are illustrated in Figure 13. It is seen that Ca^{2+} (of larger radius than Fe^{2+}) causes an increase in the lattice constant, while Al^{3+} (of smaller size) causes a decrease. Both Mg^{2+} and Si^{4+} cause very little change, suggesting less possibility of solid solution [110, 117].

C. Promoted Iron Catalysts (Reduced Form)

1. Changes During Reduction

The reduction of iron oxide to metal can be carried out above 623 K by the use of hydrogen or H_2—N_2 synthesis gas. The reduced catalyst is made up of crystallites of α-iron. The apparent external volume of the catalyst particles does not change if the promoters are finely dispersed in the catalyst bulk [118]. Thus the removal of oxygen by reduction gives rise to a porous structure. The total surface area increases linearly with the extent of reduction as exemplified for a Fe—MgO (4.6 percent)—K_2O (0.6 percent) catalyst in Figure 14, while it reaches a maximum at 99 percent reduction. Chemisorption of CO which represents the iron surface area also increases linearly to about 85 percent reduction, but more sharply above 90 percent [118]. The increase in total surface area causes a linear increase in pore volume. The average pore diameter is practically independent of extent of reduction

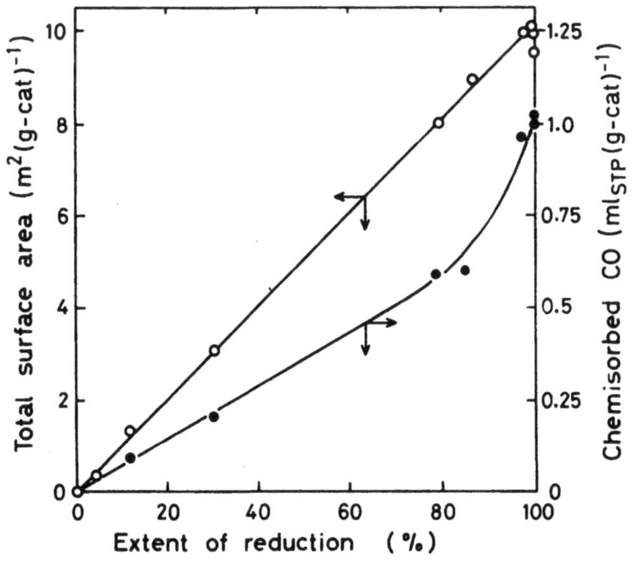

Figure 14. Variation of surface area and of CO chemisorption with extent of reduction of Fe—MgO—K_2O at 723 K [118]

from 20 to 90 percent, staying near 34 nm, but increases up to 37.1 nm as reduction approaches 100 percent [119].

2. Promoter effects on surface area

The total surface area of the iron catalyst after reduction increases on addition of alumina by about 20 times while it reaches a plateau at about 2.5 percent Al_2O_3 as illustrated in Figure 15 [120, 121]. Excess alumina above 2.5 percent seems to be liberated from the iron matrix. The state of alumina in the reduced catalyst, $Fe—Al_2O_3$, was previously shown to be γ-Al_2O_3 by dissolving the iron in acetic acid [122]. Schaeffer applied this method to $Fe—Al_2O_3$ catalysts of different Al_2O_3 contents and found that catalysts containing up to about 3 percent Al_2O_3 gave practically no γ-Al_2O_3 residue [123]. It was suggested that alumina is present in layers of 0.5–1.0 nm thickness only, and the thin layers of alumina in direct contact with the iron crystals dissolve in weak acid, thus giving no residue of γ-Al_2O_3 [123]. Such thin layers of alumina seem to be responsible for increasing the surface area. Owing to this effect of alumina, as pointed out earlier by Wyckoff and Crittenden [124], alumina has been called a structural promoter, meaning that the role of alumina is to enlarge the iron surface by inhibition of iron crystal growth. The effect is reasonably enhanced by intimate mixing of alumina with magnetite during catalyst preparation.

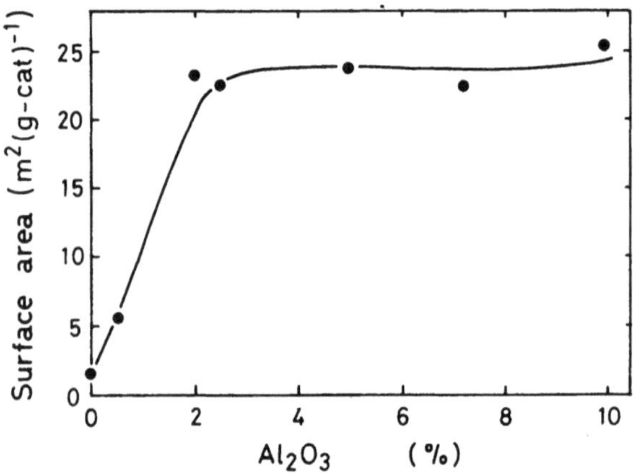

Figure 15. Specific surface area of $Fe—Al_2O_3$ catalyst as a function of Al_2O_3 content [121].

An analogous effect can be expected with other nonreducible metal oxides. According to Dry et al. [125] the effectiveness sequence of metal oxides to increase the BET area is

$$Al_2O_3 > TiO_2 > Cr_2O_3 > MgO > MnO = CaO > SiO_2 \simeq BeO .$$

Although K_2O is an important promoter, the surface area of a $Fe—Al_2O_3$ catalyst decreases on addition of K_2O [120, 126]. The average pore diameter increases from 28.4 nm for $Fe—Al_2O_3$ without K_2O to 48.4 nm for one rich in K_2O, although the pore volume remains constant at 0.15 $cm^3\ g^{-1}$ [126].

3. Surface Concentration of Promoters

One of the important roles of the promoters is definitely to create as well as to maintain the high iron surface area as described above. Such a function of the promoter is realized by its fine distribution in the catalyst bulk. Although the overall concentration of promoters is rather small, the promoter concentration in the surface of the reduced catalyst is much higher than in the bulk, as estimated from chemisorption of CO at 90 K for the iron surface, and of CO_2 at 195 K for the potash surface [87, 126, 127]. The alumina surface can be evaluated from the BET area less iron and potash surface areas, while Al_2O_3 in Fe—Al_2O_3—K_2O is mostly bound to potash forming potassium aluminate, as demonstrated by the decreased volatility of potash by the presence of alumina [127].

By this method, Brunauer and Emmett estimated the free iron surface to be 40 percent on a doubly promoted iron catalyst (1.6 percent K_2O, 1.3 percent Al_2O_3) and 45 percent on a singly promoted iron (10.2 percent Al_2O_3) [87]. A more recent reexamination using an exchange method, in which the oxide surface is estimated by equilibration with a known amount of $H_2{}^{18}O$ in H_2 at 723 K, confirmed an analogous value of 40 percent iron surface for a doubly promoted iron (1.06 percent Al_2O_3 and 0.52 percent K_2O) [128].

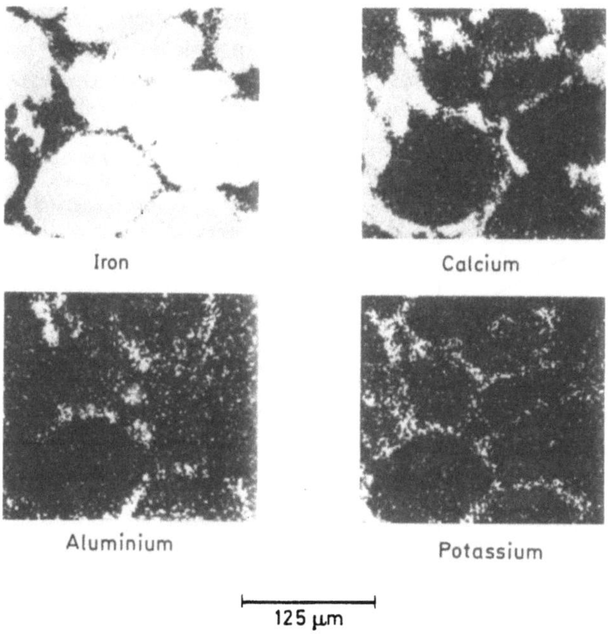

Fe_3O_4-Al_2O_3-CaO-K_2O reduced

Figure 16. Electron probe profile of a reduced iron catalyst [113]. The number of white spots is proportional to the concentration of the elements

Brunauer and Emmett also suggested that the potash promoter is finely distributed over the iron surface because CO_2 preadsorbed on a potash surface prevents subsequent adsorption of CO on iron, indicating that the adsorption sites for CO_2 and CO are neighbouring [87].

Figure 16 shows the distribution of promoters in a reduced Fe—Al_2O_3—CaO—K_2O catalyst [113]. It is clear that iron grains are surrounded by slag-like envelopes rich in promoters. It is to be noted that some amounts of promoters including potash are found in the iron grains and that the active surface should be located inside the iron grain in view of its large grain size.

4. Promoter Effects on Catalytic Activity

Peters, Schaeffer and Krabetz [121] applied the CO chemisorption method to estimate the iron surface and showed that although the surface area of Fe—Al_2O_3 increases with alumina content (cf. Figure 15), the fraction of iron surface decreases so that there exists a maximum in the total iron surface are as shown in Figure 17. It is to be noted that the yield of ammonia on these catalysts runs parallel with the iron surface area, demonstrating that the catalytic activity depends solely on free iron surface for the Fe—Al_2O_3 series. An analogous conclusion was obtained by Brill by comparing Fe and Fe—Al_2O_3 catalysts [129].

On the other hand, on addition of potash to Fe—Al_2O_3 the catalytic activity increases despite the decrease in iron surface area as shown in Figure 18, disclosing a quite different effect of the potash promoter [126]. The nature of this promoting action must be different from that of alumina

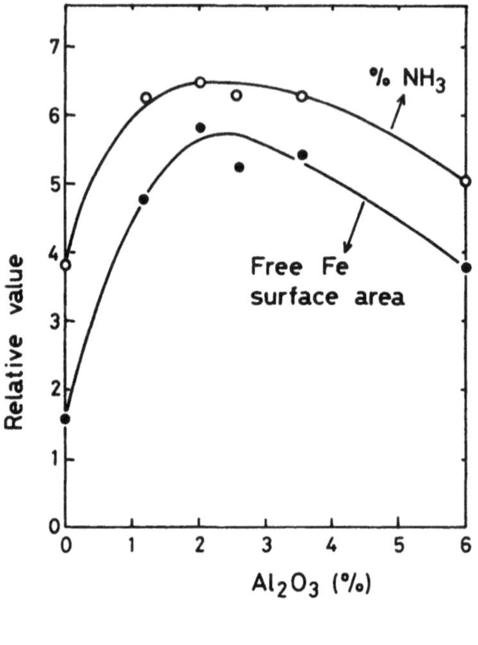

Figure 17. Variation of ammonia yield and free iron area of Fe—Al_2O_3 catalyst with Al_2O_3 content [121].

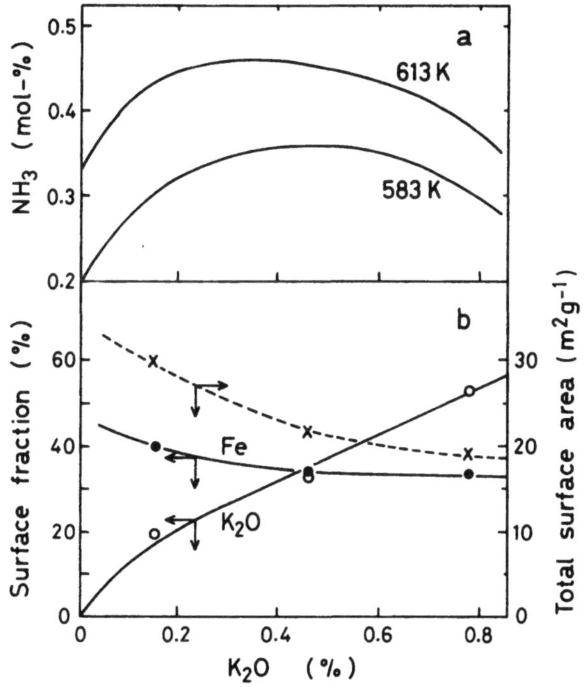

Figure 18. Tl.e effect of K_2O addition to $Fe—Al_2O_3$ catalyst, (a) on ammonia yield, (b) on surface area and surface fractions of Fe and K_2O area [126].

and we shall come back to this subject later. It is also seen from Figure 18 that too large a surface fraction of potash is unfavorable for the catalytic activity. Further studies of multicomponent iron catalysts by Krabetz and Peters gave a correlation between catalytic activity and percentage of potash surface of $Fe—Al_2O_3—K_2O—CaO(MgO)—SiO_2$ catalysts with variable SiO_2 and K_2O contents as shown in Figure 19. Maximum activity is obtained with catalysts having 25–30 percent potash surface, independent of SiO_2 and K_2O content.

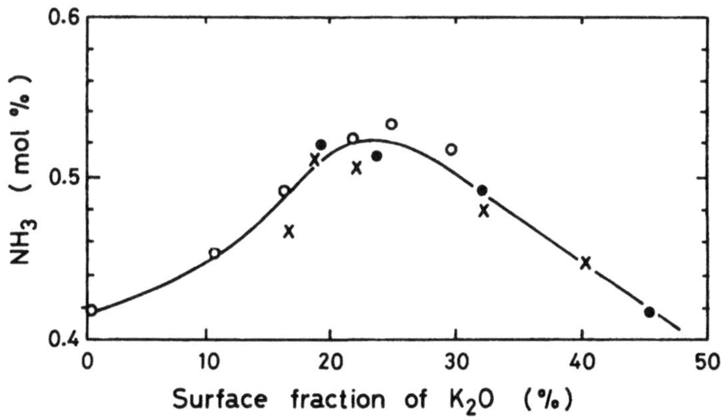

Figure 19. Variation of ammonia synthesis activity with percent free K_2O surface over total surface of $Fe—Al_2O_3—K_2O—CaO$ (MgO)$—SiO_2$ catalyst [126]
● 0% SiO_2, ○ 2.2% SiO_2, × 4.2% SiO_2

According to the results by Krabetz and Peters under atmospheric pressure, Fe—MgO is less active than Fe—Al_2O_3 as ammonia synthesis catalyst [126], whereas Uchida, Terao and Ogawa showed that Fe—MgO is essentially as active as Fe—Al_2O_3 under 10 MPa, and that the MgO-containing iron catalysts are characterized by their high heat stability at as high as 923 K, although they require a higher temperature for reduction [130].

5. *Structure of Reduced Iron*

Although alumina is homogeneously distributed in the fused unreduced catalyst, forming a solid solution with magnetite, it separates out between α-Fe crystallite on reduction. The high promoter concentration on the surface of the reduced catalyst had been understood in terms of the segregation from bulk of α-Fe. More recently it was noticed that the α-Fe in the catalyst singly promoted with Al_2O_3 has a structure different from pure α-Fe, as evidenced by an appreciable line broadening of x-ray diffraction lines. This line broadening was formerly ascribed to combined effects of small crystallite size, faulting on (211) planes and lattice strains [131]. By precise X-ray measurements of the integral reflection width, Hoseman et al. interpreted the broadening in terms of a long range disorder describable as paracrystalline lattice defects in the small crystallites (ca. 20 nm) of α-Fe [132, 134]. These defects were attributed to $FeAl_2O_4$ groups randomly dispersed in the lattice of α-Fe, replacing seven α-Fe lattice sites with a $FeAl_2O_4$ molecule [132]. The broadening caused by lattice strains [133] were ruled out because it persisted after annealing of the catalyst [135].

As the proposal by Hoseman et al. implies that an appreciable fraction of aluminum is present inside the iron crystallites, it is novel as a concept. Thus the model was examined by Mössbauer spectroscopy to find evidence for the $FeAl_2O_4$ groups incorporated in the α-Fe lattice of reduced Fe—Al_2O_3 catalysts. However no significant difference was observed in the hyperfine parameters of Fe—Al_2O_3 and of pure α-Fe [135, 136]. Since a mildly reduced catalysts exhibits a paramagnetic phase which disappears upon further reduction, it was suggested by Topsøe et al. that the aluminum is present in the form of ca. 3 nm aggregates of $FeAl_2O_4$ or Al_2O_3 depending on the extent of reduction [136], while Fagherazzi et al. suggested $FeAl_2O_4$ clusters occluded in the α-Fe crystallites [135].

According to Hoseman's theory, however, no paracrystallinity can be produced by such aggregates. The number of $FeAl_2O_4$ molecules in the cluster must be small. An identical hyperfine splitting in the Mössbauer spectra of Fe—Al_2O_3 and of pure α-Fe does not necessary mean the absence of endotactic $FeAl_2O_4$ groups. Because of the relatively few endotactic Fe atoms, a separate Mössbauer signal can be very weak. Moreover it was found from secondary ion mass spectra of a reduced Fe—Al_2O_3 catalyst that the polyatomic ions sputtered from the catalyst rule out the presence of Al_2O_3 clusters, i.e. $(AlO)^+$, $(AlO_2)^-$ and $(Fe_2Al)^+$ are observed, whereas $(Al_2)^+$, $(Al_2O)^+$, $(AlO_3)^-$, $(Al_2O_2)^+$, $(Al_2O_2)^-$ and $(Al_2Fe)^+$ are not observed, indicating that Al is coordinated with no other Al and with no more than two O, but at least two Fe [137]. This is in conformity with the proposed configuration as shown below [138].

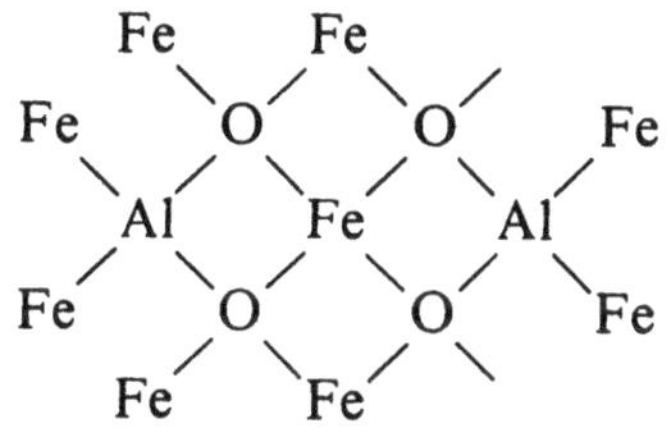

In this way the high activity of α-Fe promoted by Al_2O_3 is ascribed to paracrystallinity, and an analogous effect is expected with Sc_2O_3, Cr_2O_3 etc. [134, 138].

D. Bimetallic Catalysts

A number of bimetallic catalysts have been examined for their activity in ammonia synthesis. Fe—Mo (1/1) catalyst gives a high activity [140], although it decreases remarkably during a prolonged synthesis run unless the Mo content is higher than 80 percent [141]. Ni—Mo, and Co—Mo systems behave similarly. The catalysts are prepared by calcination of metal nitrate-ammonium molybdate mixtures, followed by reduction with N_2—H_2 gas [141]. These catalysts absorb N_2 at the beginning of operation to an extent increasing with Mo content, whereas the absorbed N_2 is gradually desorbed during operation resulting in the decrease in activity. X-ray analysis of the catalyst indicates that the desorption of N_2 is accompanied by formation of Mo-metal (Ni, Co, or Fe) mixed crystals. It appears that the real active component in this system is molybdenum. The catalytic activity of either iron or molybdenum seems to be lost on the forming mixed crystal. The effects of other metals mixed with Mo shows that Cu behaves as a simple diluent, Mn gives a favorable effect and both Cr and W are unfavorable [142].

The activity of Fe is decreased by addition of Ni [143], and a Mn—Fe alloy of 30–40 percent Mn gives a high activity [144]. The conventional promoted iron catalyst is further promoted by addition of Co, the catalysts being prepared by burning a Fe—Co alloy in O_2 followed by addition of promoters [145]. Rare earth-transition metal intermetallics and Raney Ru are described later in (5.B.2).

4. Kinetics and Mechanism of Ammonia Synthesis and Decomposition

A. Development of Temkin Theory

The kinetic investigation was started with respect to decomposition early in the 1930's. It was already known that decomposition of iron nitride Fe_2N is much slower than the nitride formation from ammonia and iron [146].

When Winter obtained a rate equation

$$V = kp_{NH_3}^{0.9}/p_{H_2}^{3/2} \tag{23}$$

for ammonia decomposition on iron foil [147], it was interpreted in terms of the partial equilibrium

$$NH_3 \rightleftharpoons 3/2H_2 + N(a) \qquad (a = \text{adsorbed})\,. \tag{24}$$

Thus, the rate of decomposition is limited by desorption of N(a), the amount of which is assumed to be determined by the partial equilibrium so that it is proportional to $p_{NH_3}/p_{H_2}^{3/2}$ as observed. Although this interpretation implies an unacceptable first order desorption of N(a), it represents an important first step to the kinetics of ammonia decomposition.

Working on promoted iron catalysts, Emmett and coworkers determined the rate of chemisorption of N_2, which was supposed to be the corresponding slow step for ammonia synthesis, and found it to be of the same magnitude as the rate of ammonia synthesis [22, 148]. The adsorption heat as well as the activation energy for N_2 chemisorption was consistent with the activation energy for N_2 desorption from Fe foil by Winter [148]. In addition, hydrogenation of iron nitride was demonstrated to take place at temperatures lower than that required for the ammonia synthesis [41], suggesting the rate of hydrogenation of adsorbed nitrogen to be fast. The rate of hydrogen adsorption as well as the rate of ammonia desorption was also thought to be fast because isotopic exchange between NH_3 and D_2 takes place on iron even at room temperature [149]. In fact, the chemisorption of hydrogen on iron catalysts was shown to be fast at 373 K [46].

Thus if one considers a scheme

$$\left.\begin{array}{l} N_2 \xrightarrow{Ia} 2N \\ H_2 \underset{Ib}{\rightleftharpoons} 2H \end{array}\right] \xrightarrow{IIa} NH \xrightarrow{IIb} NH_2 \xrightarrow{IIc} NH_3 \tag{25}$$

there was evidence towards the end of the 1930's to support the rate determining step as Ia in ammonia synthesis. Furthermore, very slow exchange between $^{15}N_2$ and $^{14}N_2$ had been demonstrated [234], although there had been no systematic explanation presented. Temkin and Pyzhev [33] first gave an overall kinetic interpretation, improving the Winter kinetic model. Namely, they introduced second order desorption of N(a) in the ammonia decomposition instead of the first order desorption in Winter's model. They introduced also the Frumkin-Slygin isotherm (equation (7)) instead of the Henry isotherm used by Winter [147], while the assumption of equilibrated decomposition of ammonia to form N(a) (equation (24)) was retained. By replacing p_{N_2} in the Frumkin-Slygin isotherm [34] with a fictitious pressure $K(p_{NH_3}/p_{H_2}^{3/2})^2$ based on the assumed equilibrium, the surface coverage of N(a), θ_N is given by

$$\theta_N = (1/f)\ln\,[a_o K(p_{NH_3}^2/p_{H_2}^3)] \tag{26}$$

where K is the equilibrium constant of the reaction

$$2\,NH_3 \rightleftarrows N_2 + 3\,H_2\,. \tag{27}$$

The rate of N_2 adsorption which is assumed to be rate-determining in the ammonia synthesis is given by introducing θ_N into the Zeldowitch equation [equation (8)]:

$$V_a = k_a p_{N_2}[a_o K p_{NH_3}^2/p_{H_2}^3]^{-g/f} \tag{28a}$$

$$= k_a' p_{N_2}[p_{H_2}^3/p_{NH_3}^2]^{\alpha}\,. \tag{28b}$$

The corresponding rate of N_2 desorption which is assumed to be rate-determining in ammonia decomposition is similarly given by

$$V_d = k_d[a_o K p_{NH_3}^2/p_{H_2}^3]^{h/f} \tag{29a}$$

$$= k_d'[p_{NH_3}^2/p_{H_2}^3]^{(1-\alpha)} \tag{29b}$$

where $\alpha = g/f = g/(g+h)$ is given by the relative change of E_a and Q, $\Delta E_a/\Delta Q$, when the heat of adsorption changes linearly with θ_N as assumed by Brunauer, Love and Keenan [37].

The Winter equation for the ammonia decomposition is given by equation (29b) when $\alpha = 0.5$. The overall rate of ammonia synthesis (V_s) is now given by

$$V_s = V_a - V_d\,. \tag{30}$$

Table 5a. Kinetic parameters of NH_3 synthesis according to the Temkin equation

Metal	Temperature/ K	Pressure/ MPa	α	ΔE_{act}/ kJ mol^{-1}	Ref.
Fe (—Al_2O_3—K_2O)	673– 723	0.1	0.5	167	[33]
	673	0.1	0.5		[112]
	590	0.1	0.6		[150]
	623	0.1	0.6		[151]
	597	0.1	0.5		[152]
	659	0.1	0.6		[152]
	606	0.1	0.8		[153]
	645	0.1	0.7		[153]
	623	1.2–3.3	0.67		[151]
	643– 723	3.3–10	0.67		[154]
	723– 773	30	0.5	176	[155]
	643– 768	15–31	0.75 (0.64)	(176)	[156]
	723	33	0.5		[112]
	773	1–50	0.5		[157]
Fe (—Al_2O_3)	645– 651	0.1	0.5–0.6		[152]
Fe (—MgO)	570– 680	0.1	0.5	(156)	[158]
Mo	720– 823	0.1	0.5	178	[159]
W	855– 951	0.1	0.5	190	[160]
	673– 773	0.1	0.5	(203)	[161]
Ru	881–1013	0.1	0.5	249	[162]
Os	673– 723	0.1	0	164	[163]
	823– 873	0.1	0.5	174	[163]

Table 5b. Kinetic parameters of NH_3 decomposition according to the equation $V = k(p_{NH_3})^m/(p_{H_2})^n$

	Temperature K	Pressure/ kPa	m	n	n/m	ΔE_{act}/ kJ mol^{-1}	Ref.
Mo	1027–1273	13.3	0	+		134–180	[164]
W	548– 873	2–20	0–1	—		205	[165]
	903–1023	4.7–20	0	—		146	[166]
	953–1153	0.1	0	0		176	[167]
	904–1214	7–26	0	0		163	[168]
	1073–1523	2–35	0	+		146–197	[164]
	950–1150	1–5	0	0		113–130	[169]
	1073	10^{-9}–10^{-6}	0.8				[170]
Re	653– 713	~100	0.53	0.89	1.68	134	[171]
	728– 843	1.3–7	0.7	1.4	2.0	205	[172]
Fe	608– 743	1–4.4	0.5	0.7	1.4	163	[173]
	653– 773	1.3–10.6	0.9–1.0	1.4–1.5	1.5	176–209	[174]
	765– 797	~100	0.9	1.5	1.67	226	[147]
Fe ($-Al_2O_3-K_2O$)	608– 703	100	0.6	0.85	1.42	192	[175]
	693	13	0.48	0.72	1.5	—	[176]
	752	13	0.75	0.38	0.5	—	[176]
Ru	543– 738	1.3–7	1.2	2.0	1.67	188	[172]
	623– 673	80–106	0.6	0.9	1.5	130	[177]
	825–1009	—	1.0	1.75	1.75	247	[162]
Co	643– 753	1.3–7	0.85	1.42	1.67	188	[172]
Rh	693– 773	1.3–7	1.35	2.45	1.81	239	[172]
Ni	573	0.2–5.3	1.0	1.5	1.5	—	[178]
	663– 773	1.3–7	0.96	1.53	1.59	180	[172]
	623– 723	4×10^{-3}–0.17	1	—			[179]
Pt	763– 833	0.001	1.5–2.0	2.2–3.0	1.5	205	[180]
	793– 883	1.3–7	0.96	1.63	1.7	247	[172]
	978	10^{-5}	0.63–0.84	0.25	0.34	17	[181]
	1045–1131	Low p_{H_2}	1	1	1	167	[182]
	1045–1131	High p_{H_2}	1	0	0	167	[182]
	1100–1485	0.03–0.5	1	1	1	184	[183]
	1100–1485	1.3–39	1.4	2.3	1.64	419	[183]
	1206–1488	13–26	1	1	1	586	[168]
	1173–1623	~10^{-3}	1	0	θ	21	[184]
Cu	768– 893	5–67	1	1	1	192	[185]
	993–1183	100	1	1.5	1.5	243–259	[186]
V	673– 753	~100 (Low p_{H_2})	0.5	0	0	138	[187]
	673– 753	High p_{H_2}	1.0	1.5	1.5	138	[187]

The Temkin-Pyzhev equation derived above was in agreement with a number of kinetic measurements made on various catalysts as summarized in Table 5. One characteristic feature of the ammonia synthesis rate is the retardation by the product ammonia, and this is beautifully explained by the Temkin theory. Thus the basic assumption of rate-determining step, N_2 chemisorption, was further supported, while it remained a subject of many later investigations (cf. 4.C.1).

Although equation (28b) was successful in describing a number of experimental data, the experimental values of α are not always the same as shown in Table 5, ranging from 0.5 to 0.8 and depending on temperature. This is partly natural because equation (28) is valid only for an intermediate range of coverage ($\theta_N = 0.2$–0.8). Thus when p_{NH_3} in the gas mixture is very low or very high, the results tend to give lower or higher value of α [191, 192]. Even with moderate coverage, the rate constants given by the Temkin equation are dependent on the H_2/N_2 ratio, and/or total pressure [112, 154, 193, 152]. When the heat of adsorption decreases linearly with an increase in surface coverage as postulated, α is equal to $\Delta E_a/\Delta Q$. As is clear from Figure 4, the ratio $\Delta E_a/\Delta Q$ depends on θ_N. Thus the value of α obtained from rate data can be an average value under the experimental conditions, and is not a constant as postulated in the original theory. Nielsen et al. reexamined the rate data on a triply promoted iron catalyst at high pressures above 643 K, and obtained a best value of $\alpha = 0.75$ [156].

B. Rate Equations Based on a Langmuir Type Isotherm

As described earlier in section 2.C, the Frumkin-Slygin isotherm can be regarded an approximation from a complete isotherm for a non-uniform surface:

$$\theta_N = (1/f) \ln [(1 + a_o p)/(1 + a_o p \exp(-f))] . \tag{12}$$

The invalidity of equation (28) at low coverages arises from the assumption that $a_o p \gg 1 \gg a_o p \exp(-f)$. In order to be valid for low coverages, the assumption $a_o p \gg 1$ should be abandoned, while the assumption $1 \gg a_o p \exp(-f)$ can be retained. That is,

$$\theta_N = (1/f) \ln (1 + a_o p) . \tag{31}$$

Referring to adsorption of N_2, f (for dissociative adsorption of NH_3) is replaced by $f/2$, and p by $Kp_{NH_3}/p_{H_2}^{3/2}$. Thus,

$$\theta_N = (2/f) \ln [1 + a_o K p_{NH_3}/p_{H_2}^{3/2}] . \tag{32}$$

Substitution of equation (32) in the Zeldowitch equation (8) gives a modified Temkin equation, applicable to a wider range of θ_N:

$$V_a = k_a p_{N_2}/[1 + a_o K p_{NH_3}/p_{H_2}^{3/2}]^{2\alpha} \tag{33}$$

The validity of this equation for high pressure synthesis was shown by Nielsen et al. with an α value of 0.75 [156].

When $\alpha = 1$, equation (33) is the rate equation on a uniform surface. Although the nonuniformity of iron surfaces has been demonstrated [16], the catalyst surface under working condition can be apparently less nonuniform than the surface at very low coverage as seen in Figure 4. In fact, the rate equation on a uniform surface

$$V_{a} = k_{a}p_{N_2}/[1 + K_{a}p_{NH_3}/p_{H_2}^{3/2}]^2 \tag{34}$$

has been applied successfully to some results including those under high pressures as summarized in Table 6. It was shown by Kubota and Shindo [194] that the results by Emmett and Kummer and others [154, 155, 198, 199, 200, 201] which were previously adapted to the Temkin equation are also described by equation (34).

Table 6. List of kinetic results on iron adapted to the Langmuir type-rate equation

Equation	Main adsorbed species	References		
		Fe	Fe—Al_2O_3	Fe—Al_2O_3—K_2O
33	N			156
34	N	195	196, 197	153, 154[a], 155[a], 192, 198[a], 199[a], 200[a], 201[a]
35	NH		153, 201[a]	191
38	N_2		197, 201[a]	

[a] plots by Kubota and Shindo [194]

The kinetics of synthesis runs on two different doubly promoted iron catalysts at 491–578 K and atmospheric or reduced pressures are best described by the rate equation

$$V_a = k_a p_{N_2}/[1 + K_a p_{NH_3}/p_{H_2}]^2 \tag{35}$$

where K_a is proportional to the equilibrium constant of the reaction [191]

$$NH_3 \rightleftarrows NH(a) + H_2 \,. \tag{36}$$

This implies that the main adsorbed species in NH(a) instead of N(a) as hitherto generally assumed, which is supported by the deuterium isotope effect on K_a as disscussed in 4.D.1. Later measurements on unpromoted iron [195] or singly promoted iron [196] as well as on doubly promoted iron at high temperatures, 673 to 748 K [202], are in conformity with the main species of N(a). Krabetz and Peters [153] also examined the main species determined by the rate equation in the temperature range 553–643 K using a flow system and "well reduced" catalysts in view of the rather low reduction temperature in the experiments by Ozaki et al. [191]. They found N(a) [equation (34)] on doubly promoted iron and NH(a) [equation (35)] on singly promoted iron, and ascribed the apparent formation of NH(a) to incomplete reduction of catalysts, because the reduction of singly promoted iron (Fe—Al_2O_3) is known to be more difficult than that of doubly promoted iron.

Another possible reason for the formation of NH(a) instead of N(a) would be the reaction temperature. It appears that lower temperatures on doubly promoted iron cause the transformation. In fact Nielsen et al. [156] noted that the Arrhenius plot of rate constant at high pressure gives a break somewhere between 603 and 643 K, suggesting a change in mechanism. Since N(a) is thermodynamically more stable than NH(a), the formation of NH(a) must be caused by a kinetic reason [191]. One possibility is a hydrogen-assisted cleavage of the dinitrogen molecule to form NH(a). We shall come back to this subject in connection with isotopic equilibration in dinitrogen.

Apart from the main species, the rate data at low temperatures and pressures which fit to the equation for a uniform surface [equation (34)] are also represented by the Temkin-Pyzhev equation giving α values ranging from 0.4 to 0.8 depending on temperature [191]. Since the rate constant in the Temkin-Pyzhev equation is given by

$$k_a' = k_a(K_s/a_o)^\alpha \tag{37}$$

k_a must be separated from $(K_s/a_o)^\alpha$ to obtain the activation energy for the reaction, whereas it is impossible when α changes with temperature. In this respect the equation (34) or (35) is more useful for the evaluation of activation energy since it can give k_a directly as noted by Brill and Tauster [192], particularly at low temperature and pressure.

Observed values of the activation energy based on k_a on iron catalysts are in a range from 60 to 100 kJ mol^{-1} when equation (34) or (35) is applied, and depend on the catalyst as summarized in Table 7. Equation (28) gives larger values. The activation energies on other metals have not been so extensively determined. At atmospheric or lower pressures, the values on Ru [203], $Fe-Al_2O_3-K_2O$ [191], and Mo_2N [204] catalysts were found to

Table 7. Activation energy of rate constant of N_2 chemisorption over iron catalyst

Equation	Sample	ΔE_{act}/ kJ mol^{-1}	Ref.
28	$Fe-Al_2O_3-K_2O$	121[a]	[33]
	$Fe-Al_2O_3-K_2O$	130[a]	[155]
	$Fe-MgO$	110	[158]
34	$Fe-Al_2O_3-K_2O-CaO$	76	[201][b]
	$Fe-Al_2O_3-K_2O$	77, 84, 94, 95	[199][b]
	$Fe-Al_2O_3-K_2O$	95	[154][b]
	$Fe-Al_2O_3-K_2O$	73	[192]
	$Fe-Al_2O_3-K_2O$	59	[153]
35	$Fe-Al_2O_3$	84–105	[153]
	$Fe-Al_2O_3-K_2O$	84	[191]

[a] Estimated by equation (37)
[b] Re-plotted by Kubota and Shindo [194]

be 100, 84, and 50 kJ mol^{-1} respectively. It appears that the activation energy increases with a decrease in metal-nitrogen affinity (cf. Figure 3).

It is expected that the adsorption constant K_a in the rate equation (34) decreases as the metal-nitrogen affinity decreases. Thus when K_a is very small, the rate equation is reduced to

$$V_a = k_a p_{N_2} \,. \tag{38}$$

In fact on Ru [203] and Os [163] for which the metal-nitrogen affinity is expected to be very small, the rate data are in agreement with equation (38). A similar situation may be expected when p_{NH_3} is very small as is the case with a very high space velocity. However it is not always so, as discussed later in 4.C.2.

Finally reference should be made to the state of adsorbed dinitrogen. The foregoing discussions are always referred to a scheme by which dinitrogen is dissociated into two nitrogen atoms in the rate-determining adsorption. However as far as the kinetics are concerned, a rate equation for molecular adsorption:

$$V_a = k_a p_{N_2}/(1 + K p_{NH_3}^2/p_{H_2}^3) \tag{39}$$

can describe the results, as pointed out by Brill [197] and also by Kubota and Shindo [194]. Similarly the Temkin-Pyzhev equation is also consistent with the molecular adsorption [205]. Thus mere kinetics cannot be proof of a mechanism.

C. The Rate-Determining Step

1. Confirmation of Rate-Determining Step

Although the rate-determining step of ammonia synthesis had been accepted as chemisorption of N_2 since the theory by Temkin-Pyzhev, a strong debate was raised by Horiuti and coworkers in terms of the "stoichiometric number" of the rate-determining step [206]. The stoichiometric number, ν, is defined as the number of occurrences of an elementary step required for the completion of an overall reaction. For instance, in ammonia synthesis, the values would be, 1 for chemisorption of N_2, 2 for hydrogenation of N(a) and 3 for dissociation of H_2. Since the free energy decrease ($-\Delta G$) corresponding to the overall reaction is made at the rate-determining step r,

$$-\Delta G = \nu_r(-\Delta G_r) \tag{40}$$

where subscript r denotes the value for the rate-determining step. The steady rate of reaction, V_s, is given by

$$V_s = (\vec{V}_r - \overleftarrow{V}_r) = \vec{V}_r[1 - \exp(\Delta G/\nu_r RT)] \tag{41}$$

where $\vec{V}_r$ and $\overleftarrow{V}_r$ are the forward and reverse rates of step r, respectively.

Under the reaction conditions,

$$-\Delta G = RT \ln [K_s p_{N_2} p^3_{H_2} / p^2_{NH_3}] \tag{42}$$

where K_s is the equilibrium constant for the synthesis. Thus in principle v_r can be evaluated from equation (41) together with the observed value of $\vec{V}_r$ as determined by a tracer method.

Early determinations gave a v_r of 2 [207, 208], which means the hydrogenation of N(a) is the rate-determining step, strongly conflicting with the generally accepted mechanism. This striking result gave rise to three further experiments. Bokhoven et al. made extensive and careful measurement of the stoichiometric number, finding $v_r = 1$ [209]. Ozaki, Taylor and Boudart made a search for a deuterium isotope effect in ammonia synthesis, and found no kinetic isotope effect (cf. 4.D.1.). [191] Tamaru made an adsorption measurement during ammonia synthesis, leading to a postulation of double cascade [217] (cf. 4.E.1).

In view of above contradictions, the method for the confirmation of the rate-determining step was improved by Tanaka to give a final conclusion [210–214]. In this method the rate-determining step is identified from the non-equivalence of the isotopic composition of the reactant and product, using a mixture of N_2 and H_2 with ^{15}N-enriched NH_3 as starting material for decomposition. If the reaction $NH_3 = N(a) + 3/2\,H_2$ is in equilibrium, the isotopic composition, Z_N, of N(a) should be identical with that of NH_3, Z_A, and if N(a) is in equilibrium with N_2, Z_N is identical with that of N_2, Z_G. During the decomposition reaction, Z_N can be estimated from the relative rates of formation of $^{30}N_2$ and $^{29}N_2$, given by

$$V_{30}/V_{29} = Z_N/2(1 - Z_N)\,. \tag{43}$$

An analogous method is applicable for the synthesis reaction.

The rate-determining step is identified by comparing Z_N with Z_A and Z_G. Measurements on doubly promoted iron at 703 K [210], and on singly promoted iron at 578–613 K [211], 613–663 K [214], and 678–713 K [213] are all in conformity with N_2 chemisorption and desorption as the rate-determining steps for synthesis and decomposition, respectively. Similar results are obtained for ammonia decomposition on nickel [215].

As an extension of the above method, isotopic mixing in unreacted dinitrogen during the synthesis reaction is also useful, and is an easier method for the confirmation of the rate-determining chemisorption of N_2. No isotopic mixing is detected during the synthesis over Mo_2N at 603 K for 122 hr [204], and on Ru—K—AC* at 505 K for 43 hr [216], when the synthesis run is carried out in a circulating system with a liquid nitrogen trap (cf. also 4.E.3).

* AC = active charcoal.

2. Change of the Rate-Determining Step

In the foregoing discussion a significant amount of ammonia exists in the reaction system, giving rise to retardation of the rate of N_2 chemisorption. The situation may be different when the synthesis is carried out far from equilibrium as realized by increasing the space velocity or by decreasing the reaction temperature. Under a condition where the reverse rate is negligible, the overall rate (V_s) is equal to the rate of chemisorption ($\vec{V}_a$), and possibly to the rate of hydrogenation of adsorbed nitrogen ($\vec{V}_h$). If $\vec{V}_a$ and $\vec{V}_h$ are given by

$$\vec{V}_a = k_a p_{N_2}(1 - \theta_N)^2 \simeq k_a p_{N_2}(1 - 2\theta_N) \quad (44)$$

$$\vec{V}_h = k_h p_{H_2}\theta_N \quad (45)$$

and if $V_s = \vec{V}_a = \vec{V}_h$
V_s is given by eliminating θ_N

$$V_s = k_a k_h P_{N_2} P_{H_2}/[2k_a P_{N_2} + k_h P_{H_2}]$$
$$\approx k' P_{N_2}^m P_{H_2}^m \quad (46)$$

where m and n are apparent orders of reaction with $m < 1$, $n < 1$, and $m + n = 1$.

Temkin et al. found kinetics corresponding to equation (46) with $m = n = 0.5$ under reaction conditions far from equilibrium with very high space velocities of about $10^6\ h^{-1}$ at 627–723 K [205]. In agreement with this, Tanaka [213] estimated, by using ^{15}N, the changes of $\vec{V}_a$ and $\vec{V}_h$ with departure from equilibrium as illustrated in Figure 20. As $-\Delta G$ increases, $\vec{V}_a$

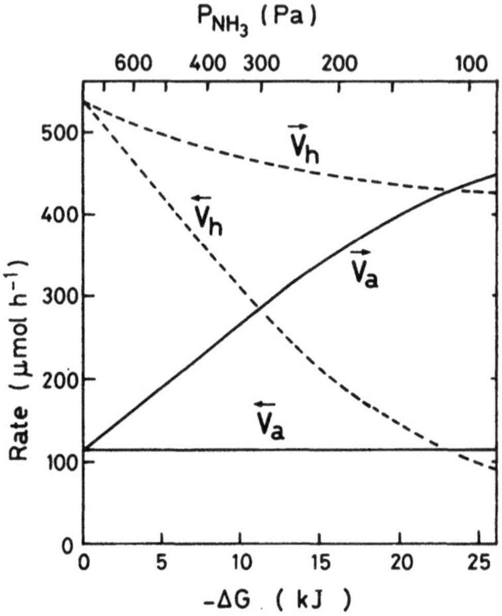

Figure 20. Variation of elementary rates with departure from equilibrium [213] (mol denotes mole)

increases steeply, and $\vec{V}_h$ decreases slowly, attaining ultimately an equal rate of $\vec{V}_a$ and $\vec{V}_h$.

Such changes in $\vec{V}_a$ with a decrease in p_{NH_3} are expected also from surface heterogeneity, since a decrease in p_{NH_3} results in a decrease in θ_N. It is reasonable that an increase in heat of adsorption of nitrogen resulting from the decrease in θ_N enhances $\vec{V}_a$ while decreases $\vec{V}_h$. Thus it is very likely that a single rate-determining step does not exist under such conditions.

For the same reason, at a sufficiently low pressure of H_2 and high temperature, the rate of N_2 chemisorption may exceed the rate of hydrogenation. The kinetics of ammonia decomposition on platinum wire under pressures below 1.3 Pa are found to conform to rate-determining dissociation of NH_3 at around 1273 K, while to desorption of N_2 at around 773 K [180, 181].

Takezawa and Toyoshima claimed a change in the rate-determining step of ammonia decomposition from N_2 desorption at 693 K to dissociation of NH_2(a) at 753 K on the basis of kinetic results [176]. They also claimed a similar change with decrease in the potash content of the iron catalyst [218]. The reason for the anomaly is not clear.

Another reason for change in the rate-determining step may come from the nature of metal in the catalyst. The nitride-forming metals such as vanadium and molybdenum, in particular, seemingly give rise to a faster rate of N_2 chemisorption and a slower hydrogenation rate for the adsorbed nitrogen. In fact, the initial rate of N_2 chemisorption on pure molybdenum is very fast even at 583 K [219]. However the rate decreases rapidly with increase in N_2 uptake, or extent of nitride formation. Thus the rate of N_2 chemisorption on Mo_2N is much slower than on Mo. Even with iron, the rate of ammonia decomposition on iron nitride (Fe_4N) is reportedly two orders of magnitude slower than on pure iron [220]. Since both vanadium and molybdenum are transformed into bulk nitride under the reaction condition of ammonia synthesis, the mechanism should be referred to a nitride surface.

Sebba and coworkers investigated ammonia decomposition over the metal nitrides TiN, VN and Cr_3N_2, finding VN the most active [221]. The kinetics of ammonia synthesis [222] and decomposition [187] on VN at around 773 K follow the Temkin-Pyzhev equation, while the VN catalyst was contaminated with oxygen. The kinetics of ammonia synthesis over uranium nitride (U_2N_3—UN_2) at 648–823 K and 3 MPa [223], as well as over molybdenum nitride (Mo_2N) at 720–828 K [159], also obey the Temkin-Pyzhev equation.

Thus chemisorption of N_2 appears to remain rate-determining over nitrides. As described before, it was confirmed by observing the absence of isotopic mixing in unreacted N_2 over Mo_2N [204]. In this experiment, the ^{15}N content of product ammonia collected in a trap was always lower than that in N_2, indicating that the bulk nitrogen in Mo_2N undergoes an exchange reaction with adsorbed nitrogen or ammonia, in conformity with an equilibrated step of hydrogenation. Moreover, the synthesis rate is as slow as expected [204] from the rate of isotopic exchange of N_2 on Mo_2N [224]. The rate-determining step of ammonia decomposition over iron nitride (Fe_4N) formed on iron films was also ascribed to the desorption of N(a) which is in equilibrium with ammonia [173].

The decomposition on tungsten is quite different from the above. Tungsten is transformed to nitride in the initial stage so that the decomposition proceeds over layers of tungsten nitride [165]. Since the rate of N_2 desorption increases, while the rate of NH_3 dissociation decreases, with increase in the nitrogen uptake, as illustrated in Figure 21, the thickness of the nitride layer is determined by a dynamic balance of two reaction rates, $2NH_3 \rightarrow 2N(a) + 3H_2$, and $2N(a) \rightarrow N_2$. Lower temperatures and higher ammonia pressures result in thicker nitride layers [225].

Although zero order kinetics with respect to ammonia have been reported (cf. Table 5), the reaction order increases at higher temperatures and low ammonia pressures; 2/3 order at 773 K, and 4/5 order at 973–1273 K and $p_{NH_3} = 10^{-6} - 10^{-3}$ Pa [170, 225]. Since no effect of hydrogen is observed on the rate of decomposition, the thermodynamic control of the amount of N(a) by NH_3 is ruled out. Thus the kinetics are in accordance with the rate-determining dissociation of NH_3 to form N(a) at least at high temperatures and low pressures. The isotope effect observed with NH_3/ND_3 is consistent with the above idea [170] (cf. 4.D.2). It appears that the rate-determining step changes from the N_2 desorption at low temperatures to the dissociation of NH_3 at high temperatures because of the larger activation energy associated with N_2 desorption.

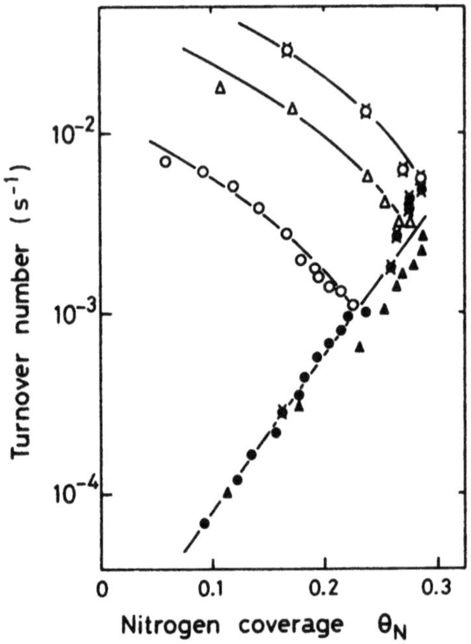

Figure 21. Rates of ammonia uptake (open symbol) and N_2 desorption (solid symbol) on tungsten (partly covered with oxygen) as a function of nitrogen coverage at 1073 K [225].
p_{NH_3}: 4.1×10^{-5} Pa (¤, ✖);
1.2×10^{-5} Pa (△, ▲);
4.0×10^{-6} Pa (○, ●)

D. Isotope Effect

1. Isotope Effect in Ammonia Synthesis

The argument on the rate-determining step made in the 1950's induced the study of a deuterium isotope effect in ammonia synthesis on two different doubly promoted iron catalysts at 491–575 K [191], leading to an unexpected discovery of an inverse isotope effect in which D_2 reacts faster than H_2 with N_2. The kinetics are in conformity with equation (35) where NH, instead of N, is assumed as the main adsorbed species. The kinetic analysis with equation (35) discloses that the observed inverse isotope effect is ascribed to a thermodynamic isotope effect in K_a, $K_H/K_D = 2.7$ (at 523–573 K), giving rise to no kinetic isotope effect in k_a or $(k_a)_H = (k_a)_D$. Since K_a is referred to dissociative adsorption of NH_3 to give NH(a) and H_2, the ratio K_H/K_D is given in terms of the partition function Q by

$$K_H/K_D = Q_{ND_3}Q_{H_2}Q_{NH(a)}/Q_{NH_3}Q_{D_2}Q_{ND(a)}\,. \tag{47}$$

The product $(Q_{ND_3}/Q_{NH_3})\,(Q_{H_2}/Q_{D_2})$ is readily calculated. $Q_{NH(a)}/Q_{ND(a)}$ may be estimated from the zero point energy difference between NH(a) and ND(a), 10.3 kJ mol^{-1} as evaluated from reasonable values of the vibrational frequencies of adsorbed imine radicals. Thus K_H/K_D is calculated as 2.7 at 523 K and 2.4 at 573 K, in excellent agreement with the observed value of 2.7. If the main adsorbed species is N, as in equation (34), calculated values of K_H/K_D are 4.3 at 573 K and 5.4 at 523 K, substantially differing from the observed value. Hence the main adsorbed species of NH is supported by the extent of the isotope effect as well as kinetics on the doubly promoted iron catalysts [191].

The results on unpromoted iron [195] and a singly promoted iron at 578 K [196], however, disclose a larger isotope effect which conforms to the main adsorbed species of N(a) as in equation (34). When H_2—D_2 mixture is used in place of D_2, the isotope effect changes as expected for the reaction of isotopically equilibrated hydrogen, with N being the main adsorbed species [195]. Temkin and coworker [202] confirmed the inverse isotope effect on doubly promoted iron at 673–748 K, the temperature being much higher than the previous work by Ozaki, Taylor and Boudart [191] and the results were interpreted by the Temkin-Pyzhev equation. Since the rate constant of the Temkin-Pyzhev equation is given by $k_a' = k_a(K_s/a_o)^\alpha$ (equation (37)), the isotope effect on k_a', i.e., k_H'/k_D' is equivalent $(K_H/K_D)^\alpha$ because the rate constant k_a and adsorption constant a_o for N_2 chemisorption should be common to both systems. Obtaining an α value of 0.5, they showed that the observed value of k_H'/k_D' is in agreement with the calculated value of $(K_H/K_D)^{0.5}$ where K refer to the synthesis reaction ($N_2 + 3\,H_2 \rightleftharpoons 2\,NH_3$). This means that the main adsorbed species is N at high temperatures even on doubly promoted iron.

Since the inverse isotope effect arises from the adsorption term of the rate equation, its extent should depend on the magnitude of the adsorption constant. The relative magnitude of the adsorption constant is expected to decrease in the order Mo > Fe > Ru [203]. In fact no isotope effect is observed on Ru for which the rate equation is devoid of the adsorption term [203]. The isotope effect on Mo_2N is in conformity with equation (34), indicating the basic mechanism to be the same as on iron [204].

2. Isotope Effect in Ammonia Decomposition

The deuterium isotope effect on the rate of ammonia decomposition was first observed in the 1930's by Jungers and Taylor [166] and also by Barrer [167] on tungsten filaments at around 953 K. Although they worked at different pressures (4–20 kPa and 0.13–13 Pa), both obtained a normal isotope effect of $V_H/V_D = 1.5$–1.6. These results were naturally interpreted as indicating rate-determining dissociation of adsorbed ammonia, since the kinetics of decomposition on W were known to be zero order in ammonia pressure without any retardation by H_2 or N_2 [167] suggesting full coverage by ammonia. If the dissociation of adsorbed ammonia determines the rate of decomposition as supposed, there should be significant amounts of hydrogen (in the form of NH_3 or NH_2) on the catalyst surface during decomposition. But working on adsorption measurements during ammonia decomposition on W powder, Tamaru observed little hydrogen remained on the surface above 773 K [165].

Thus there have been arguments on this point. Field emission studies by Dawson and Hansen [227] suggest a surface species of W_2NNH_2 and LEED studies by Estrup and Anderson [228] and also by May et al. [229] show a similar species. Peng and Dawson reconfirmed the coexistence of hydrogen with adsorbed nitrogen at the decomposition temperature [230].

Very recently, Tamaru and coworkers reinvestigated the isotope effect in ammonia decomposition on W foil [170], showing that NH_3 gives a nitrided layer of tungsten thicker than ND_3 does, and that the rate of N_2 formation is controlled by the thickness of nitrided layer, thus giving rise to a normal isotope effect. Since the isotope effect on thickness is caused by a reactivity difference, the isotope effect is kinetic in nature, whereas the situation might be different at lower temperatures as suggested before (cf. 4.C.2). If the isotope effect is observed in the very beginning of decomposition as is likely in the classic work, the predominant reaction is the formation of N(a) and H_2 so that the normal isotope effect is natural.

E. Elementary Steps

1. Hydrogenation of Preadsorbed Nitrogen

The theory of ammonia synthesis so far discussed implies that the hydrogenation of adsorbed nitrogen is fast enough to establish an equilibrium between NH_3, H_2 and adsorbed nitrogen species, N or NH. Tamaru made an attempt to measure the rate of hydrogenation of preadsorbed nitrogen on

$Fe—Al_2O_3—K_2O$ and found an inverse isotope effect in the rate, i.e. $V_D/V_H = 3 - 4$ at 453 to 593 K [217, 231]. A similar inverse isotope effect was also found for hydrogenation of molybdenum nitride, but the apparent "rate" was proportional to the flow rate of hydrogen, demonstrating that the measured "rate" did not represent the real rate of hydrogenation [204].

If the rate of hydrogenation (V_h) of preadsorbed nitrogen is measured by determining the amount of ammonia in the effluent stream of hydrogen, V_h is given by equation (48) [232]

$$V_h = C\vec{V}_h/[1 + C\vec{V}_h/FK_A p_{H_2}^{3/2} A_N] \approx (C\vec{V}_h)^n (FK_A p_{H_2}^{3/2} A_N)^{1-n} \quad (48)$$

where C ($\geqq 1$) is a correction factor, $\vec{V}_h$ forward rate of hydrogenation, F flow rate of hydrogen, K_A equilibrium constant for the reaction N(a) + 3/2 H_2 = NH_3, and A_N a function of the amount of preadsorbed nitrogen. The power n varies from unity to zero as $\vec{V}_h/F$ increases. The factor C approaches unity as n approaches unity. Thus when $\vec{V}_h$ is very large as accepted, and F is rather small to give $n = 0$, V_h is expressed by equation (49).

$$V_h \approx FK_A p_{H_2}^{3/2} A_N \,. \quad (49)$$

The observation on Mo_2N that $V_h \propto F$ demonstrates equation (49) to be valid under these conditions.

On the basis of equation (49), the isotope effect on V_h is given by

$$V_{hD}/V_{hH} = K_{AD}/K_{AH} \,. \quad (50)$$

In fact, the observed isotope effect ($V_{hD}/V_{hH} = 3.6$ on Mo_2N at 603 K) is very close to the expected value of 3.8 [204]. Similarly the extent of the isotope effect observed by Tamaru [217, 231] is in agreement with this prediction at higher temperatures, and decreases as the temperature is reduced: at 593, 523 and 453 K V_{hD}/V_{hH} was 4.2, 4.0 and 2.7 respectively, while K_{AD}/K_{AH} is 4.0, 5.5 and 8.1 respectively. This is reasonable on the basis of equation (48), since the reduction in $\vec{V}_h$ at lower temperatures should give rise to an increase in the value of n, which reduces the relative weight of K_A in determining the isotope effect. Since these experiments were carried out in circulating systems, the flow rate could not be high enough to realize $n = 1$. The space velocity was 400 hr^{-1} at most.

More recently Takezawa studied the rate of hydrogenation of adsorbed nitrogen on $Fe—Al_2O_3—K_2O$ using a flow reactor with a much higher space velocity of hydrogen, $2\text{–}3 \times 10^4$ hr^{-1}, and found that V_h depends significantly on the temperature of chemisorption [233]. The species chemisorbed at above 673 K (H type nitrogen) exhibit a normal isotope effect $V_{hH}/V_{hD} = 1.25$ at 435 K without any dependence on the flow rate F, while the species chemisorbed at around 673 K (L-type nitrogen) show an inverse isotope effect $V_{hD}/V_{hH} = 1.35$ at 435 K with V_h being proportional to the

0.75 power of F, indicating an n value of 0.25. Although the high space velocity as well as the low temperature in this experiment gave rise to a high value of n, it is still far from unity, demonstrating that $\vec{V}_h$ is extremely fast with respect to L type nitrogen.

The rapid establishment of equilibrium between the adsorbed nitrogen and NH_3—H_2 gas was confirmed very recently by Rambeau and Amariglio on a doubly promoted iron [273] and also on pure ruthenium [274] using an efficient method to determine the amount of ammonia formed. In the case of Ru, the initial rate of ammonia formation in the hydrogenation of preadsorbed N_2 is proportional to flow rate of hydrogen at above 373 K, and becomes independent of it as low as 248 K and below, demonstrating the rapid rate of hydrogenation.

2. *Isotopic Equilibration of Dinitrogen*

If the chemisorption or dissociation of dinitrogen is rate-determining in the ammonia synthesis as has been demonstrated, the isotopic equilibration of dinitrogen

$$^{15}N_2 + {}^{14}N_2 \rightleftharpoons 2\,{}^{15}N^{14}N \qquad (51)$$

must be a slow process. This was shown first by Joris and Taylor on iron catalysts [234], although the abnormally slow rates of exchange were later ascribed to incomplete reduction of the catalysts [171, 235]. Kummer and Emmett confirmed that the rate of equilibration is comparable with the rate of nitrogen desorption from the same catalyst at 773 K [235]. Accordingly, catalysts which are active for nitrogen equilibration are also active for ammonia synthesis. Known active metals include Fe [90, 236, 237], Os [238], Co [239] and Raney Ni [239].

The equilibration on Os is measurable at as low as 473 K, and is rapid at 573 K [238].

The equilibration requires higher temperatures over Re [171], W [234, 240], and Mo [54, 224, 241]. Recently a novel and highly active catalyst system, transition metal/active carbon/alkali metal, has been found for ammonia synthesis [110, 111, 242]. The relative activities of transition metals on carbon promoted by potassium run parallel for ammonia synthesis [112] and for isotopic equilibration [243]. (cf. Figure 12)

When an isotopically enriched dinitrogen is introduced on adsorbed nitrogen, the displacement of adsorbed nitrogen takes place in addition to the isotopic equilibration [90]. The rate of displacement (V) can be measured by following the variation with time of the atomic fraction (f_g) of ^{15}N in N_2 introduced without disturbing the preestablished adsorption equilibrium [90]. The result on unpromoted iron at 653 K discloses an initial rapid displacement followed by a slower one, and more than 60 percent of the adsorbed nitrogen is displaced within the initial stage.

The rate of isotopic equilibration (R) is measured by following the variation of the mole fraction (X_g) of $^{29}N_2$ or $^{30}N_2$ with time. The value of R on the unpromoted iron is about one fifth of V in the initial stage, which agrees well with V in the later stage. This shows that the displacement is accompanied by isotopic mixing in the later stage, whereas it is unaccompanied by isotopic mixing in the initial stage. Displacement without accompanying isotopic mixing demonstrates that the portion of adsorbed nitrogen (more than 60 percent at 653 K) on the unpromoted iron is undissociated. The results on promoted iron catalysts [91, 92, 242, 244] however, show that V agrees with R throughout the displacement process, suggesting that dinitrogen, once adsorbed, is rapidly equilibrated with adsorbed nitrogen before desorption. In view of this, the role of promoters, Al_2O_3 or K_2O, is not a mere increase in the surface area.

The rate of isotopic nitrogen equilibration is enhanced by the presence of hydrogen on potash-promoted iron [91], while not on unpromoted iron [90]. This subject will be discussed in the subsequent section.

The kinetics of the equilibration have been studied on Fe and Ru. A reaction order of one half with respect to nitrogen pressure is obtained on pure iron at 623 K [237], doubly promoted iron at 703 K [245] and at 773 K [249] and Ru/AC/K at 553 K [246]. Schulz gave an explanation for the half order kinetics on pure iron in terms of molecular adsorption on a surface mostly covered by N atoms [237]. However, since V is much larger than R on pure iron as indicated above, the surface exchange process is probably a slow step. The reaction order on Ru catalysts increases with temperature as follows: from 0.26 at 553 K to 0.45 at 593 K on Ru—Al_2O_3—K [247], from 0.44 at 593 K to 0.79 at at 653 K on Ru—K [247], and first order on Ru—Al_2O_3 at 653–713 K [247]. The kinetics on Ru—K and Ru—Al_2O_3—K are represented by a Langmuir type equation for dissociative adsorption;

$$R = kp_{N_2}/(1 + \sqrt{Kp_{N_2}})^2 \qquad (52)$$

in conformity with a Bonhoeffer-Farkas type mechanism [248]. The heat of chemisorption of nitrogen on Ru—K is estimated from the adsorption constant in the above equation to be as high as 167 kJ mol^{-1}.

3. *Effects of Hydrogen on Activation of Dinitrogen*

It was known previously that isotopic nitrogen equilibration is accelerated in the presence of hydrogen over promoted iron catalysts (Fe—Al_2O_3—K_2O [234] Fe—Al_2O_3 [235]). Thus it was remarkable that no enhancement by hydrogen was found on pure iron in two different laboratories [90, 236]. Subsequent study on doubly-promoted iron at 598–643 K confirmed the enhancement by hydrogen, the extent of which is more than 5-fold, with a decrease in the activation energy [91], while the effect of hydrogen on singly-promoted iron is complex [244]. That is, the presence of hydrogen increases the activation energy so that the rate of nitrogen equilibration is decreased at low temperatures, but increased at high temperatures [244]. These results are illustrated in Figure 22 as Arrhenius plots of the rates. However Boreskova et al. claimed no enhancement by hydrogen over doubly

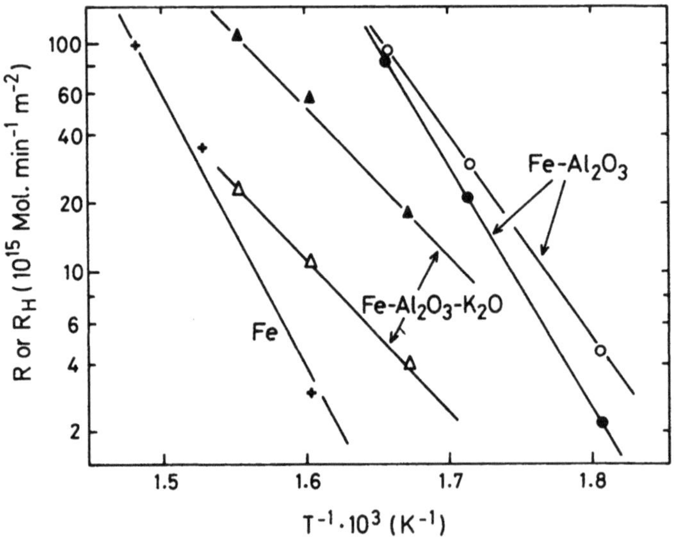

Figüre 22. Rates of N_2 equilibration over iron catalysts in the absence (open symbols) and presence (filled symbols) of hydrogen. The H_2/N_2 ratios were about 1/5 (●) and 1/10 (▲) [244] (Mol. denotes molecules)

promoted iron, and ascribed the enhancement to incomplete reduction of catalyst [92]. Takezawa and Toyoshima observed enhancement by hydrogen at 623 K, whereas none at 703 K, on a doubly-promoted iron [250]. The reason of discrepancy is not clear.

On the other hand Tamaru found that the rate of nitrogen uptake by doubly-promoted iron at 523 K is enhanced by hydrogen [188, 189]. An analogous result was obtained by Sastri and Srikant at elevated pressures up to 3.5 MPa and temperatures up to 473 K on doubly promoted iron [190]. However Scholten observed no enhancement by hydrogen on singly promoted iron [226]. Chesnokova et al. examined the effect of promoters on the enhancement by hydrogen and found that the initial rate of N_2 chemisorption at 748 K is enhanced by preadsorbed hydrogen by a factor of 6 on Fe—K_2O and Fe—K_2O—Al_2O_3, while no enhancement was found on two different samples of Fe—Al_2O_3 [245, 251].

Summarizing the above results, it appears that both isotopic equilibration and N_2 chemisorption are enhanced by hydrogen only on potash-promoted iron catalysts. In fact, hydrogen-enhanced equilibration is observed on iron after the addition of potash [244]. Thus it has been suggested that hydrogen is chemically involved in the dissociation of the N_2 molecule [91, 191] via

$$N_2 + 2\,H(a) \rightleftharpoons 2\,NH(a)\,. \qquad (53)$$

The main adsorbed species of NH(a) observed during ammonia synthesis on doubly promoted iron at lower temperatures is in agreement with this scheme.

Although the rate of isotopic equilibration is somewhat enhanced by hydrogen on Ru at 723 K, the effect of hydrogen is reversed on Ru—K, Ru—K—Al_2O_3 and Ru—K—AC [216] as illustrated in Figure 23 for a Ru—K catalyst. The equilibration rate decreases as the ratio H_2/N_2 increases, with the effect being larger when ammonia is removed by trapping (shown by arrow). The rate of equilibration with ammonia trapped finally reaches zero at about $H_2/N_2 = 3$. This is reasonable if N_2 chemisorption is rate-determining in the ammonia synthesis since the chemisorbed nitrogen is rapidly hydrogenated to ammonia and thus never returns to dinitrogen. The steady rate of ammonia formation (V_s) is given by the forward and

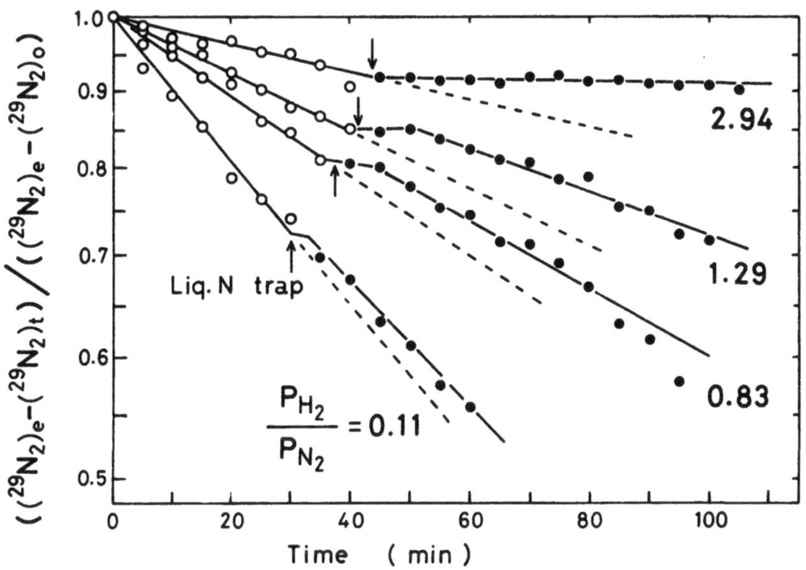

Figure 23. Isotopic equilibration of N_2 in the presence of hydrogen on Ru-K at 593 K, p_{N_2} = 8.0 kPa [216]

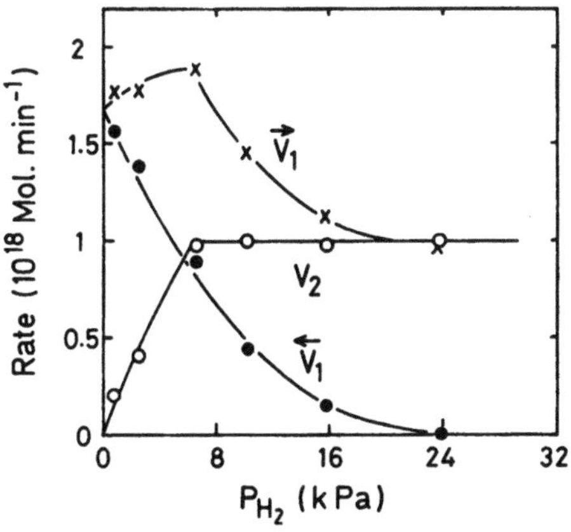

Figure 24. Relative rates of elementary steps of the ammonia synthesis as function of hydrogen pressure.
(Ru—K, 593 K, p_{N_2} = 8.0 kPa)
(Mol. denotes molecule)

backward rates of N_2 chemisorption, $\vec{V}$ and $\overleftarrow{V}$, respectively. $\overleftarrow{V}$ is given by the rate of equilibration (R). From the experimental values of V_s and R, $\vec{V}$ and $\overleftarrow{V}$ are evaluated as shown in Figure 24 as functions of hydrogen partial pressure. $\vec{V}$ passes a maximum as the hydrogen partial pressure increases, presumably because of an increase in available surface caused by the removal of adsorbed nitrogen, while the adsorption site is taken over by hydrogen at higher pressures.

4. Instrumental Studies of Adsorbed Species During the Reaction

Since the adsorbed species of NH(a) instead of N(a) was concluded from kinetics as well as the isotope effect of ammonia synthesis [191], the nature of adsorbed species has been the subject of instrumental studies. Nakata and Matsushita treated an iron on silica with N_2—H_2 mixture at 773 K and observed IR spectra of adsorbed species. The spectra at room temperature give three bands assignable to NH_2 at 3380, 3290 and 1610 cm^{-1} and another band assignable to NH at 3200 cm^{-1} [252], while the relative intensity of the 3200 cm^{-1} band increases with an increase in the temperature of IR measurement [252]. Similar room temperature measurements by Brill et al. on Fe—MgO pretreated with N_2—H_2 at 673 K gave many more bands (3270, 3200, 1610, 1525, 1225, 1140, 975, 780, 815 and 750 cm^{-1}) [80]. Although ammonia adsorbed on the same catalyst did not give so many bands, adsorbed hydrazine gave similar bands, suggesting an N_2H_x species. When Fe—MgO was treated with ammonia at 673 K and cooled to room temperature, two band were observed at 2200 and 2050 cm^{-1}, which were assigned to two kind of molecular nitrogen [79a]. Very recently similar bands were found on pure MgO treated with NH_3 at 773 K. [79b]

Ohkawa et al. observed a band at 500 cm^{-1} ascribable to NH_x on evaporated iron films treated with NH_3 at 673 K [253].

Melton and Emmett studied the transient species evolved from the catalyst surface during decomposition of NH_3 at very low pressure over platinum and iron filaments by means of a direct inlet mass spectrometer. It was found that three radicals, N, NH and NH_2 are desorbed from the surface when the filament temperature is suddenly changed from 348 to 1273 K [254]. Interestingly the relative amounts of the species are strikingly different from platinum to iron, namely, $NH_2 > NH > N$ over Pt but $N > NH \gg NH_2$ over Fe. A recent SIMS study by Weiss et al. on the system NH_3/Fe (110) shows NH(a) as the predominant adsorbed species above room temperatures [255].

Shvachko et al. adopted bombardment with argon ions to knock out adsorbed species on iron during both synthesis and decomposition of ammonia, identifying the species with mass spectrometry [256]. The adsorbed species are distinguished from the decomposition products by the temperature dependence. Since the appearence of NH^+ is quite different from others when the temperature is raised in a N_2—H_2 mixture, NH^+ seems to come from adsorbed NH species, while other species such as NH_2^+ or N^+ are probably formed by secondary decomposition of NH_3. The ion beam intensities of FeN_2^+ and NH_3^+ run parallel to each other reaching a maximum at 673 K. Thus it apears that NH_3 is formed from Fe—N_2, suggesting a pathway:

$$N_2 \rightarrow Fe\text{-}N_2 \xrightarrow{H_2} Fe{<}\begin{matrix}NH\\NH\end{matrix} \xrightarrow{H_2} NH_3\,. \tag{54}$$

Schmidt used the field emission technique to examine the adsorbed species on iron in contact with a N_2—H_2 mixture and found only H_2^+, N_2^+ and N_2H^+, but no N^+ even at 473 K [257].

Ertl and coworkers made XPS measurements of adsorbed species on doubly promoted iron treated with a N_2—H_2 mixture at 623 K and cooled in the gas mixture, finding one peak ascribable to NH(a) and/or NH_2(a) on Fe, whereas it disappeared by evacuation at 473 K [97].

5. *Dinitrogen Intermediate*

It has been widely accepted that ammonia synthesis proceeds through dissociation of dinitrogen followed by hydrogenation. Brill proposed an alternative mechanism in which hydrogenation to form hydrazine is followed dissociation [197, 258]. As mentioned earlier, the kinetics of ammonia synthesis may be interpreted in terms of either mechanism. The proposal is based on an IR observation on Fe—MgO ascribable to adsorbed hydrazine as described before (4.E.4) [80].

Block and Schulz-Ekloff examined the decomposition of hydrazine on Fe—MgO using mixtures of $^{15}N_2H_4$ and $^{14}N_2H_4$ [259] confirming that the formation of $^{14}N^{15}N$ is negligible at 299–638 K. The molecular identity of hydrazine is preserved in the nitrogen formed. An analogous result was obtained by Davis and Sayer on Rh—Al_2O_3 at 273–323 K [260]. These results exclude the recombination of a mononitrogen intermediate to form dinitrogen:

$$N_2H_4 \rightarrow 2\,NH_2 \rightarrow 2\,N \rightarrow N_2\,. \tag{55}$$

However, it is also consistent with the following mechanism for hydrazine decomposition [261]

$$\begin{aligned}&N_2H_4 \longrightarrow 2NH_2(a)\\&N_2H_4 + NH_2(a) \rightarrow N_2H_3(a) + NH_3 \dashrightarrow N_2\,.\end{aligned} \tag{56}$$

This scheme accounts for the simultaneous formation of NH_3 and is consistent with the preservation of molecular identity in dinitrogen [260].

Since hydrazine is readily decomposed to give N_2 and NH_3 on most metals below 523 K, the hydrazine intermediate is very unlikely except for very low temperatures [261].

However it is to be noted that the L type of adsorbed nitrogen which seems to be molecular nitrogen [88] is much more active than atomic nitrogen (H type) for hydrogenation [233]. It is very likely that the molecularly adsorbed N_2 is on the reaction path to form ammonia as suggested by equation (54).

5. Nature of the Active Surface

A. Structural Aspect

1. Importance of (111) Plane

The heterogeneity of iron surfaces has been demonstrated in a number of ways. One of the reasons for the heterogeneity may arise from a difference in activity of exposed faces. The (111) plane is the most loosely packed simple plane of α-Fe and thus it has been supposed to possess the highest surface energy and the highest catalytic activity among the planes [87]. The change in chemisorption of CO or H_2 per surface area upon heat treatment was interpreted in terms of a decrease in the relative exposure of the (111) plane [262]. In fact Brill et al. found by field electron microscopy that, when an iron tip with physisorbed N_2 is heated for about 3 sec above 673 K, chemisorbed N_2 appears on the (111) plane, accompanied by a rise in the work function [258]. This indicates N_2 to be selectively chemisorbed on the (111) plane, stabilizing this thermodynamically unstable plane. More recently it was confirmed that the rate of N_2 chemisorption is much faster on Fe (111) than on Fe (100) or (110) [19, 20]. Thus the ammonia synthesis activity of iron catalysts has been discussed in terms of relative exposure of Fe (111). According to Brill et al. [258], metallic iron from magnetite is more active after reduction with a 3 H_2/N_2 mixture than after reduction with H_2, presumably because of a larger fraction of Fe (111) exposed. More recently, McAllister and Hansen showed that the rate of decomposition of ammonia on single crystals of tungsten is significantly higher on (111) plane than on (100) and (110) planes [263].

2. Activation of Iron Surface with N_2—H_2 Mixture

The importance of reduction by a N_2—H_2 mixture was noted earlier. During the treatment with 3 H_2/N_2 gas at the relatively low temperature of 578 K after a reduction with H_2 at 673 K, the catalytic activity for ammonia synthesis increases slowly to a steady value for about 40 hr [191, 195]. These findings led Amariglio and Rambeau to a thorough investigation of transient activity in ammonia synthesis on iron catalysts [264]. Figure 25 illustrates the effect of various treatments on the rate of ammonia production (R_s) at a fixed temperature of 556 K using a stoichiometric mixture. Curve 1 shows that the synthesis run at higher temperatures up to 773 K using the stoichiometric mixture increases the initial value of R_s at 556 K, with the extent increasing with the temperature of runs. Curve 2 shows the effect

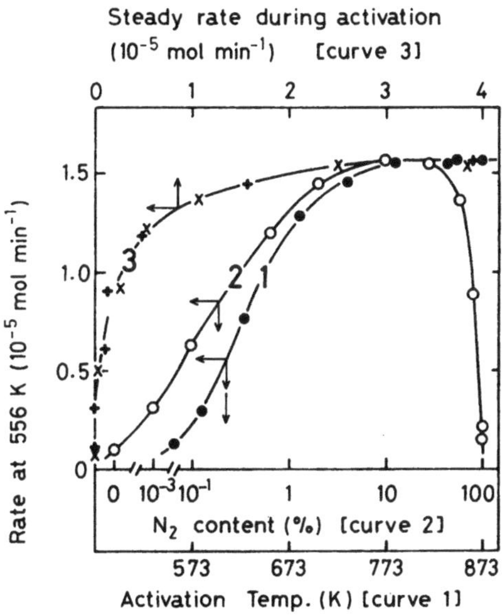

Figure 25. Rates of ammonia production from stoichiometric mixture on an iron catalyst (lg) at 556 K following activations by: the stoichiometric mixture at various temperatures (curve 1); different mixtures of varying H_2/N_2 ratio at 782 K, plotted against N_2 content (curve 2); stoichiometric mixture with different flow rates at 782 K, plotted against the steady rate during the activation period (curve 3).

of synthesis run at 782 using H_2—N_2 mixtures of different compositions. It is clear that R_s at 556 K strikingly increases with the N_2 content in the 0–1 percent range, passes a broad maximum at around 10–20 percent N_2, and decreases to a value of little effect at 100 percent N_2, showing that neither pure H_2 nor pure N_2 is effective in increasing R_s. Curve 3 shows that the initial value of R_s at 556 K is well correlated with the steady rate at 782 K using different compositions of N_2—H_2 mixtures at a fixed flow rate, or using a stoichiometric mixture at different flow rates.

All the results shown in Figure 25 demonstrate that the catalyst surface is modified to give a higher activity by treatment with a H_2—N_2 mixture at higher temperatures, although a prolonged run at 556 K results in a decrease in R_s to a steady value depending on the reaction conditions. One might suspect the activation to be caused by some impurity accompanied in the reactant. However this has been reasonably ruled out. The activation can be ascribed to the ammonia synthesis reaction itself. It appears that the structure of the catalyst surface is transformed so as to adjust itself to the atmosphere under the reaction condition [264].

The activation of an iron surface by a N_2—H_2 mixture as described above seems to lead to the reduction of iron catalysts at relatively low temperatures [195]. Although catalyst reduction at temperatures lower than 770 K has been regarded as "incomplete reduction", Ertl and Thiele recently showed using XPS that treatment of doubly promoted iron catalysts in a N_2—H_2 mixture (10^5 Pa) at 623–673 K causes complete reduction of iron [97].

3. *Structure of Activated Surface*

Boudart and coworkers observed an analogous activation of an iron surface [158, 265]. When small metallic iron particles between 1.5 and 4.0 nm dispersed on MgO are treated with ammonia at 680 K to give iron nitride followed by decomposition in a H_2—N_2 mixture, the iron particles are activated to give a higher activity in terms of turn over number, while the high activity returns to its original lower value after treatment of the catalyst in pure H_2. Such activation by NH_3 and H_2—N_2 treatments is less extensive with larger iron particles of 30 nm for which the turn over number is an order of magnitude larger than that for smaller particles. It appears that the surface of larger particles can be readily tranformed to the active form by the treatment with N_2—H_2 mixtures, while that of smaller particles requires treatment with NH_3 to be activated. Since the ammonia treatment gives rise to a higher virtual pressure of N_2, the nature of activation seems to be a nitrogen induced surface reconstruction.

One significant change in the surface property caused by the ammonia treatment is found in a decrease in CO chemisorption by the Fe—MgO despite the increase in catalytic activity, suggesting a surface reconstruction to give a structure which is unfavorable for the chemisorption of CO [266]. Since the cross-sectional size of the dumbbell molecule of CO is larger than the size of an iron atom, an adsorbed CO on a surface atom would inhibit adsorption on the nearest neighbor atoms in the surface. It is accordinghy expected that the adsorption of CO per surface area decreases as the number of nearest neighbor atoms of the adsorption sites increases. The decrease in CO chemisorption after the ammonia treatment suggests a surface reconstruction to give adsorption sites with more nearest neighbors [266].

Now the adsorption sites, C_i, are defined by the number, i, of nearest neighbor iron atoms. As shown in Figure 26, the number of nearest neighbor atoms in the *surface* increases from 0 for a C_4 atom to 4 for a C_6 atom. Since the C_7 atom is located inside the outermost three atoms, it cannot chemisorb a CO molecule which is larger than an iron atom. If the surface is reconstructed to change the (100) plane to the (110) or (111) plane, the decrease in CO chemisorption is reasonable, since the number of C_6 or C_7 atoms increases in place of C_4 atoms.

Further information from Mössbauer spectroscopy also gives evidence for the reconstruction. The metallic iron spectral area decreases upon ammonia treatment and is restored to the original value by hydrogen treatment, as is the case with the catalytic activity [266]. The reversibility of the change sug-

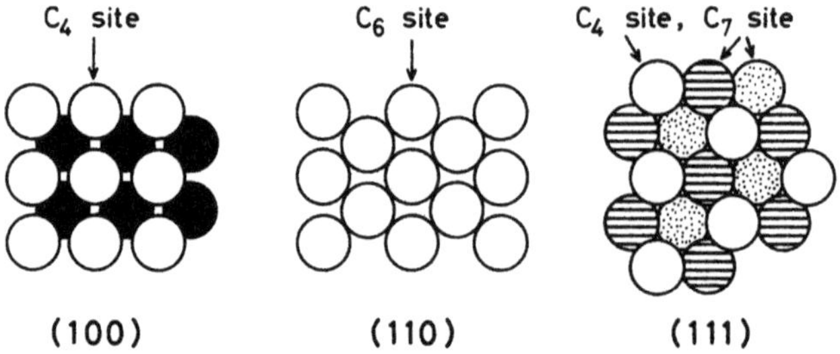

Figure 26. Low index bcc surface planes and associated surface atoms

gests that the decrease in the spectral area is due to a decrease in the magnetic anisotropy of the surface, which is likely more extensive with the C_7 sites than with the C_6 site. Thus it appears that C_7 sites are created by the ammonia treatment to give the more active surface. Since many C_7 iron atoms are available on the (111) plane, the conclusion is in agreement with the previous findings.

The structure-sensitive character of the ammonia synthesis reaction as suggested by the higher specific activity (turn over number) for large iron particles than for small ones [158] is now further supported by the enhanced activity of the special sites.

B. Electronic Aspects

1. The Promoter Action of Potash in Iron Catalysts

In the foregoing sections significant promoter effects of potash are noted for iron catalysts. They are summarized as:

a) Increase in the specific activity per surface iron atom.
b) Induction of H_2-enhanced chemisorption and isotopic equilibration of N_2 which seemingly gives rise to the formation of NH on doubly-promoted iron.

One of the roles of potash promoter was formerly ascribed to neutralization of acidic alumina to promote desorption of product ammonia [267]. But this does not explain above findings.

Electronic effects of potash promoter have been noted by Russian workers. Dmitrenko et al. [268] showed a significant effect of electric polarization of iron catalysts on their catalytic activity for ammonia synthesis. The catalyst powder was pasted on the wall of a glass reactor by use of water glass, forming an electrode, and a nickel electrode was placed on the other side of the wall, with both electrodes in contact with the stoichiometric $3\,H_2—N_2$ mixture. A direct current of 300 V was applied across the two electrodes at 623 K. The catalytic activity is increased by cathodic and decreased by anodic polarization of the iron electrode. With different iron catalysts, the more active the catalyst, the smaller the effect of cathodic polarization and the larger the effect of anodic polarization [269]. It is obvious that an electron-rich state of iron is associated with high activity, which seems to be reasonable in view of the electronegative nature of N_2 to be chemisorbed.

The above finding indicates the importance of the electronic state of the catalyst. According to Enikeev et al. [270, 271] the addition of potash results in a decrease in the electron work function of iron and the ammonia synthesis activity increases with decrease in the work function, accompanying a decrease in the activation energy for the synthesis reaction. Ivanov et al. confirmed that addition of Al_2O_3 increases, while potash decreases, the work function of iron [271, 272].

To sum up, the rate-determining chemisorption of N_2 is promoted by a decrease in work function or an increase in the electron concentration of iron. Indeed, the rate of isotopic equilibration based on the iron surface area is three to four times higher on Fe—K_2O—Al_2O_3 than on pure iron, even in the absence of H_2 [91].

2. Chemical Evidence for the Electronic Effect

If the observed effects of potash are really produced by electron donation to iron, metallic potassium would be more effective due to its much lower electronegativity. On this idea a new catalyst system promoted by metallic potassium was designed, with active carbon (AC) as support which was expected to act as a medium for electron transfer [106, 139]. The synthesis activities of transition metals in this system have already been shown in Figure 12. Figure 27 shows the effect of alkali metals on the rate of ammonia synthesis on Ru, which is most active in this system [107, 139]. The activity increases with increase in both the amount and the electropositivity of the alkali metal, demonstrating the electron donation to be effective. It is remarkable that the transition metals, known to be active for ammonia synthesis, lose their activities when supported on active carbon [106, 139], suggesting an inhibitory effect of carbon ascribable to its electron-accepting properties. In fact, iron atoms intercalated into graphite carry some positive charge even after reduction treatments, as demonstrated by Mössbauer spectroscopy [275]. The situation is reversed on addition of alkali metal resulting in a remarkable gain in activity, while potassium metal added to active carbon alone gives no activity [106, 139, 242].

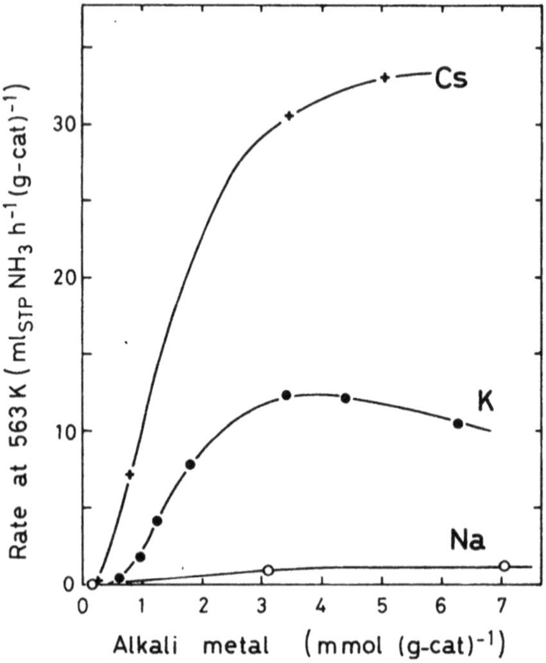

Figure 27. Rate of ammonia synthesis on Ru—AC—alkali metal catalyst as a function of alkali metal content at 563 K [106].

The effect of added metallic potassium is similarly observed with respect to isotopic equilibration of N_2 on Ru—AC [246] as well as on pure ruthenium [247], as shown in Figure 28 as Arrhenius plots of the rate of isotopic equilibration. The addition of metallic potassium results in about a 500 fold increase in the activity of Ru at 673 K, while the increase in the ammonia synthesis rate is less extensive (25 fold), due to the inhibition by coexistent hydrogen (cf. 4.E.3) [216]. The inhibition by hydrogen is accounted for by competitive strong adsorption of hydrogen as well as by a decreased adsorption constant for N_2 [216].

An analogous catalyst system including graphite has been reported by Icuikawa et al. [276]. It is known that transition metal chlorides are intercalated into layered lattice of graphite [277]. Volpin et al. showed that on reduction of these compounds, transition metals remain intercalated as metal atoms giving no activity, although they become activated for ammonia synthesis by the addition of metallic potassium [275]. In the case of iron-graphite, the activity increases with an increase in the K—Fe ratio, and passes through a maximum at around K/Fe = 6, at which the isomer shift in the Mössbauer spectra for iron passes a minimum, as shown in Figure 29. The decrease in isomer shift is ascribable to an increase in the *s*-electron density on the iron atoms, giving direct evidence for electronic promotion. The effect of a further increase in the K/Fe ratio is ascribed to an increase in the *d*-electron density. There might be an optimum in *d*-electron density to give the optimum activity.

Transition metals are also activated by alloying with electropositive metals. Raney Ru prepared from Ru—Al alloy is more active than pure Ru in terms of specific activity per surface Ru, suggesting promotion by residual Al

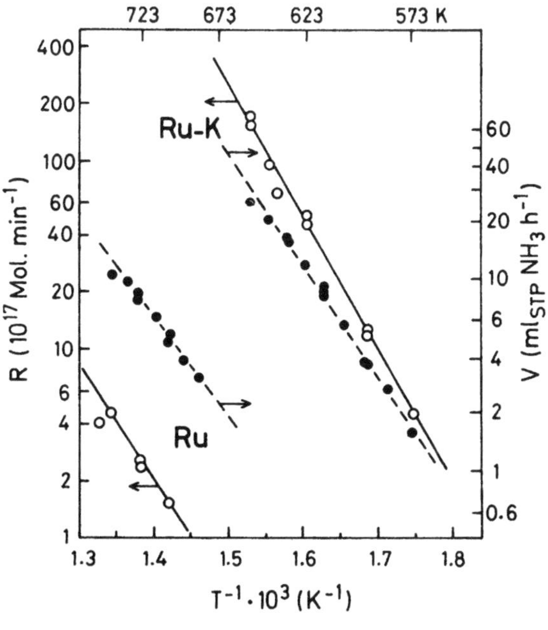

Figure 28. Enhancement of rates of isotopic equilibration (○) (p_{N_2} = 20 kPa) and of ammonia synthesis (●) (pressure of $N_2 + 3\,H_2$ = 80 kPa) by addition of potassium to ruthenium (1.6 g) [216]. (Mol. denotes molecules)

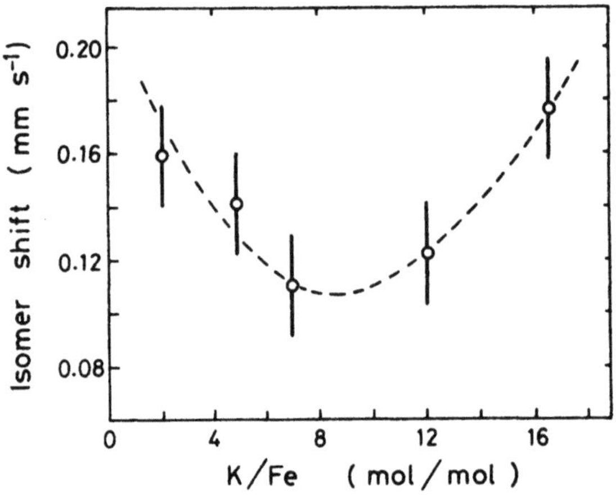

Figure 29. Effect of K/Fe ratio in graphite-$FeCl_3$-K on the electronic state of iron [275].

remaining after leaching, while it is further activated by addition of metallic potassium to give a highly active catalyst which works even at 373 K [278]. Rare earth-transition metal intermetallics such as $CeCo_3$, $CeCo_5$, $CeRu_2$, and $CeFe_2$ are claimed to be more active than doubly promoted iron catalyst on a BET area basis for ammonia synthesis under elevated pressures and at temperatures up to 873 K, while the intermetallics are decomposed into finely dispersed transition metal and cerium nitrides during reaction [279]. The high specific activities observed suggest an electron donation from the rare earths [279].

Not only alkali metals, but also alkali metal oxides act as efficient promoters. Toyoshima and Suzuki showed that N_2 chemisorption is detectable only above 523 K on cobalt, but above room temperature on potash-promoted cobalt [11]. Although the chemisorbed nitrogen is readily hydrogenated

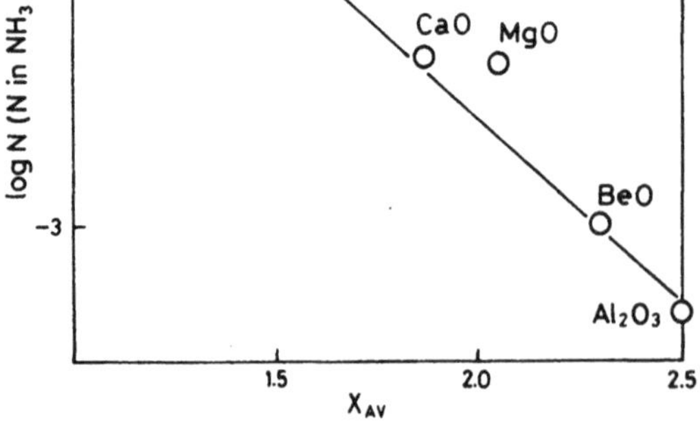

Figure 30. Specific activity per surface ruthenium at 667 K as a function of electronegativity (X_{Av}) of support oxide [280].

to NH_3 at above 323 K, the catalytic synthesis from a N_2—H_2 mixture requires much higher temperatures, because the N_2 chemisorption is strongly retarded by prechemisorbed hydrogen.

The ammonia synthesis activities of Ru promoted with basic oxides are given in Figure 30 as a function of the electronegativity of the basic oxides as estimated by the geometric mean of Pauling's value for the elements. It is clear that the activity increases with decrease in electronegativity, in conformity with the activation by electron donation [280].

3. The Nature of Promotion by Alkali Metal

If the promotion by alkali metal is caused solely by electron donation to the transition metal as discussed above, it is natural to expect a decrease in the activation energy. Rudnitsky et al. developed a theory of promotion in which the potash-promotion of iron is ascribed solely to a decrease in the activation energy caused by the Fe—K dipole on the surface [281]. It appears from Figure 28, however, that the extensive promotion by potassium metal for the equilibration reaction on ruthenium is mostly ascribable to the pre-exponential factor. This requires further insight. The kinetic analysis of the isotopic equilibration based on a Langmuir type equation [equation (52)] gives values for the activation energy (E_0) associated with k and the heat of adsorption (Q) of N_2 during the reaction to be 59 $\pm$ 13 and 167 $\pm$ 25 kJ mol^{-1}, respectively, on a Ru—K catalyst [216]. Since no comparable values are available on unpromoted Ru, the extent of decrease in E_0 is uncertain. However the value of Q is remarkably high as a heat of nitrogen chemisorption on ruthenium (cf. 2.B), disclosing the effect of potassium.

It is known that the heat of chemisorption of CO on iron is increased by addition of potash, presumably due to an increase in electron concentration which strengthens the Fe—C bond simultaneously weakening the C—O bond [282]. A similar effect can be expected for the chemisorption of N_2, strenthening the Ru—N bond and weakening the N≡N bond.

On the other hand it has been noted that the catalytic activity of Ru—K is quite stable for a long period, while Fe—K is less stable, suggesting a strong interaction between Ru and K. Although a possibility first considered to explain the high stability was the formation of a compound between Ru and K as is the case for CsAu [283], electric resistance measurements disclosed the compound CsRu to be unlikely [284]. It was noted, however, that the surface area of Ru—K increased by about 50 percent after the ammonia synthesis runs, suggesting a "corrosive chemisorption" of N_2 [216], and that the amount of N_2 uptake by Ru—K—Al_2O_3 far exceeds the number of surface ruthenium atoms, suggesting an absorption of nitrogen into the bulk [285, 287]. Indeed, Ru—K absorbs a large amount of N_2 to give a complex $(K—N_2—Ru)_n$ as evidenced by the following observations:

1) The amounts of potassium stably held by ruthenium powder after evacuation at 623 K are equal to the N_2 uptake on a mole basis [284, 286].
2) The N_2 uptake by Ru—K dispersed on Al_2O_3 or MgO is close to a ratio N_2/Ru of unity [285, 287].

3) The absorbed N_2 results in an IR bands at 2020 cm^{-1} and 520 cm^{-1} which are assigned to $N\equiv N$ and Ru—N, repectively, as confirmed by ^{15}N labelling [81, 82].
4) The ruthenium powder treated with N_2 and K gives hydrazine and ammonia on hydrolysis [284].

Thus the real surface of the Ru—K catalyst during the synthesis must be formed on the complex $(K—N_2—Ru)_n$ in which Ru assumes an anionic character. Further addition of potassium on this surface gives rise to an increase in the activity [284].

6. Activation of Dinitrogen by Metal Complexes

A. Metal Dinitrogen Complexes

Since the incidental discovery of the metal dinitrogen complex, $[Ru(NH_3)_5N_2]^{2+}$, during the synthesis of $[Ru(NH_3)_6]^{2+}$ by the reaction of N_2H_4 with $RuCl_3$ in water [288], the capability of the dinitrogen molecule as a ligand has attracted much attention, giving rise to attempted syntheses of a number of metal dinitrogen complexes including those directly from dinitrogen. Up to now more than one hundred dinitrogen complexes are known for the metals shown below [289, 290, 291].

Ti	Cr	Mn	Fe	Co	Ni
Zr	Mo		Ru	Rh	Pd
	W	Re	Os	Ir	Pt

Some of the complexes are mononuclear with the N_2 molecule being coordinated by an "end-on" type arrangement, i.e. M—N—N, as evidenced by x-ray work. Other complexes are binuclear or trinuclear, with a linear M—N—N—M′ arrangement.

1. Mononuclear Dinitrogen Complexes

The preparation of mononuclear dinitrogen complexes may be divided into two methods: a) reaction of a coordinated nitrogen compound such as N_3, NH_3, NO or N_2H_4; b) direct coordination of N_2 with or without assistance by a reducing agent such as Na/Hg or Me_3Al. Typical examples of b) are [292, 293]:

$$[CoH_3(PPh_3)_3] + N_2 \rightarrow [CoH(PPh_3)_3N_2] + H_2$$

$$MoCl_4(PMe_2Ph)_2 + N_2 \xrightarrow{Na/Hg} Mo(N_2)_2\,(PMe_2Ph)_4$$

with some exceptions, the dinitrogen ligand in those complexes is generally lost as N_2 on interaction with another ligand such as CO, H_2O or even H_2, on heating, or on oxidation. In general, the behaviour resembles that of weakly chemisorbed dinitrogen on metals. Nevertheless, the nature of bonding between N_2 and metal atom is of interest, even on the ground of its relevence to the heterogeneous activation of dinitrogen.

It has been well established that the end-on coordination of dinitrogen is similar to metal carbonyls, that is, an occupied σ-orbital of the dinitrogen molecule interacts with an empty $d\sigma$ orbital of the metal, and electron back-donation takes place via an occupied $d\pi$ orbital of the metal and one of the empty antibonding orbitals of dinitrogen as depicted in Figure 31. The back donation gives rise to weakening of the bond between two nitrogen atoms as evidenced by the N—N bond length (mostly 110–112 pm for mononuclear complexes) longer than in the free dinitrogen molecule (109.8 pm). The weakening of the N—N bond is also reflected in the lowering of the N—N stretching frequency as compared to 2331 cm^{-1} for free dinitrogen. Since the N—N stretching frequency is readily observed by infrared absorption even with unstable complexes, more information is available on this value than on the N—N bond length.

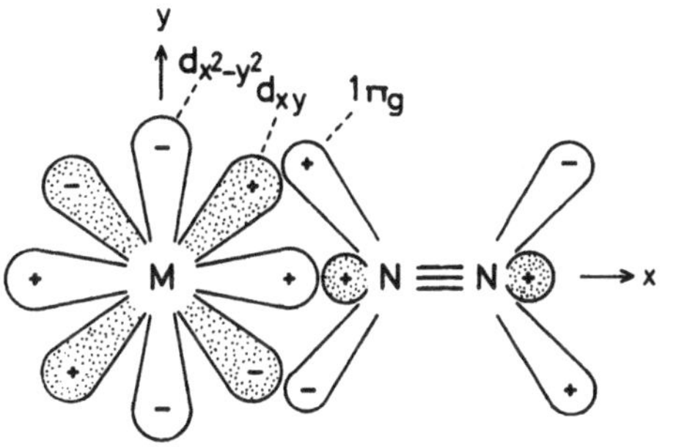

Figure 31. Scheme for the metal-dinitrogen bond (occupied orbital shaded)

Table 8 summarizes the effects of metals, ligands, and oxidation states on the N—N stretching frequency for some dinitrogen complexes which are relevant for the study of the above effects. It is obvious that the weakening of the N—N bond as indicated by a decrease in the stretching frequency advances with the change from 3*d* or 4*d* to 5*d* metals, except for Fe. This seems to be reasonable since the π-bonding ability or π-basicity of the metal atom increases as one goes down in the transition series, $3d < 4d < 5d$, or as the polarizability increases. In view of this, the extensive decrease observed with the iron complexes is noticeable, suggesting a high efficiency of the iron atom in transferring electrons from donor ligand to dinitrogen ligand.

Ligand effects given in Table 8 are in conformity with the above concept. Less electron accepting phosphines such as PBu_3 or $PPhMe_2$ give rise to more extensive decreases in the stretching frequency. The effect of oxidation state is also in line with this. The lower the oxidation state, the lower the frequency because of the high electron density on the metal. It can be concluded that any effect which increases the electron density at the metal atom increases the metal-nitrogen bond strength by increasing back donation, resulting in a corresponding decrease in the N—N bond strength.

Table 8. The N—N stretching frequency (cm^{-1}) in some dinitrogen complexes

A) Effect of metal	3*d*	4*d*	5*d*		Ref.
M =	Fe	Ru	Os		
$[HM(N_2)(depe)_2]^+BPh_4^-$	2090	2163	2136		[294]
$H_2M(N_2)(PPh_2Et)_3$	2055	2147	2085		[295]
$[M(N_2)(NH_3)_5]Br_2$		2118	2028		[288, 296]
M =		Rh	Ir		
$M(N_2)Cl(PPh_3)_2$		2152	2105		[297]
M =	Mn		Re		
$M(N_2)Cl(depe)_2$	2100		1980		[298, 299]
$M(N_2)(C_5H_5)(CO)_2$	2169		2141		[300, 301]
M =		Mo	W		
trans-$M(N_2)_2(dppe)_2$		1970	1953		[293, 302]
B) Ligand effect					**Ref.**
X =	I	Br	Cl		
$Ir(N_2)X(PPh_3)_2$	2113	2107	2105		[304]
R_3 =	Ph_3	Ph_2Et	$PhMe_2$		
$Ir(N_2)Cl(PR_3)_2$	2105	2093	2051		[304]
R_3 =	Ph_3	Ph_2Me	Ph_2Et	Bu_3	
$CoH(N_2)(PR_3)_3$	2090	2077	2060	2050	[305]
C) Oxidation state (Z)					**Ref.**
Z =	+3		+2		
$[Os(N_2)(NH_3)_5]^{+Z}$	2140		2010		[296, 306]
Z =	+1		0		
$[trans\text{-}Mo(N_2)_2(dppe)_2]^{+Z}$	2047		1976, 1970		[307, 302]
Z =	0		−1		
$[Co(N_2)(PPh_3)_3]^{+Z}$	2093		1845		[308, 309]

dppe: $Ph_2P(CH_2)_2PPh_2$, depe: $Et_2P(CH_2)_2PEt_2$

2. *Binuclear Dinitrogen Complexes*

In the above mononuclear complexes, the extent of weakening of the N—N bond is not large enough to give chemical reactivity to the dinitrogen ligand. The stretching frequency remains mostly above 2000 cm^{-1}. Since the weakening is caused by coordination to a metal atom, it may be expected that coordination to two metal atoms may give rise to a further weakening. But this is so only in particular cases, if the two metals are same. For homo-binuclear dinitrogen complexes of Fe, Ru, Rh, Ni and Pt, the stretching frequencies are above 2000 cm^{-1} and only observed by Raman spectroscopy. Molecular orbital interpretations for this fact has been given by Sellman [310].

There are, however, several dinitrogen complexes that show exceptionally low frequencies as summarized in Table 9. Although they are homo-binuclear, the Ti, Zr and Mo complexes exhibit very low frequencies. Different oxidation state of two Fe atoms in *5* (Table 9) seems to give rise to a low frequency.

The hetero-binuclear complexes generally give low frequencies as are shown. The $[CoN_2(PPh_3)_3]^{-1}$ complex shown in Table 8 has Na^+ as a counter cation so that it is a sort of heterobinuclear complex, giving a low frequency [308]. Similarly a further weakening of the N—N bond is made possible by applying organo-alkali ligands as in *8–10* (Table 9). Thus $[(PhLi)_6Ni_2N_2(Et_2O)_2]$ gives an N—N bond length as large as 135 pm [321].

The complex *10* (Table 9) is particularly of interest in view of the compound $[KN_2Ru]_n$ suggested in heterogeneous system. The structure of the complex *10* has been determined, as shown in Figure 32.

3. Matrix-Isolated Dinitrogen Complexes

There has been found another type of dinitrogen complexes which are prepared by direct reaction of metal atoms with dinitrogen molecules. The metal atom vapor and gaseous dinitrogen are simultaneously condensed onto a plate made of inert material such as sodium chloride and cooled to 4.2–10 K. The dinitrogen for the preparation is usually diluted with argon. Complex formation by this method may be identified by vibrational spectra at low temperatures (10–40 K), as summarized in Table 10. The complexes

Table 9. Some of multinuclear dinitrogen complexes

No	Complex	ν_{N-N}/cm^{-1}	Ref.
1	$[(C_5Me_5)_2ZrN_2]_2N_2$	1556	[311]
2	$[(C_6H_6)(PPh_3)_2Mo]_2N_2$	1910 (Raman)	[312]
3	$[(Cp)_2Ti]_2N_2$	1280	[303, 313]
4	$[(Cp)_2Ti]_2N_2MgCl$	1255	[314]
5	$(PPh_3)_2(OEt_2)HFe—N_2—Fe(OEt_2)(PPh_3)$	1761	[315]
6	$(PPhMe_2)_4ClRe—N_2—CrCl_3(THF)$	1890	[316]
7	$(PPhMe_2)_4ClRe—N_2—MoCl_4(OMe)$	1660	[317]
8	$[Co(N_2)(PPh_3)_3]_2Mg(THF)_4$	1840	[318]
9	$[Co(N_2)(PPh_3)_3]Li(THF)_3$	1890	[319]
10	$[KN_2Co(PMe_3)_3]_6$	1795	[320]

THF: tetrahydrofuran, C_5H_8O
Cp: cyclopentadienyl group, C_5H_5

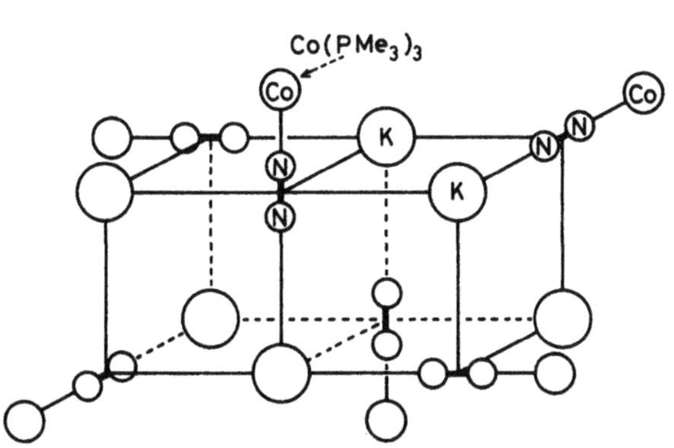

Figure 32. Crystal structure of $K[(PMe_3)_3CoN_2]$ [320]

Table 10. Infrared spectra (cm^{-1}) for matrix-isolated dinitrogen complexes

	$n = 1$	$n = 2$	$n = 3$	$n = 4$	Ref.
CrN_2	2127				[322]
	2134				
CoN_2	2101				[323]
$Ni(N_2)_n$	2088	2106	2134	2174	[324]
$Ni(O_2)(N_2)_n$	2242	2260			[325]
		2283			
$Ni(CO)_{4-n}(N_2)_n$	2264	2270	2258		[326]
	2266	2240	2210		[327]
$Rh(N_2)_n$	2154	2188	2196	2203	[328]
				2179	
$Pd(N_2)_n$	2215	2234	2241		[324]
	2211				
$Pt(N_2)_n$	2173	2150	2217		[329]
	2168	2198	2206		[330]
	2170				
LiN_2	1800				[331]

can involve up to four N_2 molecules per metal atom (n in Table 10). The number n may be controlled by the N_2/Ar ratio in the reacting gas mixture. In $Ni(N_2)_4$, the N_2 ligands are tetrahedrally coordinated to the Ni atom as is the case with $Ni(CO)_4$. Thus, a mixed N_2—CO complex [326, 327], as well as N_2—O_2 [325], and N_2—X_2 [332] complexes of nickel have also been prepared. The isotope shifts of the N—N stretching frequency disclose that the nature of the coordination bonds is mostly end-on type. Coordinated $^{29}N_2$ gives two absorptions corresponding to M—$^{14}N{\equiv}^{15}N$ and M—$^{15}N{\equiv}^{14}N$ for NiN_2 [324] and RhN_2 [328]. For CoN_2, however, only one absorption is observed with $^{29}N_2$ suggesting side-on (or edge-on) type coordination [323]. $Pt(N_2)_2$ has also been claimed to be of a side-on type [329]. The N—N stretching frequency of LiN_2 is exceptionally low at 1800 cm^{-1}, suggesting an ionic structure $Li^+N_2^-$ or a supernitride.

Although such complexes are interesting species from the theoretical view point because of their simplicity in structure, they can not stand normal

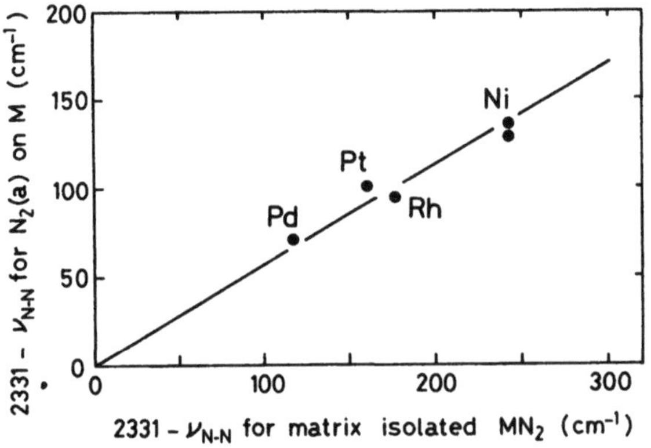

Figure 33. Correlation in N—N stretching frequency shift between adsorbed species on metals and matrix isolated metal-dinitrogen complex.

temperatures so that no information is available for their reactivity with other molecules such as hydrogen. However there is a good correlation between changes of stretching frequency with different transition metals for the matrix-isolated complexes, and for the chemisorbed species on metal solids as illustrated in Figure 33, where the wave number differences from the free dinitrogen molecule for the both systems are plotted. Thus the matrix-isolated complex can be regarded as a model for the chemisorbed species.

B. Reaction of Dinitrogen in Solution

1. Reduction of Coordinated Dinitrogen

The extensive weakening of the N—N bond as realized in the Ti and Zr complexes renders the coordinated dinitrogen reactive. In fact, the complexes *1* and *2* (Table 9) give N_2H_4 on reaction with HCl (*1*) or with organo-alkali compounds (*2*). The complexes *5*, *8* and *9* (Table 9) also give N_2H_4 on reaction with acid. However the reaction of complex *6* or *7* (Table 9) to give N_2H_4 or NH_3 has not been reported despite the very low frequency, suggesting that the N—N stretching frequency does not always reflect the reactivity of coordinated dinitrogen. It appears that effective polarization in dinitrogen is another factor to determine the reactivity.

Recently Chatt and coworkers were successful in a dinitrogen reaction of a mono-nuclear Mo complex for which the N—N stretching frequency is not very low, 1970 cm^{-1}. That is the reactions

$$\textit{trans}\text{-Mo(N}_2)_2(\text{dppe})_2 \xrightarrow{\text{HX}} \text{MoX}_2(\text{N}_2\text{H}_2)(\text{dppe})_2 + \text{N}_2$$

$$\textit{cis}\text{-Mo(N}_2)_2(\text{PPhMe}_2)_4 \xrightarrow[\text{MeOH}]{\text{HX}} \text{MoX}_2(=\text{N}-\text{NH}_2)(\text{PPhMe}_2)_3 + \text{N}_2$$

were found to proceed by careful adjustment of reaction conditions [333, 334]. The latter product gave NH_3 on reaction with H_2SO_4—MeOH. Analogous reactions with W complexes were also found giving nearly quantitative amount of NH_3 [334]. The formation of NH_3 suggests that the N_2 molecule is dissociated in the reaction.

Chatt and coworkers also found that the same complex of Mo or W gives M—N=N—COR on reaction with RCOCl [335] and M—N=N—R on reaction with RX under illumination [336]. Hidai et al. also succeeded in reactions of the Mo complex with aldehyde or ketone to give a complex with hydrazone type N_2CRR' ligand [337]. In this way the reaction of coordinated dinitrogen has been extended to give a C—N bond.

2. Activation of Dinitrogen in Solution

Prior to the discovery of the dinitrogen complexes, Volpin and Shur discovered a low temperature fixation of dinitrogen in solution to form metal compounds that give ammonia on hydrolysis [338]. This unique discovery gave rise to extensive studies of nitrogen fixation in solution. The Volpin system is generally made up of a transition metal compound (chloride or acetylacetonate of Ti, V, Nb, Cr, Mo, W, Mn, Fe, Co, or Ni) as catalyst, and a free metal or metal organic compound of IA to IIIA Group as reducing agent in an organic solvent. Dinitrogen is introduced into this system mostly at an elevated pressure up to 152 bar, and kept standing at 298–403 K for several hours. $TiCl_4$ or $(Cp)_2TiCl_2$ gives generally a better yield of NH_3 than the others. Basic solvents such as ether or THF are favorable. Dihydrogen usually inhibits the reaction [339]. Addition of $AlBr_3$, a strong Lewis acid, into the system brings about a catalytic reduction of dinitrogen, thus greatly enhancing the yield of NH_3 per catalyst metal [340].

The product of the N_2 reaction appears to be a nitride. For example, $[TiCl_3(THF)_3]$—Mg absorbs N_2 to give a black diamagnetic powder $[TiNMg_2Cl_2(THF)]$ which forms NH_3 on reaction with HCl [341]. The nitride formed in $(Cp)_2TiCl_2$—Mg—N_2 system reacts with ketone as well as acid chloride to give amine and nitrile, respectively [342]. In order that the reaction be catalytic, however, the nitride formed should not involve the transition metal. Thus in the catalytic cycle with $AlBr_3$ added, the product is the nitride of the reducing agent. In some cases, the reaction product gives hydrazine on hydrolysis, suggesting a dinitrogen complex in the product [343, 344].

In summary, the Volpin reaction is a reduction of dinitrogen by a strong reductant promoted by a transition metal compound. It is likely that the transition metal is first reduced to a low oxidation state which gives rise to an increased affinity to dinitrogen. The kinetics of N_2 reduction in the Cp_2TiCl_2—EtMgBr system is zero order with respect to N_2 pressure above 1 bar, is zero order with respect to EtMgBr concentration, and is second order in Ti concentration, suggesting the rate-determining formation of a binuclear complex of lowvalent Ti [345]. An analogous conclusion is reached for the $FeCl_3$—LiPh—N_2 system [346, 347].

More recently, dinitrogen fixation has been extended by Shilov's school to protic media in addition to the aprotic one described above. Fixation in protic solvents takes place in the presence of Mo(III) complexes with $Ti(OH)_3$, $Cr(OH)_2$ or Na acting as reductant [348, 349]. The N_2 reaction in protic solvent gives rise to direct formation of N_2H_4 and NH_3, making the system catalytic based on Mo, although the reducing agent, $Ti(OH)_3$, is an expensive material and the system including solid is heterogeneous.

In the same year, Schrauzer et al. also reported an analogous reduction of N_2 to NH_3 by $NaBH_4$ in the presence, in water, of Mo complexes containing SH groups [350]: the yield of NH_3 is much less than that by $Ti(OH)_3$. Another type of reduction is made in protic media by $V(OH)_2$ which acts as promoter as well as reducing agent [351, 352]. Addition of $Mg(OH)_2$ greatly enhances the yield of N_2H_4. Finally, an efficient system which works in homogeneous solution is found by combining a V(II) complex with aromatic diols. By using catechol in methanol at room temperature and 1 bar of N_2, the yield of NH_3 reaches 75 percent with respect to V(II) [353]. In this way, many efforts have been made in the field of dinitrogen fixation in solution, mostly aiming at a chemical reproduction of the function of "nitrogenase".

References

1. Skinner, K. J.: Chem. Eng. News *54*, Oct. 4, 22 (1976)
2. Weast, R. C. (ed.): Handbook of Chemistry and Physics, 55th ed. Cleveland: CRC Press 1974
3. Borodko, Yu. G.; Shilov, A. E.: Russ. Chem. Rev. *35*, 355 (1969), Usp. Khim. *38*, 761 (1969)
4. Leigh, G. J.: In: The Chemistry and Biochemistry of Nitrogen Fixation (Postgate, J. R., ed.), London: Plenum Press 1971, pp. 45
5. Henrici-Olivé, G.; Olivé, S.: The dinitrogen molecule. In: A treatise on Dinitrogen Fixation Sec. I and II. (Hardy, R. W. F. et al., ed.), New York: J. Wiley 1979, pp. 3—29
6. Jolly, W. L.: The Inorganic Chemistry of Nitrogen. New York, Amsterdam: Benjamin 1964, pp. 36
7. Sanderson, R. T.: Chemical Periodicity. New York: Reinhold 1960, pp. 201
8. Trapnell, B. M. W.: Proc. Roy. Soc. A *218*, 566 (1953)
9. Bond, G. C.: Catalysis by metals. London—New York: Academic Press 1962
10. Schlier, E.; Farnsworth, H. E.: Phys. Rev. *78*, 316 (1950)
11. Toyoshima, I.; Takezawa, N.; Suzuki, H.: JCS Chem. Commun. *1973*, 270; Proc. 6th Int. Congr. Catal. (Bond, G. C. et al., ed.) London: Chem. Soc. 1977, pp. 708
12. Ozaki, A.; Aika, K.: The synthesis of ammonia by heterogeneous catalysis. In: A treatise on Dinitrogen Fixation Sec. I and II (Hardy, R. W. F. et al., ed.) New York: John Wiley & Sons 1979, pp. 169–247
13. Beeck, O.: Advan. Catal. *2*, 151 (1950)
14. Bagg, J., Tomkins, F. C.: Trans. Faraday. Soc. *51*, 1071 (1955)
15. Kwan, T.: J. Res. Inst. Catal., Hokkaido Univ. *3*, 109 (1955), J. Phys. Chem. *60*, 1033 (1956)
16. Scholton, J. J. F., Zwietering, P., Konvalinka, J. A., De Boer, J. H.: Trans. Faraday Soc. *5*, 2166 (1959)
17. Beeck, O.; Colle, W. A.; Wheeler, A.: Discuss Faraday Soc. *8*, 314 (1950)
18. Grabke, H. J.: Ber. Bunsenges. Phys. Chem. *72*, 541 (1968)
19. Bozso, F.; Ertl, G.; Weiss, M.: J. Catal. *49*, 18 (1977)
20. Bozso, F.; Ertl, G.; Weiss, M.: J. Catal. *50*, 519 (1977)
21. Oguri, T.: J. Phys. Soc. Jap. *19*, 77 (1964)
22. Emmett, P. H.; Brunauer, S.: J. Amer. Chem. Soc. *56*, 35 (1934)
23. Romanushkina, A. E.; Kiperman, C. L.; Temkin, M. I.: Zh. Fiz. Khim. *27*, 1181 (1953)
24. Davis, R. T.: J. Amer. Chem. Soc. *68*, 1395 (1946)
25. Kisliuk, P. J.: J. Chem. Phys. *30*, 174 (1959)
26. Joyner, R. W.; Rickman, J.; Roberts, M. W.: J. Chem. Soc. Faraday Trans. I *70*, 1825 (1974)
27. Eley, D. D.: Discuss. Faraday Soc. *8*, 34 (1950)
28. Trapnell, B. M. W.: Chemisorption. London: Butterworths 1955
29. Miyazaki, E.; Yasumori, I.: Surface Sci. *55*, 747 (1976)
30. Sachtler, W. M. H.; Van Reijen, L. L.: J. Res. Inst. Catal., Hokkaido Univ. *10*, 87 (1962)
31. Tanaka, K.; Tamaru, K.: J. Catal. *2*, 366 (1963)
32. Mimeault, V. J.; Hansen, R. S.: J. Phys. Chem. *70*, 3001 (1966)
33. Temkin, M. I.; Pyzhev, V. M.: Acta Phisicochim. (USSR) *12*, 327 (1940)
34. Frumkin, A.; Slygin, A.: ibid. *3*, 791 (1935)
35. Zeldowitsch, Ya.: ibid. *1*, 449 (1934)
36 Langmuir, I.: J. Amer. Chem. Soc. *54*, 2798 (1932)
37. Brunaur, S.; Love, K. S.; Keenan, R. G.: ibid. *64*, 751 (1942)
38. Halsey, G.; Taylor, H. S.: J. Chem. Phys. *15*, 624 (1947)
39. Zwietering, P.; Roukens, J. J.: Trans. Faraday Soc. *50*, 178 (1954)
40. Oh-kita, M.; Midorikawa, H.; Aika, K.; Urabe, K.; Ozaki, A.: Proc. 7th Intern. Congr. Catal. (Seiyama, T. et al., ed.) Tokyo-Amsterdam: Kodansha-Elsevier 1981, pp. 1494
41. Emmett, P. H.; Love, K. S.: J. Amer. Chem. Soc. *55*, 4043 (1933)
42. Grabke, H. J.: Ber. Bunsenges. Phys. Chem.: *72*, 533 (1968)

43. Grabke, H. J.: Z. Phys. Chem. (N. F.) *100*, 185 (1976)
44. Ehrlich, G.: J. Phys. Chem. *60*, 1388 (1956)
45. Ehrlich, G.: J. Chem. Phys. *34*, 29 (1961); Advan. Catal. *14*, 255 (1963)
46. Emmett, P. H.; Harkness, R. W.: J. Amer. Chem. Soc. *57*, 1631 (1935)
47. Rigby, L. J.: Can. J. Phys. *43*, 532 (1965)
48. Yates Jr., J. T.; Madey, T. E.: J. Chem. Phys. *43*, 1055 (1965)
49. Oguri, T.: J. Phys. Soc. Jap. *18*, 1280 (1963)
50. Delchar, T. A.; Ehrlich, G.: J. Chem. Phys. *42*, 2686 (1965)
51. Madey, T. E.; Yates, Jr., J. T.: J. Chem. Phys. *44*, 1675 (1966)
52. Goymour, C. G.; King, D. A.: J. Chem. Soc. Faraday Trans. I *69*, 749 (1973)
53. Pasternak, R. A.; Endow, N.; Bergsnov-Hansen, B.: J. Phys. Chem. *70*, 1304 (1966)
54. Gasser, R. P. H.; Hale, A.; Marsay, C. J.: Trans. Faraday Soc. *62*, 1789 (1967)
55. Han, H. R.; Schmidt, L. D.: J. Phys. Chem. *75*, 227 (1971)
56. Mahnig, M.; Schmidt, L. D.: Z. Phys. Chem. (N. F.) *80*, 71 (1972)
57. Kunimori, K.; Kawai, T.; Kondow, T.; Onishi, T.; Tamaru, K.: Surface Sci. *54*, 525 (1976)
58. Yates Jr., J. T.; Madey, T. E.: J. Chem. Phys. *51*, 334 (1969)
59. Ponec, V.; Knor, Z.: J. Catal. *10*, 73 (1968)
60. Wedler, G.; Borgmann, D.; Geuss, K.-P.: Surface Sci. *47*, 592 (1975)
61. Wedler, G.; Borgmann, D.: J. Catal. *44*, 139 (1976)
62. Yates, J. T.; Klein, R.; Madey, T. E.: Surface Sci. *58*, 469 (1976)
63. Fuggle, J. C.; Menzel, D.: Vak. Tech. *27*, 130 (1978)
64. Wilf, M.; Folman, M.: Surface Sci. *52*, 10 (1975)
65. Adams, D. L.; Germer, L. H.: Surface Sci. *26*, 109 (1971)
66. King, D. A.; Wells, M. G.: Proc. Roy. Soc. (London) Ser. A *339*, 245 (1974)
67. Singh-Boparai, S. P.; Bowker, M.; King, D. A.: Surface Sci. *53*, 55 (1975)
68. Tamm, P. W.; Schmidt, L. D.: ibid. *26*, 286 (1971)
69. Adams, D. L.; Germer, L. H.: ibid. *27*, 21 (1971)
70. Singh-Boparai, S. P.; King, D. A.: ibid. *61*, 275 (1976)
71. Eischens, R. P.; Jacknow, J.: Proc. 3rd Intern. Congr. Catal. (Sachtler, W. M. H. et al., ed.) Amsterdam: North Holland Pub. 1965, pp. 627
72. Hardeveld, R. V.; Montfoort, A. V.: Surface Sci. *4*, 396 (1966)
73. Hardeveld, R. V.; Montfoort, A. V.: Surface Sci. *17*, 90 (1969)
74. Borodko, Yu. G.; Lyutov, V. S.: Kinet. Katal. *12*, 238 (1971)
75. Egerton, T. A.; Sheppard, N.: J. Chem. Soc. Faraday Trans. I *70*, 1357 (1974)
76. Eischens, R. P.: Acc. Chem. Res. *5*, 74 (1972)
77. Chang, C. C.; Kokes, R. J.: J. Phys. Chem. *77*, 2640 (1973)
78. Cohen de Lara, E.; Delaval, Y.: J. Chem. Soc. Faraday Trans. II *74*, 790 (1978)
79a. Okawa, T.; Onishi, T.; Tamaru, K.: Z. Phys. Chem. (N. F.) *107*, 239 (1977)
79b. Kagami, S.; Naito, S.; Onishi, T.; Tamaru, K.: 41st Ann. Meet. Chem. Soc. Japan, 1980, (Paper No. 2T08)
80. Brill, R.; Jiru, P.; Schulz, G.: Z. Phys. Chem. (N. F.) *64*, 215 (1969)
81. Oh-kita, M.; Aika, K.; Urabe, K.; Ozaki, A.: J. Catal. *44*, 460 (1976); Chem. Commun. *1975*, 147
82. Oh-kita, M.; Urabe, K.; Ozaki, A.: J. Catal. *52*, 432 (1978)
83. Ravi, A.; King, D. A.; Sheppard, N.: Trans. Faraday Soc. *64*, 3359 (1968)
84. Bradshaw, A. M.; Pritchard, J.: Surface Sci. *19*, 198 (1970)
85. Rissmann, E. F.; Parry, J. M.: J. Phys. Chem. *79*, 1975 (1975)
86. Nakata, T.; Matsushita, S.: J. Catal. *4*, 631 (1965)
87. Brunauer, S.; Emmett, P. H.: J. Amer. Chem. Soc. *62*, 1732 (1940)
88. Takezawa, N.; Emmett, P. H.: J. Catal. *11*, 131 (1968)
89. Huang, Y.-Y.; Emmett, P. H.: ibid. *24*, 101 (1972)
90. Morikawa, Y.; Ozaki, A.: ibid. *12*, 145 (1968)
91. Morikawa, Y.; Ozaki, A.: ibid. *23*, 97 (1971)
92. Boreskova, E. G.; Kuchaev, V. L.; Temkin, M. I.: Kinet. Katal. *18*, 973 (1977)
93. Amenomiya, Y.; Pleizier, G.: J. Catal. *28*, 442 (1973)
94. Madey, T.; Yates Jr., J. T.; Erickson, N. E.: Surface Sci. *43*, 526 (1974)

95. Brundle, C. R.: J. Vac. Sci. Tech. *13*, 301 (1976)
96. Kishi, K.; Roberts, M. W.: Surface Sci. *62*, 252 (1977)
97. Ertl, G.; Thiele, N.: Appl. Surface Sci. *3*, 99 (1979)
98. Perman, E. P.: Proc. Roy. Soc. (London) *76*, 167 (1905)
99. Haber, F.; van Oordt, G.: Z. Anorg. Allg. Chem. *43*, 111 (1905)
100. Nernst, W.: Z. Elektrochem. *13*, 521 (1907)
101. German Patent 235421 (1908), Haber, F.: Z. Elektrochem. *16*, 244 (1910)
102. Mittasch, A.: Geschichte der Ammoniak-Synthese. Verlag Chemie, Weinheim, 1951
103. Mittasch, A.: Advan. Catal. *2*, 81 (1950)
104. Larson, A. T.; Tour, R. S.: Chem. Met. Eng. *26*. 493, 555, 588, 647 (1922), Larson, A. T.: ibid. *16*, 1002 (1924)
105. Tanaka, K.; Tamaru, K.: Kinet. Katal. *7*, 242 (1966)
106. Aika, K.; Hori, H.; Ozaki, A.: J. Catal. *27*, 424 (1972)
107. Aika, K.; Yamaguchi, J.; Ozaki, A.: Chem. Lett. (Jap.) *1973*, 161
108. Westrik, R.: J. Chem. Phys. *21*, 2049 (1953)
109. Verwey, E. J. W.; Heilmann, E. L.: J. Chem. Phys. *15*, 174 (1947)
110. Dry, M. E.; Ferreira, L. C.: J. Catal. *7*, 352 (1967)
111. Michel, A.: Ann. Chim. *8*, 317 (1937)
112. Nielsen, A.: An Investigation on Promoted Iron Catalysis for the Synthesis of Ammonia. Copenhagen: Jul. Gjellerups Forlag 1968
113. Chen, H. C.; Anderson, R. B.: J. Colloid. Interface Sci. *38*, 535 (1972); J. Catal. *28*, 161 (1973)
114. Wilchinsky, Z. W.: Anal. Chem. *21*, 1188 (1949)
115. Sasaki, S.; Osumi, Y.: Nippon Kagaku Zasshi (Jap.) *73*, 808 (1952); Osumi, Y.: Bull. Chem. Soc. Jap. *26*, 519 (1953)
116. Kobayashi, H.; Nishijima, O.: Kogyo Kagaku Zasshi (Jap.) *57*, 189 (1954)
117. Uchida, H.; Terao, I.; Ogawa, K.; Oba, M.: Bull. Chem. Soc. Jap. *37*, 1508 (1964)
118. Hall, W. K.; Tarn, W. H.: Anderson, R. B.: J. Amer. Chem. Soc: *72*, 5436 (1950)
119. Emmett, P. H.; De Witt, T. W.: ibid. *65*, 1253 (1943)
120. Emmett, P. H.; Brunauer, S.: ibid. *59*, 1553 (1937)
121. Peters, C.; Schäfer, K.; Krabetz, R.: Z. Elektrochem. *64*, 1194 (1960)
122. Brill, R.: ibid. *38*, 669 (1932)
123. Schäfer, K.: ibid. *64*, 1191 (1960)
124. Wyckoff, R. W. G.; Crittenden, E. D.: J. Amer. Chem. Soc. *47*, 2866 (1925)
125. Dry, M. E.; Du Plessis, J. A. K.; Leuteritz, G. M.: J. Catal. *6*, 194 (1966)
126. Krabetz, R.; Peters, C.: Angew. Chem. **77**, 333 (1965)
127. Emmett, P. H.; Brunauer, S.: J. Amer. Chem. Soc. *59*, 310 (1937)
128. Solbakken, V.; Solbakken, A.; Emmett, P. H.: J. Catal. *15*, 90 (1969)
129. Brill, R.; Tauster, S.: Ber. Bunsen ges. Phys. Chem. *67*, 390 (1963); Brill, R.: J. Catal. *19*, 236 (1970)
130. Uchida, H.; Terao, I.; Ogawa, K.: Bull. Chem. Soc. Jap. *37*, 653 (1964)
131. Herbstein, F. H.; Smuts, J.: J. Catal. *2*, 69 (1963)
132. Hosemann, R.; Preisinger, A.; Vogel, W.: Ber. Bunsenges. Phys. Chem. *70*, 796 (1966); Chem. Ing. Tech. *42*, 1252 (1970)
133. Schultz, J. M.: J. Catal. *27*, 64 (1972)
134. Pernicone, N. et al.: Proc. 5th Intern. Congr. Catal. Amsterdam: North Holland Publ. 1973, pp. 1241
135. Fagherazzi, G.; Galante, F.; Garbassi, F.; Pernicone, N.: J. Catal. *26*, 344 (1972)
136. Topsφe, H.; Dumesic, J. A.; Boudart, M.: J. Catal. *28*, 477 (1972)
137. Buhl, R.; Preisinger, A.: Surface Sci. *47*, 344 (1975)
138. Ludwiczek, H. et al.: J. Catal. *51*, 326 (1978)
139. Ozaki, A.; Aika, K.; Hori, H.: Bull. Chem. Soc. Jap. *44*, 3216 (1971)
140. Larson, A. T.; Brooks, A. P.: Ind. Eng. Chem. *18*, 1305 (1926)
141. Keunecke, E.: Z. Elektrochem. *36*, 690 (1930)
142. Mittasch, A.; Keunecke, E.: Z. Phys. Chem. *574* (1931)

143. Brill, R.; Osumi, Y.: Bull. Chem. Soc. Jap. *39*, 1678 (1966); Zabuga, V. Ya.; Markova, G. P.: Kataliz i Katalizatory, Akad. Nauk. Ukr. SSR, Resp. Mezhved. sb. *1965*, 110; CA *64*, 4312d (1966)
144. Chalenko, V. G.; Tovbin, M. V.: Ukr. Khim. Zh. *30*, 1128 (1964)
145. Artyukh, Yu. N.; Rusov, M. T.; Boldyreva, N. A.: Kinet. Katal. *8*, 1319 (1967)
146. Mittash, A.; Kuss, E.; Emert, O.: Z. Elektrochem. *34*, 829 (1928)
147. Winter, E.: Z. Phys. Chem. *B13*, 401 (1931)
148. Emmett, P. H.; Brunauer, S.: J. Amer. Chem. Soc. *55*, 1738 (1933)
149. Taylor, H. S.; Jungers, J. C.: ibid. *57*, 660 (1935)
150. Brill, R.: J. Chem. Phys. *19*, 1047 (1951)
151. Bokhoven, C.; van Heerden, C.; Westrik, R.; Zwietering, P.: in Catalysis (Emmett, P. H., ed.) New York: Reinhold 1955, Vol 3, pp. 265
152. Peters, C.; Krabetz, R.: Z. Elektrochem. *60*, 859 (1956)
153. Krabetz, R.; Peters, C.: Ber. Bunsenges. Phys. Chem. *67*, 381 (1963)
154. Emmett, P. H.; Kummer, J. T.: Ind. Eng. Chem. *35*, 677 (1943)
155. Sidrov, I. P.; Livshits, V. D.: Zh. Fiz. Khim. *21*, 1177 (1947)
156. Nielsen, A.; Kjaer, J.; Hansen, B.: J. Catal. *3*, 68 (1964)
157. Sidrov, I. P.; Livshits, V. D.: Zh. Fiz. Khim. *26*, 538 (1952)
158. Dumesic, J. A.; Topsøe, H.; Khammouma, S.; Boudart, M.: J. Catal. *37*, 503 (1975)
159. Kiperman, S.; Temkin, M. I.: Acta. Physiochim. *21*, 267 (1946)
160. Kiperman, S.; Temkin, M. I.: Zh. Fiz. Khim. *20*, 623 (1946); CA *41*, 332
161. Schultz-Ekloff, G.; Baresel, D.; Sahholz, W.: J. Catal. *43*, 353 (1976)
162. Kiperman, S.: Zh. Fiz. Khim. *21*, 1435 (1947); CA *42*, 5320
163. Kiperman, S.; Granovskaya, V.: Zh. Fiz. Khim. *25*, 557 (1951); CA *45*, 9346
164. Kunsman, C. H.: J. Amer. Chem. Soc. *50*, 2100 (1928)
165. Tamaru, K.: Trans. Faraday Soc. *57*, 1410 (1961)
166. Jungers, J. C.; Taylor, H. S.: J. Amer. Chem. Soc. *57*, 679 (1935)
167. Barrer, R. M.: Trans. Faraday Soc. *32*, 490 (1936)
168. Hinshelwood, C. N.; Burk, R. E.: J. Chem. Soc. (London) *127*, 1105 (1925)
169. Hailes, H. R.: Trans. Faraday Soc. *27*, 601 (1931)
170. Shindo, H.; Egawa, C.; Onishi, T.; Tamaru, K.: Z. Naturforsch. *34a*, 96 (1979)
171. McGeer, J. P.; Taylor, H. S.: J. Amer. Chem. Soc. *73*, 2743 (1951)
172. Logan, S. R.; Kemball, C.: Trans. Faraday Soc. *56*, 144 (1960)
173. Logan, S. R.; Moss, R. L.; Kemball, C.: ibid. *54*, 922 (1958)
174. Schwab, G. M.; Krabetz, R.: Z. Elektrochem. *60*, 855 (1956)
175. Love, K. S.; Emmett, P. H.: J. Amer. Chem. Soc. *63*, 3297 (1941)
176. Takezawa, N.; Toyoshima, I.: J. Phys. Chem. *70*, 594 (1966)
177. Amano, A.; Taylor, H.: J. Amer. Chem. Soc. *76*, 4201 (1954)
178. Tamaru, K.; Tanaka, K.; Fukasaku, S.; Ishida, S.: Trans. Faraday Soc. *61*, 765 (1965)
179. Friedlander, A. G.; Courty, Ph. R.; Montarnal, R. E.: J. Catal. *48*, 322 (1977)
180. Apel'baum, L. O.; Temkin, M. I.: Zh. Fiz. Khim. *33*, 2697 (1959); CA *55*, 12008.
181. Robertson, A. J. B.; Willhoft, E. M. A.: Trans. Faraday Soc. *63*, 476 (1967)
182. Dixon, J. K.: J. Amer. Chem. Soc. *53*, 2071 (1931)
183. Schwab, G. M.; Schmidt, F. H.: Z. Elektrochem. *35*, 605 (1929)
184. Elöd, E.; Banholzer, W.: ibid. *32*, 555 (1926)
185. Dixon, J. K.: J. Amer. Chem. Soc. *53*, 1763 (1931)
186. Rusov, M. I.; Pevsner, Ts. V.: Zh. Fiz. Khim. *28*, 1765 (1954)
187. McGill, W. J.; Sebba, F.: J. Catal. *2*, 104 (1963)
188. Tamaru, K.: Prep. 2nd Intern. Congr. Catal. Paris, 1960, No. 8
189. Tamaru, K.: Trans. Faraday Soc. *59*, 979 (1963)
190. Sastri, M. V. C.; Srikant, H.: J. Sci. Ind. Res. (India) *20D*, 321 (1961)
191. Ozaki, A.; Taylor, H. S.; Boudart, M.: Proc. Roy. Soc. (London) *258*, 47 (1960)
192. Brill, R.; Tauster, S.: J. Chem. Phys. *36*, 2100 (1962)
193. Bokhoven, C.; van Raayen, W.: J. Phys. Chem. *58*, 471 (1954)
194. Kubota, H.; Shindo, M.: Kagaku Kogaku (Jap.) *23*, 242 (1959)
195. Aika, K.; Ozaki, A.: J. Catal. *13*, 232 (1969)
196. Aika, K.; Ozaki, A.: ibid. *19*, 350 (1970)

197. Brill, R.: ibid. *16*, 16 (1970)
198. Emmett, P. H.: Fixed Nitrogen. 1932, pp. 227
199. Kobayashi, H.; Kubota. H.: Rep. Faculty Engng. Hokkaido Univ. *3*, 136 (1949)
200. Comings, E. W.; Adams, R. M.: Chem. Eng. Progr. *49*, 359 (1953)
201. Uchida, H.; Kuraishi, M.: Bull. Chem. Soc. Jap. *28*, 106 (1955)
202. Shapatina, E. N.; Kutaev, B. L.; Temkin, M. I.: Kinet. Katal. *12*, 1476 (1971)
203. Aika, K.; Ozaki, A.: J. Catal. *16*, 97 (1970)
204. Aika, K.; Ozaki, A.: ibid. *14*, 311 (1969)
205. Temkin, M. I.; Morozov, N. M.; Shapatina, E. N.: Kinet. Katal. *4*, 260 (1963)
206. Horiuti, J.: Proc. Jap. Acad. *29*, 100 (1953) J. Res. Inst. Catal. Hokkaido Univ. *5*, 1 (1957)
207. Enomoto, S.; Horiuti, J.: ibid. *2*, 87 (1953)
208. Enomoto, S.; Horiuti, J.; Kobayashi, H.: ibid. *3*, 185 (1955)
209. Bokhoven, C.; Gorgels, M. J.; Mars, P.: Trans. Faraday Soc. *55*, 315 (1959)
210. Tanaka, K.; Yamamoto, O.; Matsuyama, A.: Proc. 3rd Intern. Congr. Catal. (Sachtler, W. M. H. et al., ed.) Amsterdam: North Holland Pub. 1965, pp. 676
211. Tanaka, K.: J. Res. Inst. Catal., Hokkaido Univ. *13*, 119 (1965)
212. Tanaka, K.: ibid. *14*, 153 (1966)
213. Tanaka, K.: Prepr. 4th Intern. Congr. Catal. Moscow, 1968, No. 9
214. Tanaka, K.; Matsuyama, A.: J. Res. Inst. Catal., Hokkaido Univ. *19*, 63 (1971)
215. Kazusaka, A.: ibid. *19*, 42 (1971)
216. Urabe, K.; Aika, K.; Ozaki, A.: J. Catal. *42*, 197 (1976)
217. Tamaru, K.: Proc. 3rd Intern. Congr. Catal. Amsterdam: North Holland Pub. 1965, pp. 665
218. Takezawa, N.; Toyoshima, I.: J. Catal. *6*, 145 (1966)
219. Hill, M. R.; Kemball, C.; Robert, M. W.: Trans. Faraday Soc. *62*, 3570 (1966)
220. Khrizman, I. A.; Korneyiuchuk, G.: Acta Physicochim. (USSR) *18*, 420 (1943); CA *38*, 5719
221. Lotz, C. R.; Sebba, F.: Trans. Faraday Soc. *53*, 1246 (1957)
222. King, D. A.; Sebba, F.: J. Catal. *4*, 253 (1965)
223. Segal, N.; Sebba, F.: ibid. *8*, 105 (1967)
224. Boreskov, G. K. et al.: Kinet. Katal. *16*, 1218 (1975)
225. Shindo, H.; Egawa, C.; Onishi, T.; Tamaru, K.: J. Chem. Soc. Faraday Trans. I 76, 280 (1980)
226. Scholten, J. J. F.; Konvalinka, J. A.; Zwietering, P.: Trans. Faraday Soc. *56*, 262 (1960)
227. Dawson, P. T.; Hansen, R. S.: J. Chem. Phys. *45*, 3148 (1966)
228. Estrup, P. J.; Anderson, A.: ibid. *49*, 523 (1968)
229. May, J. W.; Szostak, R. J.; Germer, L. H.: Surface Sci. *15*, 37 (1969)
230. Peng, Y. K.; Dawson, P. T.: J. Chem. Phys. *54*, 950 (1971)
231. Tamaru, K.: Bull. Chem. Soc. Jap. *37*, 771 (1964)
232. Ozaki, A.: J. Catal. *45*, 370 (1976)
233. Takezawa, N.: ibid. *24*, 417 (1972)
234. Joris, G. G.; Taylor, H. S.: J. Chem. Phys. *7*, 893 (1939)
235. Kummer, J. T.; Emmett, P. H.: ibid. *19*, 289 (1951)
236. Schulz, G.; Schaefer, H.: Z. Phys. Chem. (N. F.) *64*, 333 (1969)
237. Schulz-Ekloff, G.: Ber. Bunsenges. Phys. Chem. *75*, 110 (1971)
238. Guyer, W. R. F.; Joris, G. G.; Taylor, H. S.: J. Chem. Phys. *9*, 287 (1941)
239. Gorbunov, A. I.; Boreskov, G. K.: In: Problemy Kinet. Katal. Akad. Nauk. (USSR) 1960, No. 10, pp. 192
240. Gasser, R. P. H.; Lowrence, C. P.; Newman, D. G.: Trans. Faraday Soc. *61*, 1771 (1965)
241. Moore, G. E.; Unterwald, F. C.: J. Chem. Phys. *48*, 5393 (1968)
242. Ozaki, A.; Aika, K.; Morikawa, Y.: Proc. 5th Intern. Congr. Catal. (Hightower, J. W., ed.) Amsterdam-London: North-Holland Pub. 1973, pp. 1251
243. Urabe, K.; Oh-ya, A.; Ozaki, A.: J. Catal. *54*, 436 (1978)
244. Morikawa, Y.; Naka, Y.; Ozaki, A.: Nippon Kagaku Zasshi *1972*, 1023
245. Chesnokova, R. V., et al.: Kinet. Katal. *11*, 1486 (1970)
246. Urabe, K.; Aika, K.; Ozaki, A.: J. Catal. *32*, 108 (1974)
247. Urabe, K.; Aika, K.; Ozaki, A.: ibid. *38*, 430 (1975)

248. Ozaki, A.: Isotopic Studies of Heterogeneous Catalysis. Tokyo-New York: Kodansha-Academic Press 1977
249. Boreskov, G. K.; Gonbunov, A. I.; Masanov, O. L.: Dokl. Akad. Nauk. SSSR *123*, 90 (1958)
250. Takezawa, N.: Toyoshima, I.: J. Catal. *19*, 271 (1970)
251. Grunze, M.; Bozso, F.; Ertl, G.; Weiss, M.: Appl. Surface Sci. *1*, 241 (1978)
252. Nakata, T.; Matsushita, S.: J. Phys. Chem. *72*, 458 (1968); Ann. Mtg. Jap. Chem. Soc. 1966 Abstr. IZA012
253. Okawa, T.; Onishi, T.; Tamaru, K.: Chem. Lett. (Jap.) *1977*, 1077
254. Melton, C. E.; Emmett, P. H.: J. Phys. Chem. *68*, 3318 (1964)
255. Weiss, M.; Ertl, G.; Nitschke, F.: Appl. Surface Sci. *2*, 614 (1979)
256. Shvachko, V. I.; Fogel, Ya. M.; Kolot, V. Ya.: Kinet. Katal. *7*, 834 (1966)
257. Schmidt, W. A.: Angew. Chem. *80*, 151 (1968)
258. Brill, R.: Ber. Bunsenges. Phys. Chem. *75*, 455 (1971)
259. Block, J.; Schulz-Ekloff, G.: J. Catal. *30*, 327 (1973)
260. Davis, K. M. C.; Sayer, C. F.: J. Chem. Soc. Faraday Trans. I *68*, 1884 (1972)
261. Aika, K.; Ohhata, T.; Ozaki, A.: J. Catal. *19*, 140 (1970)
262. Westrik, R.; Zwietering, P.: Proc. Kon. Ned. Akad. Wetenschappen *B56*, 492 (1953)
263. McAllister, J.; Hansen, R. S.: J. Chem. Phys. *59*, 414 (1973)
264. Amariglio, H.; Rambeau, G.: Proc. 6th Intern. Congr. Catal. (Bond, G. C. et al., ed.) London: Chem. Soc. 1977, pp. 1113
265. Boudart, M. et al.: J. Catal. *37*, 486 (1975)
266. Dumesic, J. A.; Topsøe, H.; Boudart, M.: J. Catal. *37*, 513 (1975)
267. Frankenburg, W. G.: In: Catalysis (Emmett, P. H., ed.) New York: Reinhold 1955, Vol 3, pp. 234
268. Dmitrenko, L. M.; Lachinov, S. S.; Sibyakova, R. F.: Kinet. Katal. *1*, 379 (1960)
269. Dmitrenko, L. M.; Lachinov, S. S.; Sivyakova, R. F.: Kinet. Katal. *6*, 121 (1965)
270. Enikeev, E. Kh. et al.: Dokl. Akad. Nauk SSSR *131*, 1126 (1960); CA *57*, 7965
271. Krylova, A. V.; Kuznetsov, L. D.; Konyukhova, I. N.: Kinet. Katal. *5*, 948 (1964)
272. Ivanov, M. M. et al.: Kinet. Katal. *9*, 1239 (1968)
273. Rambeau, G.; Amariglio, H.: J. Chim. Phys. *75*, 333 (1978)
274. Amariglio, H.: Private communication
275. Volpin, M. E. et al.: Z. Anorg. Allg. Chem. *428*, 231 (1977)
276. Ichikawa, M. et al.: Chem. Commun. *1972*, 176
277. Croft, R. C.: Austr. J. Chem. *9*, 184, 194, 201, 206 (1956)
278. Urabe, K.; Yoshioka, T.; Ozaki, A.: J. Catal. *54*, 52 (1978)
279. Takeshita, T.; Wallace, W. E.; Craig, R. S.: ibid. *44*, 236 (1976)
280. Ohya, A.; Aika, K.; Ozaki, A.: to be published
281. Rudnitsky, L. A.; Berengarten, M. G.: Kinet. Katal. *13*, 115 (1972); J. Catal. *30*, 444 (1973)
282. Jones, A.; McNicol, B. D.: J. Catal. *47*, 384 (1977)
283. Sommer, A.: Nature (London) *152*, 215 (1943); Spicer, W. E., Sommer, A. H., White, J. G.: Phys. Rev. *115*, 57 (1959)
284. Ohya, A.; Urabe, K.; Aika, K.; Ozaki, A.: J. Catal. *58*, 313 (1979)
285. Urabe, K.; Shiratori, K.; Ozaki, A.: ibid. *55*, 71 (1978)
286. Ohya, A.; Urabe, K.; Ozaki, A.: Chem. Lett. (Jap.) *1978*, 233
287. Aika, K.; Ozaki, A.: J. Catal. *35*, 61 (1974)
288. Allen, A. D.; Senoff, C. V.: Chem. Commun. *1965*, 621
289. Chatt, J.; Richards, R. L.: In: The Chemistry and Biochemistry of Nitrogen Fixation (Postgate, J. R., ed.) London: Plenum Press 1971, pp. 57
290. Taqui Khan, M. M.; Martell, A. E.: In: Homogeneous Catalysis by Metal Complexes. New York: Academic Press 1974, pp. 181—292
291. Bottomley, F.: In: A Treatise on Dinitrogen Fixation, sec. I and II (Hardy, R. W. F. et al., ed.) New York: John Wiley & Sons 1979, pp. 109—167
292. Yamamoto, A.; Kitazume, S., Ikeda, S.: J. Amer. Chem. Soc. *90*, 1089 (1968)
293. Bell, B.; Chatt, J.; Leigh, G. J.: Chem. Commun. *1970*, 842
294. Bancroft, G. M. et al.: J. Chem. Soc. A *1970*, 2146

295. Bell, B.; Chatt, J.; Leigh, G. J.: Chem. Commun. *1970*, 576
296. Allen, A. D.; Stevens, J. R.: Chem. Commun. *1967*, 1147
297. Collman, J. P. et al.: J. Amer. Chem. Soc. *90*, 5430 (1968)
298. Chatt, J.; Dilworth, J. R.; Leigh, G. J.: Chem. Commun. *1969*, 687
299. Taqui Khan, M. M.; Anwaruddin, Q.: Unpublished results in reference 290, pp. 240
300. Sellman, D.: J. Organometal. Chem. *36*, C27 (1972)
301. Sellmann, D.: Angew. Chem. Int. Ed. *10*, 919 (1971)
302. Hidai, M.; Tominari, K.; Uchida, Y.; Misono, A.: Chem. Commun. *1969*, 1392
303. Marvich, R. H.; Brinzinger, H. H.: J. Amer. Chem. Soc. *93*, 2046 (1971)
304. Chatt, J.; Melville, D. P.; Richards, R. L.: J. Chem. Soc. A *1969*, 2841
305. Rossi, M.; Sacco, A.: Chem. Commun. *1969*, 471
306. Chatt, J. et al.: Chem. Commun. *1970*, 90
307. George, T. A.; Seibold, C. D.: J. Amer. Chem. Soc. *94*, 6859 (1972)
308. Aresta, M. et al.: Chem. Commun. *1971*, 781
309. Speier, C.; Marko, L.: Inorg. Chim. Acta. *3*, 126 (1969)
310. Sellman, D.: Angew. Chem. Int. Ed. *13*, 639 (1974)
311. Manriquez, J. M.; Bercaw, J. E.: J. Amer. Chem. Soc. *96*, 6229 (1974); Manriquez, J. M.; Sanner, R. D.; Marsh, R. E.; Bercaw, J. E.: J. Amer. Chem. Soc. *98*, 3042 (1976)
312. Green, M. L. H.; Silverthorn, W. E.: Chem. Commun. *1971*, 557; J. Chem. Soc. Dalton *1973*, 301
313. Borodko, Yu. G. et al.: Chem. Commun. *1972*, 1178
314. Borodko, Yu. G. et al.: Chem. Commun. *1973*, 169
315. Borodko, Yu. G. et al.: Chem. Commun. *1971*, 1185
316. Chatt, J. et al.: Chem Comm. *1970*, 955, Chatt, J. et al.: J. Chem. Soc. Dalton *1973*, 1167
317. Mercer, M.; Crabtree, R. H.; Richards, R. L.: Chem. Commun. *1973*, 808
318. Miura, Y.; Yamamoto, A.: Chem. Lett. (Jap.) *1978*, 937
319. Ito, T. et al.: Abst. 26th Symposium Organometal. Chem. Jap. 1979, A205
320. Hammer, R. et al.: Angew. Chem. Int. Ed. *16*, 485 (1977)
321. Krüger, C.; Tsay, Y.-H.: ibid. *12*, 998 (1973)
322. Burdett, J. K.; Graham, M. A.; Turner, J. J.: J. Chem. Soc. Dalton *1972*, 1620
323. Ozin, G. A.; Vander Voet, A.: Can. J. Chem. *51*, 637 (1973)
324. Huber, H. et al.: J. Amer. Chem. Soc. *95*, 332 (1973)
325. Klotzbücher, W. E.; Ozin, G. A.: ibid. *95*, 3790 (1973)
326. Kündig, E. P.; Moskovits, M.; Ozin, G. A.: Can. J. Chem. *51*, 2737 (1973)
327. Rest, A. J.: J. Organometal. Chem. *40*, C76 (1972)
328. Ozin, G. A.; Vander Voet, A.: Can. J. Chem. *51*, 3332 (1973)
329. Kündig, E. P.; Moskovits, M.; Ozin, G. A.: ibid. *51*, 2710 (1973)
330. Green, D. W.; Thomas, J.; Gruen, D. M.: J. Chem. Phys. *58*, 5453 (1973)
331. Spiker, Jr. R. C.; Andrews, L.; Trindle, C.: J. Amer. Chem. Soc. *94*, 2401 (1972)
332. DeKocck, C. W.; Van Leirsburg, D. A.: ibid. *94*, 3235 (1972)
333. Chatt, J.; Heath, G. A.; Richards, R. L.: Chem. Commun. *1972*, 1010, J. Chem. Soc. Dalton *1974*, 2074
334. Chatt, J.; Pearmann, A. J.; Richards, R. L.: Nature *253*, 39 (1975), J. Organmometal. Chem. *101*, C45 (1975)
335. Chatt, J.; Heath, G. A.; Leigh, G. J.: Chem. Commun. *1972*, 444, Chatt, J. et al.: J. Chem. Soc. Dalton *1977*, 688
336. Diamantis, A. A. et al.: J. Organometal. Chem. *84*, C11 (1975), Iske, S. D. A.; Day, V. W.; George, T. A.: J. Amer. Chem. Soc. *97*, 4127 (1975)
337. Hidai, M.; Mizobe, Y.; Uchida, Y.: J. Amer. Chem. Soc. *98*, 7824 (1976)
338. Vol'pin, M. E.; Shur, V. B.: Dokl. Akad. Nauk. SSSR *156*, 1102 (1964), C.A. *61*, 8933a; Nature *209*, 1236 (1966)
339. Vol'pin, M. E.; Shur, V. B.: Chemical Fixation of Molecular Nitrogen. In: Organometallic Reactions, Vol. 1 (Becker, E.; Tsutsui, M., ed.) New York: John Wiley & Sons 1970, pp. 55—117
340. Vol'pin, M. E. et al.: Chem. Commun. *1968*, 1074
341. Yamamoto, A.; Ookawa, M.; Ikeda, S.: ibid. *1969*, 841

342. van Tamelen, E. E.; Rudler, H.: J. Amer. Chem. Soc., *92*, 5253 (1970)
343. Shilov, A. E.; Shilova, A. K.; Kvashina, E. E.: Kinet. Katal. *10*, 1402 (1969)
344. Shilov, A. E.; Shilova, A. K.: Zh. Fiz. Khim. *44*, 288 (1970)
345. Maskill, R.; Pratt, J. M.: J. Chem. Soc. A *1968*, 1914
346. Broitman, M. O.; Vorontsova, T. A.; Shilov, A. E.: Kinet. Katal. *13*, 61 (1972)
347. Tchoubar, B.; Shilov, A. E.; Shilova, A. K.: Kinet. Katal. *16*, 1079 (1975)
348. Denisov, N. T. et al.: Kinet. Katal. *11*, 813 (1970)
349. Shilov, A. E.: In: A Treatise on Dinitrogen Fixation Sec. I and II. (Hardy, R. W. F. et al., ed.) New York: John Wiley & Sons 1979, pp. 31—108
350. Schrauzer, G. N.; Schlesinger, G.: J. Amer. Chem. Soc. *92*, 1808 (1970), Schrauzer, G. N. et al.: J. Amer. Chem. Soc. *96*, 641 (1974)
351. Shilov, A. E. et al.: Nature *231*, 460 (1971)
352. Denisov, N. T. et al.: Zh. Fiz. Khim. *48*, 2238 (1974)
353. Nikonova, L. A. et al.: J. Molec. Catal. *1*, 367 (1975/76)

Chapter 4

The Fischer-Tropsch Synthesis

M. E. Dry

Research Department
SASOL, Sasolburg, South Africa

This review concentrates mainly on the development of the Fischer-Tropsch process from the late 1950's to 1979. During this period the Sasol plant was the only FT process in operation and hence a large part of this review deals with the information generated at Sasol. The various types of reactors are compared and discussed.

The preparation of catalysts is described with emphasis on the influence that chemical and structural promoters have on the physical and chemical properties of the catalysts as well as on their performance in the FT synthesis. The main factors influencing the loss of catalytic activity are discussed as well as the rate of carbon deposition on iron catalysts.

As the reaction products can vary from predominantly methane to predominantly high molecular mass waxes the factors controlling the spread of hydrocarbons obtained and the interrelationships that exist between the products are dealt with in detail.

The various mechanisms for the FT reaction which have been proposed are discussed and since it is concluded that the differences are more apparent than real a simple unifying scheme is suggested. The overall kinetics of the reaction is reveiwed with emphasis on the influence of the partial pressures of H_2, CO, H_2O and CO_2. For iron catalysts a relatively simple equation satisfactorily describes the reaction profile in various reactors over a wide range of conditions.

Contents

1. Introduction

Up to and during the second world war the major research effort in Fischer-Tropsch synthesis was carried out in Germany and several production plants were operated there. Initially after the war the research activity continued, especially in the USA, but with the discovery of the large oil deposits in the Middle East during the mid 1950's interest in the Fischer-Tropsch synthesis waned. Several books and reviews [1, 2, 3, 4, 5] have been written which cover the above time period and in particular those written or co-authored by R. B. Anderson contain a mass of information. Since this period the major event in the further development of the Fischer-Tropsch synthesis was the construction of Sasol I which came on line in 1955 and has been in commercial operation ever since. Research and development continued at Sasol although it had been largely discontinued elsewhere in the world, a few notable exceptions being Pichler, Kölbel and their co-workers in Germany. The present review will deal in some depth with the Sasol processes and with the research work carried out at Sasol as well as dealing with the Fischer-Tropsch synthesis overall. When no specific reference is made to information presented in this review it may be assumed that it was generated at Sasol.

Since the Middle East oil crisis of 1973 there has been a resurgence of interest in the Fischer-Tropsch process and research publications and reviews have again started appearing in the literature. A recent review [6] deals largely with the past German research effort but also includes some information of the Sasol process.

The synthesis of methane (where this is the only desired product), of methanol, of aldehydes and alcohols from olefins (i.e. the Oxo process) and of other products from mixtures of CO and H_2 will not be dealt with in any detail although they are probably closely related mechanistically to the Fischer-Tropsch process.

2. A Brief History of the Process

As early as 1902 it was observed that methane was formed from mixtures of H_2 and CO over nickel and cobalt catalysts [7]. At high pressure over cobalt catalysts BASF [8] produced liquid products. In 1923 Fischer and Tropsch [9] reported their work using alkalized iron catalyst at high pressure (10 to 15 MPa). The liquid product was very rich in oxygenated compounds. At lower pressures hydrocarbons were produced but it was found that the iron catalysts deactivated rapidly and because of this further work was concentrated on nickel and cobalt catalysts operating at low pressures. In time nickel too was discarded because of its very high tendency to produce methane and also because its activity declined due to loss of nickel from the reactor as nickel carbonyl. In 1936 the first four Fischer-Tropsch production plants were commissioned and had a total capacity of 200,000 tons of hydrocarbon per year [6]. By 1944 the potential capacity of the nine plants in Germany was about 700,000 tons per year. The catalyst used was mainly the "standard" cobalt catalyst which had the composition in relative mass units Co:ThO_2: MgO:Kieselguhr 100:5:8:200.

Development work on iron catalysts was continued and in 1936 Pichler [4] made the important discovery that when the pressure was increased from atmospheric to about 15 bar the life of the iron catalyst was markedly improved. Further development of iron as a catalyst was carried out by various German firms and this led to the comparative tests at Schwarzheide in 1943 [2]. Although the tests were apparently successful none of the catalysts were used to replace the cobalt catalyst in the production plants during the war.

After the war the two German firms Ruhrchemie and Lurgi formed an *Arbeitsgemeinschaft* (ARGE) and developed the fixed bed reactor using a precipitated iron catalyst to produce high yields of wax. In the USA several firms (e.g. Standard Oil and Hydrocarbon Research Inc.) developed fluidized bed reactors with the Kellogg Co. researching a circulating entrained catalyst version. The fluidized reactors produce high yields of gasoline. The first

Sasol plant which was commissioned in 1955 incorporated both the fixed bed Arge reactors and the entrained bed Kellogg reactors. Although the fixed bed plant operated more or less as predicted, the circulating fluid bed reactor could not be operated as a commercial unit until numerous modifications had been made, both operational and mechanical. In addition considerable further research had to be done on the catalyst before it gave an acceptable life with the required product selectivity. Thus the entrained fluid bed system, now known as the Sasol Synthol system, was converted into a highly reliable large scale industrial operation. Further R & D has also greatly improved the performance of the fixed bed units.

Although throughout this review frequent reference is made to work done at Sasol it should be appreciated that a large part of the Sasol work is of a proprietary nature and so not available for publication.

3. Types of Reactors and their Development

The laboratory reactors used by Fischer and Tropsch were simply glass tubes of about 5 mm ID with the catalyst held in a fixed position. As early as 1930 laboratory experiments with moving bed catalysts were also carried out by Fischer [10]. For larger reactors the design of the reactor is of great importance since to obtain optimum product selectivity and long catalyst life it is essential to remove the exothermic heat of reaction as effectively as possible.

A. Fixed Bed Reactors

In the original German industrial reactors of 1936 (*Lamellenöfen*) 10 m^3 catalyst was packed between perpendicular parallel metal plates spaced 7 mm apart [6]. The heat of reaction was removed by water circulating through horizontal tubes which passed through the vertical plates. The heat removal was, however, not sufficient due to the low gas linear velocities employed (fresh feed space velocities were only about 100 hr^{-1}). This resulted in localized overheating which in turn resulted in carbon deposition and in the break-up of the catalyst particles.

The next improvement was the development of the concentric tube reactor. The tubes had ID's of 21 and 24 mm and were 4.5 m long. The reactor contained about 2,000 pairs of tubes. The 10 m^3 of catalyst was placed in the annular space between the tubes. Cooling water was circulated past the outside of the outer tube and through the inside of the inner tube. The reactor operated at 1.2 MPa and at 493 K but the fresh feed space velocity was still low at about 100 hr^{-1} (m_n^3 [1] gas per m^3 catalyst per hour).

[1] m_n^3 is a cubic meter of gas measured at STP

A big improvement in the technique of operating tubular reactors was the employment of recycle gas. The reactor in Harnes consisted of 1,250 single tubes of 32 mm ID [6]. A 3 to 1 recycle ratio (recycle gas to fresh gas volume ratio) was used and the fresh feed space velocity was increased to 250 hr^{-1}. It is desirable to have a high linear gas velocity through the catalyst bed as under turbulent flow conditions there is a higher rate of heat exchange between the catalyst and the tube walls and this allows for higher fresh gas loads. These improvements led to the development of the reactors which were installed at Sasol in 1954 (cf. Figure 1). Each of these reactors has about 2,050 single tubes of 50 mm ID. The length of the tubes is about 12 m. The outside of the tubes is surrounded by boiling water, the temperature of which is controlled by regulating the pressure. The catalyst charge is about 40 m^3 and the original design fresh feed space velocity was 500 hr^{-1}. Typical operating conditions were 2.7 MPa and 493 to 523 K. As the result of changes in the mode of operation of the reactors and also the development of more active catalysts by Sasol the fresh feed load and subsequently the wax production rate has been considerably increased. The use of a high gas linear velocity through the catalyst bed ensures that the heat of reaction is removed along the length of the tubes and this results in a near-isothermal reactor. As the catalyst's activity declines the temperature can be raised to maintain conversion. The run length can exceed 350 days but normally it is the desired product selectivity which determines the run length.

In large layered catalytic bed reactors (in contrast to the tubular reactors described above) the bulk of the reaction heat must necessarily go out with the gas at the reactor exit. The result is a high differential temperature over

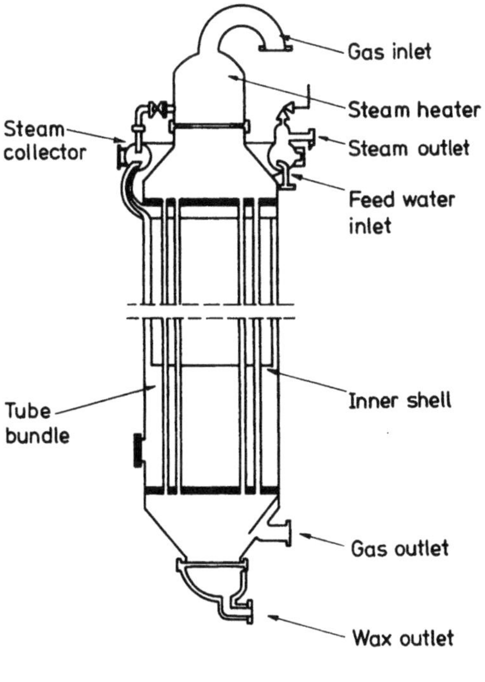

Figure 1.
Sasol fixed bed reactor. Only a few of the 5 cm ID reactor tubes are shown

the catalyst beds. To minimise the temperature rise BASF [11] utilized the concept of a very high recycle ratio (100 to 1) in a single bed reactor of catalyst volume 3.8 m^3. However, catalyst overheating still occurred. Lurgi [11] improved the situation by splitting the bed into several sections with the fresh feed also split, i.e. fed between successive beds. Their demonstration unit was charged with 90 dm^3 catalyst divided into 6 separate beds. The overall recycle to fresh feed gas ratio was ca 20 to 1 and the temperature rise per bed was about 5 K. The reactor was operated at about 543 K, 2 MPa pressure and a fresh feed space velocity of about 200. $CO + H_2$ conversion of about 85 percent was obtained.

In order to lower the pressure drop over reactors operating with a high recycle ratio, the US Bureau of Mines [12] employed catalyst beds of high voidage (about 90 percent). Thus the reactors were packed with either activated steel lathe turnings or with parallel metal plates onto which was flame sprayed either iron oxide powder or Raney nickel. Typically the reactors were operated at fresh feed space velocities of about 1,000 hr^{-1}, 593 K, 2.8 MPa and a recycle ratio of about 20. The temperature differential over the reactors was about 40 to 50 K. Work was also done on tube wall reactors [13] in which a thin layer of catalyst is attached to the outside of the tubes and coolant is passed through the inside, the objective being to improve the removal of reaction heat.

B. Slurry Bed Reactors

In these reactors gas is bubbled through a suspension of finely divided catalyst (typically $< 50\ \mu m$) in a liquid which has a low vapour pressure at the temperature being used. In Fischer-Tropsch the liquid is conveniently a cut from the product spectrum, e.g, a high boiling wax. The heat of reaction is removed by circulating the slurry through external heat exchangers or by heat exchangers immersed directly into the slurry bed. The slurry bed has an advantage over the fixed bed in that it can be used at higher temperatures as carbon deposition on the catalyst will not adversely effect its performance. Compared to "dry" fluidized bed reactors it has the advantage that it can be used at lower temperatures and/or with lower H_2/CO ratio gases. Under these conditions the wax selectivity is high and this would lead to the defluidization of the "dry" bed. The slurry bed reactor is thus potentially more flexible than either the fixed or fluidized bed reactor.

Early slurry bed studies were carried out by Fischer but it was only in 1953 that a demonstration plant was put into operation at Rheinpreussen [15]. The 1.5 m wide reactor had a working volume of 10 m^3 and operated at about 5.2 MPa. The fresh feed space velocity was only 270 h^{-1} but Kölbel [6] claims that this restriction was due to the gas compressors used and that the throughput could be raised threefold. The Fuel Research Station's unit [16] at Stevenage, England had a volume of about 270 dm^3 and was operated at about 533 K and 1.0 MPa, but again the space velocity used was only about 200 hr^{-1}. The US Bureau of Mines' oil circulation demonstration unit [17] of 3 m^3 capacity also operated at the relatively low space velocity of 345 hr^{-1}.

C. Fluidized Bed Reactors

Following the successful utilization of fluidized beds in catalytic cracking in the petroleum industry this technique was also applied to the F-T synthesis. Because of the higher density of iron catalysts relative to the $SiO_2—Al_2O_3$ used in FCC units it is more difficult to fluidize properly and critical consideration should be given to factors such as catalyst particle size and distribution, gas velocity and effective gas distribution. There are two basic types of units, (a) the fixed fluidized bed (FFB) in which the catalyst bed remains "stationary" with gas passing upwards through it, and (b) the circulating fluidized bed (CFB) in which the catalyst is entrained in the fast moving gas stream.

FFB units were developed by Hydrocarbon Research Inc [18] in the USA and by Standard Oil Co., and this led to the construction of the demonstration unit at Brownsville in 1950. The design capacity of the plant was 360,000 tons per year for two reactors. Each reactor had a width of 5 m, a height of 18 m and was loaded with about 200 tons of finely divided iron catalyst. They were operated at about 593 K, 2.7 MPa and gas velocities above 20 cm sec^{-1} were used. The heat of reaction was removed by heat exchangers immersed in the fluidized bed. The recycle to fresh gas ratio used was about 1.5 and conversions up to 96 percent were achieved at high fresh feed space velocities (2,000 to 3,000 hr^{-1}).

Initially considerable difficulties were experienced, the main problem apparently being to achieve uniform fluidization of the entire catalyst bed which is essential in order to avoid gas by-passing. The Brownsville plant was shut down in 1957 for economic reasons and it was claimed that the main operational problems had been overcome but with what degree of success has not been published. Because of the potentially high throughput, high conversion and good temperature control of this type of reactor its further industrial development is considered to be worthwhile. Such work is currently being undertaken by Sasol. Thus it has been demonstrated that the quality of fluidization is markedly improved with the addition of a relatively small amount of finely divided charcoal without this adversely effecting the catalysts' activity or selectivity [19].

The Kellogg Company [20] (USA) developed the circulating fluidized bed reactor (CFB) and this was scaled up from their 10 cm ID reactor to the 230 cm ID commercial units at Sasol I. After several mechanical and process modifications the system which is now known as the Sasol Synthol process can reliably achieve $CO + CO_2$ conversions of 85 percent at fresh feed rates of 100,000 m^3 hr^{-1}. The overall height of the reactors is about 46 m. A diagram of the reactor is given in Figure 2. The fresh feed and recycle gases are fed in at the bottom at about 22 bar where it meets a down flowing stream of the hot finely divided catalyst. The rate of flow of catalyst from the standpipe is controlled by the slide valves. The combined gas and catalyst stream

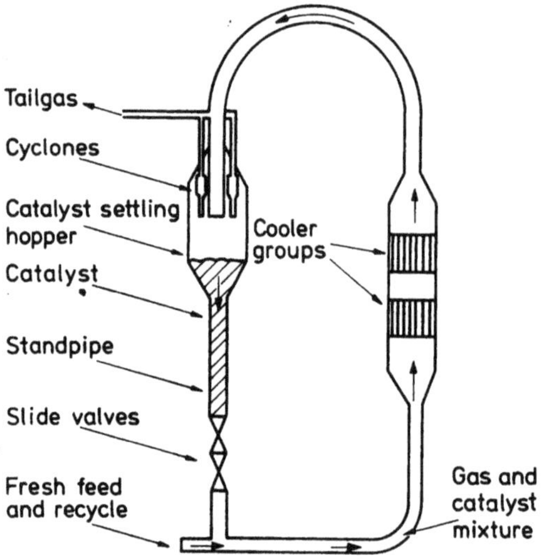

Figure 2. Sasol Synthol reactor

sweeps through the reaction zones. The two banks of heat exchangers inside the reactor remove 30 to 40 percent of the heat of reaction. The balance goes out with the recycle gas and reaction products. The exit temperature is about 593 to 633 K. The catalyst and gas disengage in the wide settling hopper above the standpipe. The gas leaves the reactor via the cyclones which remove the entrained finer catalyst particles and return them to the settling hopper.

Originally the run lengths were typically about 40 days. Process and operational changes have extended this considerably. The conversion normally declines to about 78 percent, depending on the fresh feed load. Because of the low cost of the catalyst used it is economically worthwhile to terminate the run at this stage and restart with a fresh charge of catalyst. Sasol is currently carrying out tests for on-line replacement of catalyst which achieves a higher overall level of conversion and longer on-stream times, which ideally will only be limited by maintenance requirements.

The new reactors at Secunda are modified versions of the above reactors. The heat exchangers have been improved and the reactor capacities have been increased by increasing the reactor diameter as well as the process pressure.

D. Comparison of Different Types of Reactors

The pilot plant studies of Hall [21] and co-workers showed that when normal sized catalysts were used (i.e. normal for the reactor type) the space time yield increased in the order slurry, fixed and fluidized beds. When the catalysts were of the same size the order was slurry, fluidized and fixed beds. Thus in both cases the slurry bed was the least active. This was ascribed to the retardation of the gas diffusion rate by the liquid phase.

Similar studies have been carried out at the Sasol pilot plant with the currently used commercial catalysts. The fluidized and slurry bed tests were carried out in 5 cm ID and 380 cm long tubular reactors surrounded by Dowtherm jackets and topped by wider disengaging sections. The 12 m long 5 cm ID fixed bed reactors were water jacketed. In these near-isothermal types of reactor the temperatures were controlled by regulating the pressure over the jacket fluids. Adiabatic reactor tests were carried out in a 7.5 cm ID tube surrounded by a layer of heat insulating material which in turn was heated in sections by separate electrical heating elements. Heating was so controlled that there were no temperature differences between thermocouples within the catalyst bed and within the lagging material at various heights of the reactor. Thus virtually all the heat generated within the reactor passed out with the exit gas. This contrasts with the reactors described above where virtually all the reaction heat was exchanged via the tube wall along the length of the reactors.

The results of the comparative tests are summarized in Table 1. For each set of tests the same catalysts were used except for particle size differences as required by the systems. The process conditions were also the same for each set. Cases 1 and 2 show that under the conditions employed the slurry bed had a somewhat higher conversion than the fixed bed reactor. The smallness of the catalyst particles in the case of the slurry bed (an essential requirement for good fluidization) clearly more than compensates for the lower mass of catalyst charged. As observed by Kölbel [6] and also by Hall [21] the selectivity in the case of the slurry bed is shifted towards the heavier products. This may be due to the possibility that the actual temperature of the catalyst particles in the slurry bed is lower than that of the catalyst extrudates in the fixed bed. The main factor here could be the big difference in size of the two types of particles, the smaller particles with their much larger external surface per unit mass being able to dissipate the heat of reaction faster than the larger particles. Cases 3 and 4 show that the fluidized bed has a higher activity than the slurry bed. The bed height of the fluidized catalyst was only ca half of that of the slurry bed but nevertheless contained four times the mass of catalyst. Increasing the catalyst charge to the slurry bed does not in practice increase the conversion activity as the actual gas hold-up (and hence the effectiveness of catalyst to gas contact) is adversely effected. The excess liquid present in the case of the slurry bed presumably slows up the rate of transport of the reactants from the gas phase to the catalyst surface sites. There is no discernible difference in the hydrocarbon selectivity spread between the two types of reactor. From this it may be deduced that there is little difference in the actual particle temperatures in the two cases.

Cases 5 to 9 simulate individual stages of a multistage reactor in which the fresh feed is split into roughly equal portions and fed separately to each successive catalyst bed. Cases 5 and 6 represent the first stages whereas cases 7, 8 and 9 simulate the third stage. From cases 5 and 6 and also 7 and 9 it can be seen that the adiabatic reactor, because of the higher average catalyst bed temperature, has a higher conversion. Also the product selectivity

Table 1. Comparison of fixed, slurry and fluidized bed reactors

Case	1	2	3	4	5 [b]	6 [b]	7	8	9
Bed type	Fixed	Slurry	Fluidized	Slurry	Fixed Isothermal	Fixed Adiabatic	Fixed Isothermal		Fixed Adiabatic
Catalyst type[a]	Precipitated		Fused		Precipitated ca. 2.5 mm extrudates				
Particle size	ca. 2.5 mm	40–150 μm	<70 μm	<40 μm					
Catalyst load/kg Fe	2.7	0.8	4.2	1.0					
Catalyst bed height/m	3.8	3.8	2.0	3.8	1.6	1.45	1.6		1.45
Bed Inlet Temp./K	496	508	593	593	493	493	493	523	493
Bed Outlet Temp./K	509	511	598	601	496	529	496	528	526
H_2O Exit Pressure/MPa					0.020	0.025	0.039	0.041	0.055
Recycle to Fresh Feed ratio	1.9	1.9	2.0	2.0	25	25	25	25	25
Total Gas Linear Velocity/cm sec^{-1}	36	36	45	45	34	36	34	34	36
Conversion % ($CO + H_2$) or ($CO + CO_2$)	46	49	93	79	75	88	60	75	67
Selectivity/% carbon atom									
CH_4	7	5	12	12					
C_3			14	14					
Gasoline	14	15	43	42	10	15	11	20	16
Hard wax	27	31	0	0	42	27	40	16	30
Rate of Activity decline/ % Conv. per day					0.37	0.78	0.46	0.83	1.6

[a] The preparation of the two basic types used is described fully in section 5.

[b] Cases 5 and 6 simulate the first stage of a multistage (successive catalyst beds) reactor system, while cases 7, 8 and 9 represent the third stage. Deactivation rates for cases to 4 are not given since the catalyst loadings are widely different in these cases.

has shifted to lower wax levels. Note that when the temperature of the isothermal reactor is increased (case 8) so that it approximately equals the exit temperature of the adiabatic reactor (case 9), the conversion is higher and the wax selectivity is lower since now the average temperature is higher for the isothermal case. A major drawback of the adiabatic reactor is that the rate of activity decline was markedly higher than for the isothermal reactor. Note that in case 8 where the average bed temperature of the isothermal reactor is higher than for the adiabatic case 9 the rate of activity decline is nevertheless lower.

If the objective is to make high yields of wax the isothermal fixed and slurry bed reactors are clearly the most suitable. A "dry" fluidized system cannot be used as the wax produced will rapidly result in defluidization of the bed. The adiabatic reactors, whether single- or multi-staged, are not satisfactory as the wax selectivity is lower and the rate of activity decline higher than for the other reactors. If lighter hydrocarbons are desired the fluidized "dry" bed has the highest production rate. Fixed bed reactors cannot operate at the same high temperatures at which fluidized beds operate because the carbon deposition which occurs at these higher temperatures will rapidly result in plugging of the fixed beds.

4. Sasol Commercial Plants

A. Synthesis Gas Manufacture

Lurgi gasifiers are used to produce the raw synthesis gas which has the approximate molar composition (in percent): 9 CH_4, 29 CO_2, 0.5 H_2S, 1 A + N_2 and 60 H_2 + CO [22, 23]. The coal contains 20 to 35 percent ash. The original gasifiers have an ID of 3.6 m and were designed to produce about 25,000 $m_n^3 \times hr^{-1}$ of raw gas each. Due to process improvements they are currently producing about 35,000 $m_n^3\ hr^{-1}$. The gasifiers at Secunda are larger (3.85 m ID) and each will produce about 55,000 $m_n^3\ hr^{-1}$ of raw gas. Coal is fed via lock hoppers into the top and oxygen and steam is fed in at the bottom (cf. Figure 3). The gas exit temperature is about 673 K. As the coal moves down it is progressively stripped of volatile tar products, charred and finally gasified at about 1570 K. The residual ash is cooled to about 670 K by the incoming O_2 and steam and is removed from the gasifier by a revolving grate into lock hoppers. The pressure inside the gasifiers is about 2.7 MPa. By controlling the amounts of oxygen and steam fed a heat balance is maintained between the two main reactions:

$$C + H_2O \rightarrow CO + H_2 \quad \text{(endothermic)}$$

$$C + O_2 \rightarrow CO_2 \quad \text{(exothermic).}$$

The product gas stream from the gasifier is cooled to condense out the excess water and the heavier tar oils. The water contains dissolved NH_3

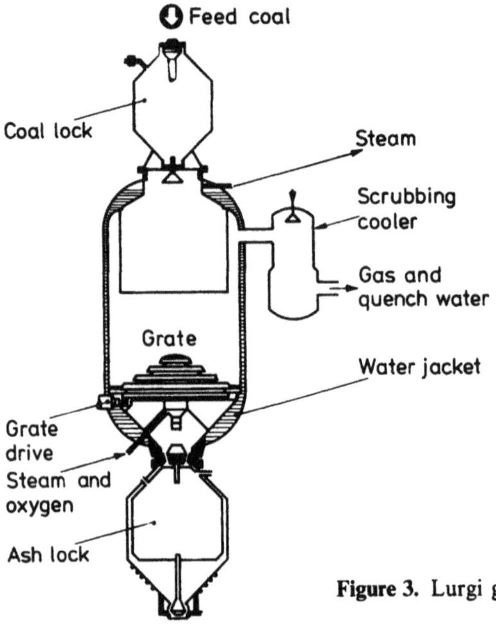

Figure 3. Lurgi gasifier

and phenols. The gas then goes to the Lurgi Rectisol purification plant where it is progressively cooled to knock out the light tar naphthas plant and finally washed with methanol at about 210 K in various stages to remove all sulfur containing gases as well as the bulk of the CO_2. The composition of S-containing gases in the raw gas is about 97 percent H_2S, 2 percent CH_3SH and 1 percent $COS + CS_2$. The molar composition of the purified gas is (percent) 13 CH_4, 1 $A + N_2$, 1 CO_2 and 85 $H_2 + CO$. The total S content is typically 0.03 mg m_n^{-3} of which about 20 percent is H_2S, 50 percent CH_3SH and the balance $COS + CS_2$. The Rectisol plant is thus extremely effective and removes all the undesired materials in a single process.

B. Sasolburg Plant Flow Scheme

Figure 4 gives the block flow diagram of the plant. Oxygen, produced in a cryogenic air separation plant, coal and steam are fed to the Lurgi gasifiers. The raw gas is cooled to remove water, NH_3 and tars. The water phase is treated in the Phenosolvan plant where the phenols are extracted with butyl acetate. The butyl acetate is recovered by distillation and recycled. The NH_3 is steam-stripped from the remaining water and is converted to $(NH_4)_2SO_4$.

The tar is distilled into various cuts, from light tar naphtha through creosotes to pitch. The lighter tar naphthas together with the Rectisol naphtha are hydrofined and the products are either blended into the gasoline pool or sold as aromatic solvents.

The Rectisol pure gas is fed to the fixed bed and to the Synthol reactors. The effluent from the F-T reactors is cooled and water and oil are condensed

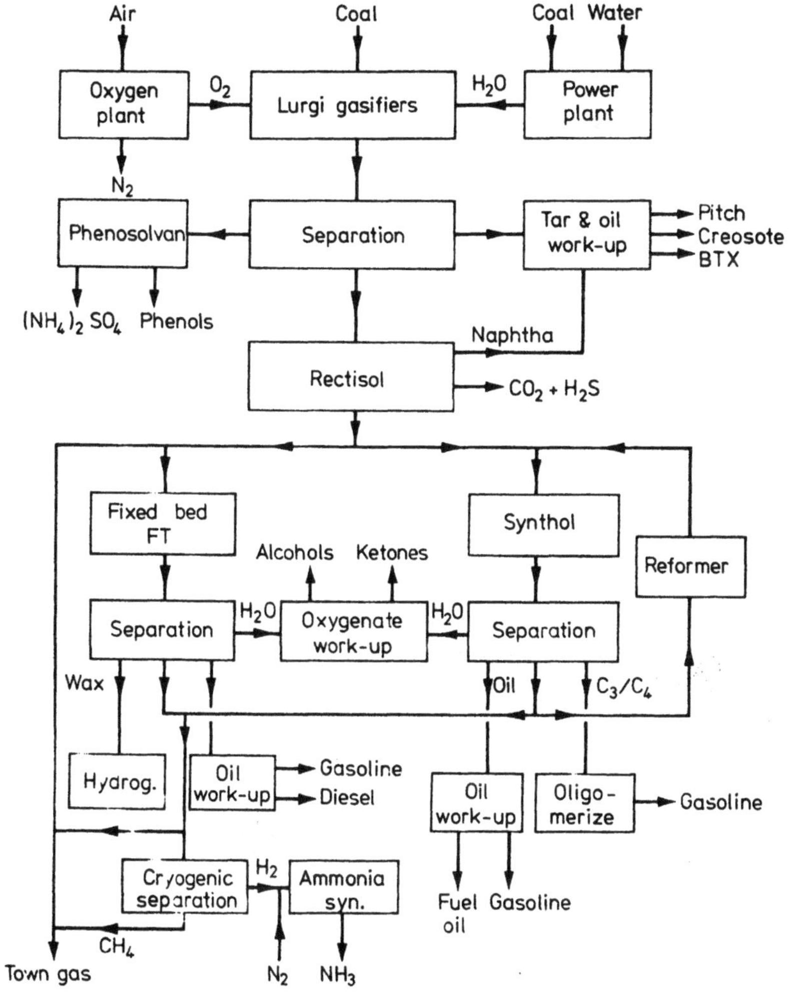

Figure 4. Sasol I flow scheme

out. The oil is further scrubbed with water to extract more oxygenates. All the water is then fed to the oxygenate recovery and refining plant.

The Synthol light oil (C_5 to about C_{12}) which is about 75 percent olefinic is treated over an acidic catalyst at about 670 K and 0.1 MPa pressure. The oxygenates are converted to olefins and the olefins are isomerised. Both double bond shift and skeletal isomerization occur. This treatment improves the octane rating of the gasoline.

The tailgas containing the uncondensed hydrocarbons is passed through an oil absorption tower in which the C_3 and heavier hydrocarbons are extracted. The $C_3 + C_4$ products from both the Synthol and fixed bed reactors are treated over phosphoric acid/kieselguhr catalyst at about 470 K and 3.8 MPa to oligomerize all the olefins to gasoline.

The condensed hydrocarbon product from the fixed bed reactors is distilled to recover the gasoline and diesel. The remaining wax is vacuum distilled to produce "medium wax" (590 to 770 K) and "hard wax" (+770 K). The waxes are then hydrofined to remove all oxygenated compounds and olefins. Nickel catalyst at about 7.0 MPa and 490 K is used in this latter process.

The tailgas from the reactors contains CH_4, ethylene, ethane and unconverted synthesis gas. It is consumed in several different ways. A portion is treated in a cryogenic unit and the extracted H_2 together with additional H_2 from the water-gas shift unit is catalytically converted to NH_3. Town gas (0.020 GJ m_n^{-3}) is produced by blending tailgas (about 0.028 GJ m_n^{-3}) with CH_4 from the cryogenic unit and with Rectisol gas (about 0.016 GJ m_n^{-3}). The remaining tailgas is catalytically reformed over nickel at about 1270 K with steam and oxygen. The approximate stochiometry of this process is

$$1.0\ CH_4 + 0.24\ H_2O + 0.62\ O_2 \rightarrow 0.52\ CO + 0.48\ CO_2 + 2.24\ H_2\ .$$

The product gas is water scrubbed to lower the CO_2 content and then recycled to the Synthol reactors.

C. The Secunda Plant Flow Scheme

Figure 5 is a simplified block flow diagram of the new Sasol plant. The synthesis gas production and associated plants are similar to those at Sasolburg. However, instead of $(NH_4)_2SO_4$, anhydrous NH_3 will be produced. Also all water soluble phenols and tar oils over and above that which the market can absorb will be hydrofined/hydrocracked to gasoline and diesel.

In the Fischer-Tropsch section only the Synthol reactors are being used. The work-up of the oxygenated compounds dissolved in the product water is similar to that used at Sasolburg.

The recovery and work-up of the hydrocarbon products at the Secunda plant is considerably different from that used at the Sasolburg plant. The work-up of the condensed oils include hydrofining, catalytic reforming (Pt—Al_2O_3), isomerization and selective hydrodewaxing. The gaseous hydrocarbons are separated into various cuts in a cryogenic unit. The CH_4 is reformed to synthesis gas and recycled to the Synthol units. The C_2's go to an ethylene plant and the C_3 and heavier olefins are oligomerized to gasoline and diesel.

A considerable degree of flexibility is possible in the product work-up. By changing the fractionation cut points and by changing the mode of operation of the hydrodewaxing, of the oligomerization and of the hydrocracking units, the overall ratio of gasoline to diesel can be varied from about 10/1 to 1/1.

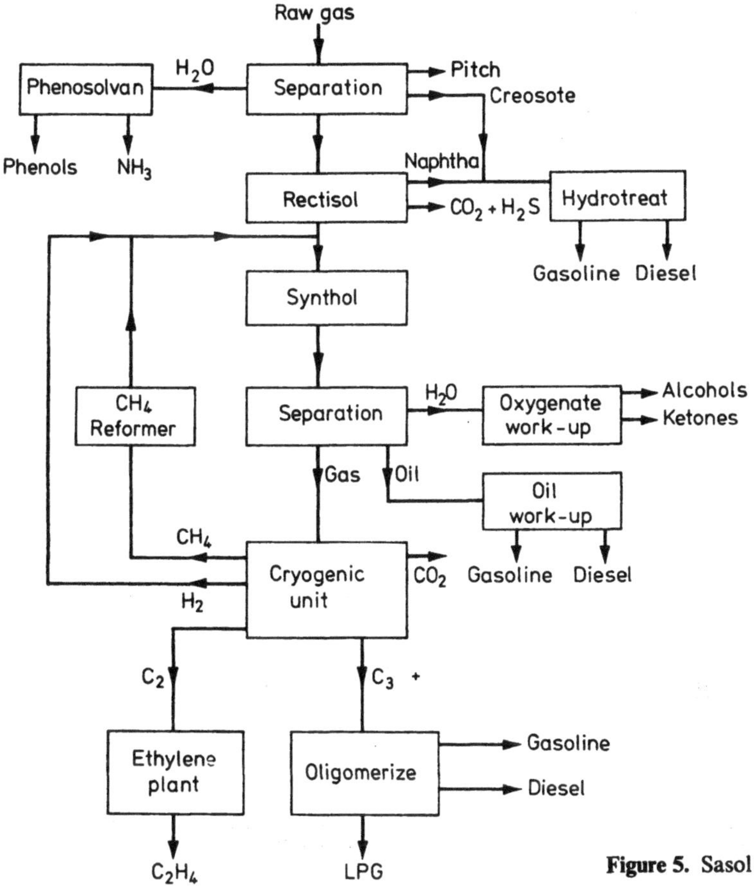

Figure 5. Sasol II flow scheme

Table 2. Product selectivities of Sasol commercial reactors

Product	Composition/% carbon atom	
	Fixed bed at 493 K	Synthol at 598 K
CH_4	2.0	10
C_2H_4	0.1	4
C_2H_6	1.8	4
C_3H_6	2.7	12
C_3H_8	1.7	2
C_4H_8	3.1	9
C_4H_{10}	1.9	2
C_5 to C_{11} (gasoline)	18	40
C_{12} to C_{18} (diesel)	14	7
C_{19} to C_{23}	7	
C_{24} to C_{35} (Medium Wax)	20	4
$>C_{35}$ (Hard Wax)	25	
Water soluble non-acid chemicals	3.0	5
Water soluble acids	0.2	1

Table 3. Composition of water soluble products from SASOL reactors

Category	Compound	Composition/wt. percent	
		Fixed bed reactors	Synthol reactors
Non Acid	CH_3CHO	~2	3
	C_2H_5CHO		1
	C_3H_7CHO	2	0.5
	CH_3COCH_3	1	10
	$C_2H_5COCH_3$		3
	$C_2H_5COC_2H_5 + C_3H_7COCH_3$	24	1
	CH_3OH	50	1
	C_2H_5OH	11	55
	n-C_3H_7OH		13
	i-C_3H_7OH	6	3
	n-C_4H_9OH		4
	i-C_4H_9OH		3
	2-C_4H_9OH		1
	$C_5H_{11}OH$	4	1
Acids	Acetic		70
	Propionic		16
	Butyric		9

Table 4. Selected properties of liquid and solid products from Sasol reactors

Product Cut	Property	Fixed Bed[a]	Synthol[a]
Gasoline C_5—C_{11}	Olefins	32%	65%
	Paraffins	60%	14%
	Aromatics	0%	7%
	Alcohols	7%	6%
	Ketones	0.6%	6%
	Acids	0.4%	2%
	n-Paraffins	95%[b]	55%[b]
	RON (Pb free)	~35	88
Diesel C_{12}—C_{18}	Olefins	25%	73%
	Paraffins	65%	10%
	Aromatics	0%	10%
	Alcohols	6%	4%
	Ketones	<1%	2%
	Acids	0.05%	1%
	% n-Paraffins	93%[b]	60%[b]
	Cetane No	75	55
Medium Wax C_{24}—C_{35}	Olefins	10%	—

[a] wt. % of cut except for RON and cetane No

[b] % of the paraffins which are straight chained

D. Product Characteristics

Table 2 compares the product spectrums obtained in the commercial fixed bed and Synthol reactors. The data is for the products as produced in the Fischer-Tropsch process. The liquid and solid products are not pure hydrocarbons but contain varying amounts of alcohols, ketones and acids. The selectivity spectrums can be widely varied in practice and this aspect will be discussed in more detail in a later section. The data given in Tables 2 to 4 merely illustrate typical results obtained in the Sasol units as currently operated. Table 3 compares the composition of the water soluble products obtained for the two Fischer-Tropsch processes and Table 4 compares a few selected properties of the liquid and solid products. As a result of the subsequent work-up processes the properties of the fuel products are considerably altered and the final gasoline and diesel fuels produced meet all the required specifications.

5. Catalysts

A. Precipitated Catalysts

1. Preparation and Promotion

The majority of the early successful catalysts developed in Germany for the fixed bed reactors were prepared by precipitation techniques. Anderson's [2] review gives detailed descriptions of many of the developments. Use of the carbonates of potassium, sodium or ammonia as precipitating agent resulted in better catalysts than when the hydroxides were used. The use of metal sulphates and chlorides resulted in catalysts of low activity.

The standard cobalt catalyst was prepared by the rapid addition of a boiling solution of the nitrates of cobalt, magnesium and thorium to a boiling solution of sodium carbonate [6]. The correct amount of Kieselguhr was then added with intensive mixing. The catalyst was then filtered and washed with hot water in a filter press. The filter cake was extruded and dried.

The early precipitated iron catalysts were promoted only with potassium salts. To improve the stability of these catalysts towards thermal sintering many supports such as zinc oxide, dolomite, kieselguhr and alumina were investigated. Silica supports appeared to be the most effective. Ruhrchemie used alkali-waterglass solution as the source of the silica instead of Kieselguhr. The catalyst developed by Ruhrchemie and used in the fixed bed reactors at Sasol for several years has been fully described elsewhere [6]. To recap briefly, the successive preparation steps were as follows. A near boiling solution of iron and copper nitrate (40 g Fe and 2 g Cu per dm^3) was poured into a hot solution of sodium carbonate with vigorous stirring over a period of several minutes until the pH reached 7 to 8. The precipitate was washed with hot distilled water in a filter press until the Na content in the effluent was very low. The washed precipitate was re-slurried with water and impregnated with a potassium waterglass solution to give 25 g SiO_2 per 100 g Fe. Sufficient nitric acid was added to remove some of the excess potassium so that after filtration the K_2O content was 5 g per 100 g Fe. The filter cake was pre-dried, extruded and finally dried to a water content of less than 10 wt. percent. In the preparation technique currently employed at Sasol several changes have been made in both the preparation process and in the type and contents of promoters which have improved the economics of the catalyst production as well as its performance in the Fischer-Tropsch synthesis.

When the solution of potassium waterglass is added to the slurry of precipitated iron oxide the silica is effectively precipitated. It presumably adsorbs onto or reacts with the high area Fe_2O_3, up to about 300 $m^2 g^{-1}$, since on thorough washing and filtering all of the potassium but none of the SiO_2 is removed. Thus the addition of HNO_3 to the Fe_2O_3/waterglass slurry is to control the K_2O content of the catalyst and not to precipitate the silica. The presence of the intimately mixed silica phase stabilizes the high area of the Fe_2O_3 by preventing crystal growth, i.e. sintering, of neighbouring Fe_2O_3 crystallites. In Table,5 the influence of the SiO_2 on the total area and pore structure is illustrated. As can be seen SiO_2 increases the area and stabilizes the presence of larger pores. The increase in area with increasing SiO_2 content has also been reported elsewhere [6]. The pore structure of the catalyst is largely determined in the very first stage of the catalyst preparation, i.e. the precipitation of the Fe_2O_3. Thus if the precipitation conditions are such that the Fe_2O_3 has narrow pores then the final catalyst will also have this feature. The nature of the pore size distribution has been found to be influenced by the concentrations of the solutions used, the time over which the precipitation occurs, the precipitating temperature and the *pH* during precipitation. Figure 6 for instance illustrates the influence of the environment in which the Fe_2O_3 precipitate is formed. Figure 6a is for Fe_2O_3 prepared by adding the iron solution to the carbonate solution while Figure 6b is for the reverse procedure, all other conditions being identical. The two cases have very similar total pore volumes and surface areas but as can be seen from the shapes of the N_2 adsorption isotherms there is a marked difference in the distribution of pore sizes. Figure 7 illustrates the result if an excess of Fe^{3+} ions is added to the carbonate solution resulting in a low final *pH*. The final *pH* is obviously a critical factor in determining the pore size and distribution. The chemical nature of the precipitating agent also has an influence on the pore structure. This influence could partly be a *pH* effect. Table 6 shows that using carbonates results in catalysts with more big pores. The SiO_2 content for the four cases in Table 6 is all the same.

The total porosity of precipitated catalysts is largely determined by the amount of shrinkage that occurs during the drying process. The presence of water appears to enhance the degree of shrinkage. When the precipitate is re-slurried in an excess of a low surface tension liquid such as acetone, and the catalyst is then dried, the pore volume is increased more than two fold. This technique has also been applied in the preparation of silica gels [24].

Table 5. Influence of SiO_2 on precipitated Fe_2O_3

SiO_2/ g (100 g Fe)$^{-1}$	Unreduced			H_2 reduced			
	Pore Vol/ $cm^3\ g^{-1}$	Area/ $m^2\ g^{-1}$	Area in pores >4.5 nm/$m^2\ g^{-1}$	Pore Vol/ $cm^3\ g^{-1}$	Area/ $m^2\ g^{-1}$	Area in Pores >4.5 nm/$m^2\ g^{-1}$	Reduction/% [a]
0	0.37	275	41	0.22	35	35	100
8	0.47	345	59	0.43	190	68	80
19	0.74	375	90	0.48	250	80	46
25	0.71	390	94	0.61	270	84	58
29	0.75	370	96	0.65	265	85	57
50	N.A.	405	N.A.	N.A.	280	N.A.	N.A.

[a] % of total Fe present in metallic state after a fixed time at a fixed temperature
N.A.: Not available

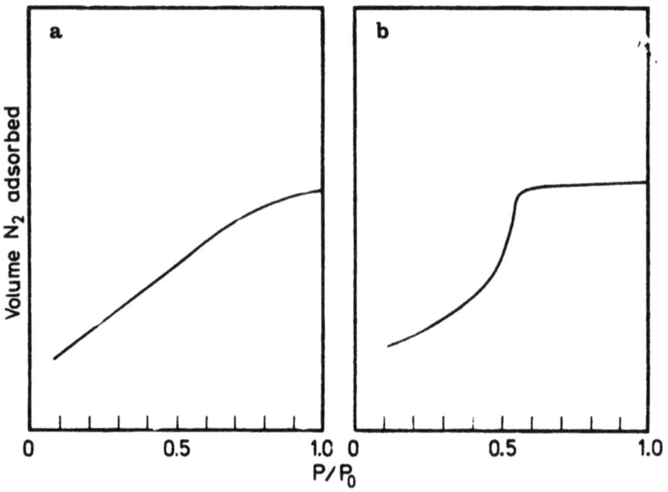

Figure 6. Plots of N_2 adsorption isotherms at liquid N_2 temperature against relative saturation pressure. Curve 6A is for the Fe_2O_3 sample prepared by adding the iron solution to the carbonate solution. 6B is for a Fe_2O_3 sample prepared by adding the carbonate solution to the iron solution

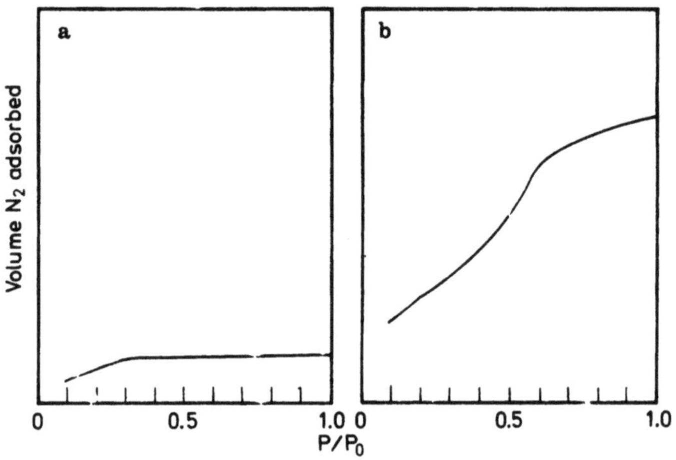

Figure 7. Plots of N_2 adsorbtion isotherms at liquid N_2 temperature against relative saturation pressure. Both Fe_2O_3 samples were prepared by adding the iron solution to the carbonate solution. For case A the final *pH* was 2.0 (ie an excess of Fe^{3+} ions was added). For case B the final *pH* was 6.9

In practice it is necessary to strike a balance between physical strength and pore volume as large pore volume catalysts are usually weaker.

2. *Reduction and Conditioning*

It has always been the practice to pre-reduce Ni, Co or Fe Fischer-Tropsch catalysts with hydrogen under relatively mild conditions so as to generate

Table 6. Influence of precipitating agent

Precipitant	Average pore size/nm	Pore volume in pores >12.5 nm/%
NaOH	2.9	1
Na_2CO_3	4.0	40
NH_4OH	2.6	1
$(NH_4)_2CO_3$	3.1	18

the high metallic surface areas required for active catalysts. Pichler [2] found that whereas unreduced precipitated iron catalysts were initially inactive in F-T synthesis, reduction at the relatively high temperature of 633 K did not improve the activity. Similarly Scheuermann [2] found that reduction at 453 to 493 K resulted in active catalysts while 573 K reduction gave poor results. Catalysts were not always pretreated with hydrogen, e.g. Pichler used CO at about 10 kPa and Ruhrchemie pretreated their test catalysts with a $H_2 + CO$ gas at atmospheric pressure. The selectivity of the catalyst in the F-T synthesis was also influenced by the reduction conditions. Thus prereduction with H_2 at about 570 K resulted in a more active catalyst but it had a lower wax selectivity than one pretreated with synthesis gas at 503 K [2]. High linear velocities of the hydrogen is desirable as this results in higher surface areas after reduction. The rapid removal of water vapour is the key factor involved.

The addition of copper to the catalyst was found to enhance the rate of reduction and this enabled reduction to be carried out at lower temperatures [1–4]. The practice was dropped in the case of cobalt catalysts as the presence of copper increased the rate of decline of the catalyst's activity in the F-T synthesis. With precipitated iron catalysts, however, no such effect was observed. If no copper is added the degree of reduction obtained at normal temperatures is very low and the activity of the catalyst is inferior although the wax selectivity is not affected. If the catalyst is reduced at higher temperatures to attain the same degree of reduction as when copper is present, the activity of the catalyst is still inferior.

The initial rate of reduction is rapid and is independent of the catalyst particle size. The initial reduction is characterized by a marked exotherm, due to the rapid reduction of Fe_2O_3 to Fe_3O_4, of KNO_3 and of CuO. Beyond this initial stage further reduction occurs very slowly in the case of SiO_2 promoted catalysts. By analogy with nickel silica precipitated catalysts [25] it is postulated that some iron silicate complex is formed during the catalyst preparation and it is this compound which reduces slowly. Note from Table 5 that as the SiO_2 content increases the degree of reduction attained in a fixed time decreases. This supports the postulate of the formation of an iron silicate. No x-ray diffraction evidence for the existence of such a silicate has however been found. Due to the smallness of the crystallites the diffraction patterns are very diffuse and only that of α-Fe_2O_3 is discernible. Even after hydrotreating in an autoclave at 553 K to encourage crystal growth, the only identifiable phases were α- and γ-Fe_2O_3.

On reduction the total surface area decreases, the average pore size increases and the pore size distribution shifts towards the larger sized pores. Iron catalysts promoted with copper have lower areas and wider pores than catalysts that do not contain copper. The effect is due to the greater degree of reduction attained in the former cases. The shift towards lower areas and wider pores continues when the partially reduced catalyst is subjected to Fischer-Tropsch synthesis and this shift too can be attributed to the further reduction of the catalyst. The changes in area and pore structure are illustrated in Table 7 for two differently prepared silica promoted iron catalysts. Note that although the changes that occur are relatively large the differences between the two catalysts remain.

It has been found that when a normally promoted catalyst is reduced under standard conditions for different lengths of time the activity of the catalyst in F-T synthesis first increases with the reduction degree and then decreases. It thus appears that although the total surface area is decreasing as reduction proceeds more active sites are being created. These sites are probably fully reduced metallic iron sites. This concept is supported by the finding that the amount of strongly chemisorbed CO, as determined by the dynamic GC pulse technique [26], increases with reduction degree. It is of interest to note that with the static volumetric technique [27], which reflects the sum of the strong as well as the weak CO chemisorption sites, the amount of CO adsorbed decreases with increasing degree of reduction. The presence of many sites capable of only weakly chemisorbing CO is also indicated by the finding that the measured heat of adsorption of CO on a reduced precipitated catalyst is only half that of a fully reduced fused iron oxide catalyst. The very low surface coverage values (<5 percent) obtained on supported precipitated catalysts is presumably due to both incomplete reduction of the surface iron as well as the presence of the high area inert silica. In keeping with this unsupported and unpromoted precipitated Fe_2O_3 reduces rapidly and completely and over 90 percent CO chemisorption coverage is obtained. It is clear from this above discussion that chemisorption data on partially reduced iron catalyst needs to be interpreted with care. The finding that excessive reduction time results in lowered synthesis activity could be due to sintering which will result in fewer surface iron sites.

Table 7. Changes in area and pore structure of precipitated iron catalysts

State	Catalyst	Total pore vol $cm^3\,g^{-1}$	Total area $m^2\,g^{-1}$	Area in pores >4.5 nm/%	Vol in pores >16.0 nm/%
Unreduced	A	0.39	355	5	0
	B	0.67	340	25	19
Partially reduced	A	0.30	195	20	1
	B	0.46	150	45	22
After some F—T	A	0.22	51	100	10
synthesis	B	0.38	47	100	44
(All wax extracted)					

On introducing synthesis gas to the partly reduced $Fe—Cu—SiO_2—K_2O$ catalyst at normal synthesis temperatures the metallic iron is very rapidly converted to Hägg carbide (Fe_5C_2). Further reduction/carbiding continues slowly for several days and the activity increases somewhat during this period due no doubt to the continued creation of more active sites. As already mentioned the surface area decreases and the pores shift to larger average diameters during this process. During the conditioning period it is advantageous to control the gas composition and the temperature in the reactor such that the catalyst attains its maximum activity as soon as possible without any permanent damage to the catalyst.

3. Influence of Promoters and Supports on Synthesis

The synthesis performance of Co and Ru catalysts is not very sensitive to the presence of promoters [2]. Thus for cobalt operating at atmospheric pressure, promotion with thoria and alkali increased wax production but when operating at 1.5 MPa and higher these promoters had little effect. Supports were incorporated mainly to increase the active metal surface area and so ensure high activity. With iron catalysts, however, promoters and supports are essential ingredients. It is well known that the stronger bases of the Group I metals, especially K_2O, are key promoter components. They markedly influence both the activity and especially the selectivity of iron catalysts. The "basicity" of the catalyst is thus of importance. Since the basic alkalis readily form chemical compounds with many of the commonly used supports such as silica or alumina, it is clear that even though these supports are added mainly to stabilize a high metal surface area they invariably also influence the overall basicity of the catalyst.

The potassium promoter is usually added as the carbonate but it has been found that the hydroxide, nitrate, fluoride, borate, oxalate, silicate and permanganate salts of potassium give equally satisfactory results. The dichromate and di-hydrogen phosphate salts appear to be somewhat less effective. The chloride, bromide, sulphate and sulphide salts result in low activity catalysts. The above results were all obtained by impregnating a potassium-free stock of standard $Fe—Cu—SiO_2$ precipitated catalyst. In general these results are in agreement with those obtained by other workers [2]. Hammer [29], however, reported that at 0.2 percent K_2CO_3 equivalent, KCl and KBr do not depress the activity of unsupported precipitated Fe—Cu catalysts.

The relative effectiveness of the Group I alkalis was also tested using the same $Fe—Cu—SiO_2$ base stock mentioned above. Equivalent amounts of Li, Na, K and Rb were added by impregnation. Promotion with K produced the highest activity. Equating the activity of the latter to 100 the activity of both the Na and Rb promoted catalysts were about 90 while that of the Li promoted catalyst was only about 40. As expected from the relative basicities of the promoters the selectivity to wax increased in the order Li, Na, K and Rb. The observation that the Rb promoted case was out of line as far as the correlation between basicity and activity was concerned is probably due to its high wax selectivity. The accumulation of the high molecular mass wax products in the catalyst pores should slow up the rate of diffusion and thus depress the F-T synthesis reaction rate. Another set of catalysts was prepared using the normal technique described in section 5.A, but using sodium waterglass instead of potassium waterglass which results in sodium promoted catalysts. The activity of the latter catalysts (at equivalent molar alkali contents) were about 25 percent lower and the wax selectivity about 20 percent lower than that of the potassium promoted catalysts. These results are similar to those obtained by Fischer [28].

The influence of K_2O promotion on the product selectivity has long been established. It shifts the product spectrum towards the longer chained molecules. The higher the alkali level the greater the shift, irrespective of whether the iron is supported or not (cf. Table 8). The acid production also increases with alkali content and is particularly high for unsupported iron catalysts. The change in activity depends on the alkali level itself. At low levels the activity increases with alkali addition but beyond a certain value it decreases again. These results agree with the findings of others, e.g. the US Bureau of Mines reported an increase in activity with alkali content and Japanese workers found that the activity went through a peak on increasing the K_2O content of precipitated Fe—Cu Kieselguhr catalyst [2].

The objective of adding supports is to create and stabilize a high active metal area but the choice of the support is usually determined by its influence on the alkali promoter. For instance, with iron catalysts the use of large amounts of kieselguhr is not recommended as it reacts with the alkali and thereby renders the alkali less effective. It lowers both the activity and the wax selectivity [2]. Kieselguhr was used as a support to prepare active cobalt and nickel catalysts but it was also found that alkali addition to these catalysts had no benefical effect on the selectivity. Cobalt-kieselguhr catalysts had a satisfactory selectivity at higher synthesis pressures and so the ineffectiveness of alkali was of less consequence than in the case of iron catalysts.

Table 8. Influence of alkali content on synthesis performance of precipitated iron catalysts

Catalyst type	K_2O Level[a]	H_2/CO Ratio of total feed	Hard wax selectivity[b]	Water analysis wt. %		Activity[a]
				Alcohols	Acids	
Unsupported	0	1.9	5		<1	26
Fe_2O_3	1.0	1.9	34		3	47
	1.6	1.9	41		8	50
	2.0	1.9	53		12	53
	3.0	1.9	63		19	40
Silica supported	12	2.2	18	2.5	0.2	112
Fe_2O_3	16	2.4	20	3.0	0.3	109
	21	2.3	30	3.4	0.3	85
	24	2.2	38	2.0	0.5	83
	32	2.4	44	2.0	0.7	75

[a] relative quantities
[b] wt. % of hydrocarbon product boiling above 773 K

Table 9. Influence of different supports on synthesis performance of precipitated iron catalysts

Catalyst	Support/ g (100 g Fe)$^{-1}$	K_2O [b]	Activity[b]	Wax selectivity[b]
Fe—Cu—SiO_2	24	10	45	34
Unsupported Fe—Cu	0	0.6	34	25
Fe—Cu—Cr_2O_3	24	2.0	31	30
Fe—Cu—MgO	32	3.6	16	13
Fe—Cu—MgO	68	3.2	28	10
Fe—Cu—Al_2O_3	23	3.0	35	10
Fe—Cu—Al_2O_3	100	12	18	34
Fe—Cu—ZnO	30	2	40	12
Fe—Cu—SiO_2—CaO	10 CaO [a]	10	38	31
Fe—Cu—SiO_2—Cr_2O_3	10 Cr_2O_3	10	42	30
Fe—Cu—SiO_2-Kieselguhr	20 Kieselguhr	10	31	29
Fe—Cu—SiO_2—Al_2O_3	10 Al_2O_3	10	41	25
Fe—Cu—SiO_2—V_2O_5	10 V_2O_5	10	41	31
Fe—Cu—SiO_2—ThO_2	10 ThO_2	10	47	31
Fe—Cu—SiO_2—MgO	10 MgO	10	30	30
Fe—Cu—SiO_2—TiO_2	10 TiO_2	10	43	30

[a] Added in addition to the standard amount of SiO_2
[b] Relative quantities

Early studies by Fischer [30] showed that for Co-kieselguhr catalysts promotion with Cr_2O_3 and Al_2O_3 lowered the activity while MgO, MnO or ZnO had little influence. Silica prepared from waterglass was used by Lurgi as a catalyst support for precipitated iron catalysts while Kölbel employed dolomite [2]. Table 9 gives some of the results obtained at Sasol on the influence of various alternative supports for precipitated iron catalysts. In one set of experiments the normal silica support was replaced with other materials and in the other set a variety of compounds was additionally incorporated into the normal Fe—Cu—SiO_2—K_2O catalyst. When altering the catalyst formulation it is usually necessary to also alter the level of the alkali promoter in order to attain optimum performance. Because of this many formulations need be carried out in practice and those given in Table 9 are not necessarily the ones with the optimum promoter compositions. As can be seen SiO_2 as a support produces the best results. Both the activity and the wax selectivity is higher than for the other cases including that with no support at all. The incorporation of other support material in addition to the normal amount of SiO_2 resulted in a lower activity and/or wax selectivity.

B. Fused Iron Catalyst

1. Preparation

The iron oxide used should preferably contain low levels of impurities such as SiO_2, TiO_2, MgO and Al_2O_3. Although Sasol has prepared satisfactory catalysts from iron ores containing several mass percent of the above impurities it becomes more difficult to accurately control the required promoter levels, especially if the impurity levels in the iron ore fluctuate. It is thus mainly for control reasons that a low impurity iron ore is preferred. At Sasol millscale from a steel works is used. It is electrically fused together with the desired amounts of promoters. The fused ingots are crushed to the particle size required for the Synthol reactors. The particle size spread must be accurately controlled as too fine a material will result in high losses through the reactor cyclones and too coarse a material will not fluidize or transport properly. Before loading the catalyst into the Synthol reactors it is reduced with hydrogen in fixed fluidized bed reactors.

Microscopic investigations of polished sections of the fused catalyst (both before and after crushing of the ingots) have been carried out. Only the very outside skin of the ingots contains Fe_2O_3. This is formed during the cooling of the molten magnetite (the Fe_2O_3 phase is not stable in contact with air at the fusion temperature of about 1770 K). Small occlusions of alkali silicates are present throughout the whole ingot, the size of the occlusions increasing towards the centre of the ingot where the rate of cooling is slower. Figure 8 is a photograph of a typical alkali silicate occlusion. Little of the added structural promoters (e.g. MgO, Al_2O_3 etc.) is present in these silicate occlusions. The structural promoters are in solid solution in the magnetite phase.

On crushing the ingots the silicate occlusions are still to be found in the larger sized particles but in the very fine particles the occlusions are absent and the silicate phase is present as separate particles. Therefore on a macro scale the alkali silicate distribution is very heterogeneous in the ground catalyst. There is evidence, however, that during reduction and synthesis the alkali does spread over the catalyst surfaces and thus on an atomic scale the alkali distribution is not as heterogeneous in the working catalyst as would appear from the investigation of the catalyst before reduction.

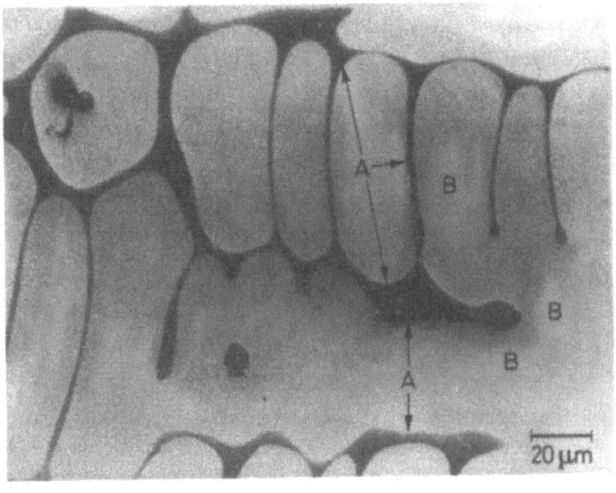

Figure 8. A polished section of promoted fused magnetite. A are the alkali silicate inclusions. B is the magnetite phase

2. Reduction

Due to the nature of its preparation fused iron catalysts have very low surface areas and pre-reduction is essential to develop the area required for high activity. In order to achieve a reasonable rate of reduction temperatures in the range 623 to 723 K are used. The actual temperature depends on the type and quantity of promoters/impurities present as these can markedly influence the rate of reduction. In the above temperature range hydrogen is a much more effective reducing agent than carbon monoxide or a mixture of CO and H_2. This is illustrated in Table 10.

Table 10. Degree of reduction at 673 K and 8 hours at fixed space velocity. Promoted fused Fe_3O_4

Gas	Reduction/%[a]	Phases present
100% H_2	80	α-Fe, Fe_3O_4
98% H_2 + 2% CO	20	Fe_5C_2, Fe_3O_4
100% CO	4	Fe_5C_2, Fe_3O_4

[a] amount of iron in the oxide form converted to either metallic or carbidic iron

The amount of water vapour present in the reduction system has a marked influence on the rate of reduction [31]. This is demonstrated in Figure 9. The observed rate of reduction is insignificant at H_2O/H_2 molar ratios of only about 20 percent of the thermodynamic equilibrium value for the equation

$$4\,H_2 + Fe_3O_4 = 3\,Fe + 4\,H_2O$$

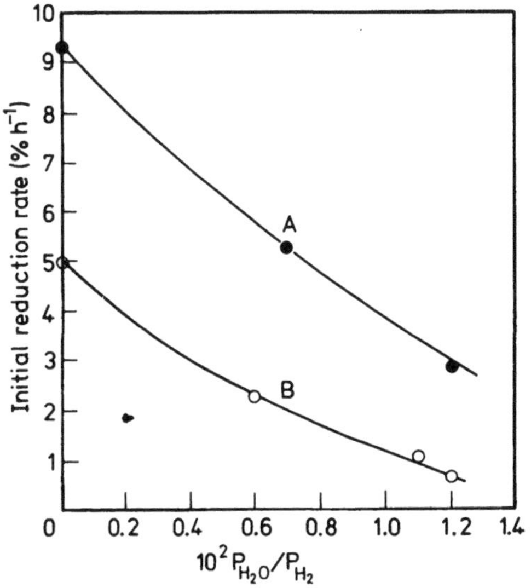

Figure 9. The retarding effect of water vapour on the initial reduction rate. Samples A and B contain 0.012 and 0.030 g atom Al/g atom Fe respectively

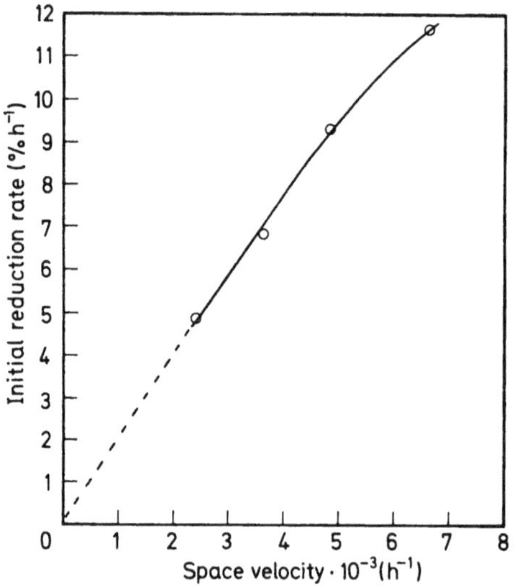

Figure 10. The effect of the hydrogen space velocity on the reduction rate

High H_2 space velocities increase the rate of reduction as is shown in Figure 10. This can be attributed to the lower H_2O/H_2 ratios attained by using high flow rates of H_2. The marked depressing effect of a small amount of CO in the H_2 (cf. Table 10) is probably partly due to the H_2O produced in the simultaneously occurring F-T synthesis.

If the reduction rate is varied, by for instance changing the H_2 flow rate, it is found that the higher the initial reduction rate the higher the surface area of the fully reduced catalyst. Reduction-oxidation-re-reduction cycles (using H_2O as oxidant) result in lower surface areas. This probably explains why high H_2 flow rates result in higher area catalysts.

Figure 11 illustrates that there is a general relationship between the rate of reduction (depicted by the H_2O/H_2 ratio of the exhaust gas) and the final

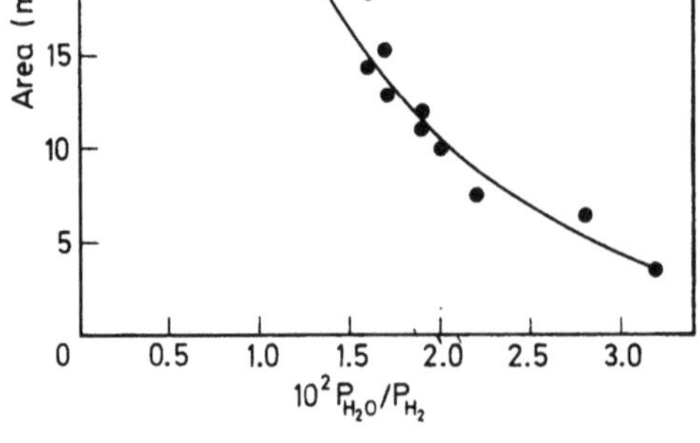

Figure 11. The relation between the final surface area developed and the initial reduction rate as depicted by the value of p_{H_2O}/p_{H_2} of the exhaust gas. The reduction temperature was the same in all cases. The samples contained different contents of various promoters

area of the catalysts. Promoters such as Al_2O_3, Cr_2O_3, TiO_2 and MgO retard the reduction rate but give high surface area catalysts. These promoters are in solid solution in the magnetite. When these catalysts are re-oxidised at about 670 K and then again re-reduced the rate of the re-reduction is much higher than that of the original fused oxide. This indicates that the promoters are not re-dissolved in the iron oxide phase which is formed on re-oxidising the reduced iron. The incorporation of copper in fused catalysts has no clear influence on the rate of reduction. This in unlike the case for precipitated iron catalysts. The relative amounts of Wustite (FeO) and of magnetite (Fe_3O_4) present in the fused catalyst also influence the final surface area. Magnetite-rich samples give higher areas. [32].

The surface area and also the pore volume of the catalyst increases linearly with the degree of reduction [33]. Depending on the reduction temperature and on the type and amount of promoter incorporated the final areas vary from about 2 to 30 $m^2\ g^{-1}$. It appears that there are no pores smaller than about 5 nm. A typical reduced catalyst will have an average pore size of about 30 nm.

3. *Influence of Promoters on Physical Properties*

If foreign cations replace either Fe^{2+} or Fe^{3+} ions in the magnetite and the ionic size of the foreign ions are different from that of the host lattice then the unit cell size, as determined by x-ray diffraction, will change. Using pure iron oxide as starting material the influence of incorporating various cations was studied [34]. Al^{3+}, Li^+ and Mg^{2+} which are smaller than the host cations result in a linear decrease in the Fe_3O_4 unit cell size as the promoter content is increased. For the larger Ca^{2+}, Mn^{2+}, Ti^{3+} and Na^+ ions the unit cell size increases progressively (cf. Figure 12). This indicates that all the above cations can go into solid solution into magnetite and thus can be homogeneously distributed throughout the fused catalyst. K^+ ions have no effect on the unit cell size presumably because they are too large to replace Fe ions in the Fe_3O_4 lattice. SiO_2 also does not go into solid solution but it does influence the solubility of basic cations such as Na^+ and Ca^{2+}. This is due to the ability of SiO_2 to combine with these promoters to form the silicates $Na_2O(SiO_2)_2$ and CaO. SiO_2 respectively. Only the excess Na^+ or Ca^{2+} ions go into solid solution [34]. SiO_2 apparently does not prevent MgO from going fully into solid solution with magnetite. Microscopic investigations of polished sections confirmed all the above deductions in that when all the cations appear to be in solid solution only one phase, Fe_3O_4, can be seen, while when K^+ ions or SiO_2 is added small occlusions of a different phase can be seen.

On reducing magnetite which has foreign cations dispersed throughout its structure one can logically except that the incorporated promoters will end up being finely dispersed between the α-Fe metal crystallites. The presence of crystals of promoters between (and thus separating) neighbouring crystallites of α-Fe should inhibit crystal growth of the α-Fe and so result in a high surface area of the reduced catalyst. This effect has been observed [27, 35]. The extent of the effect is related to the ratio ionic charge to ionic radius

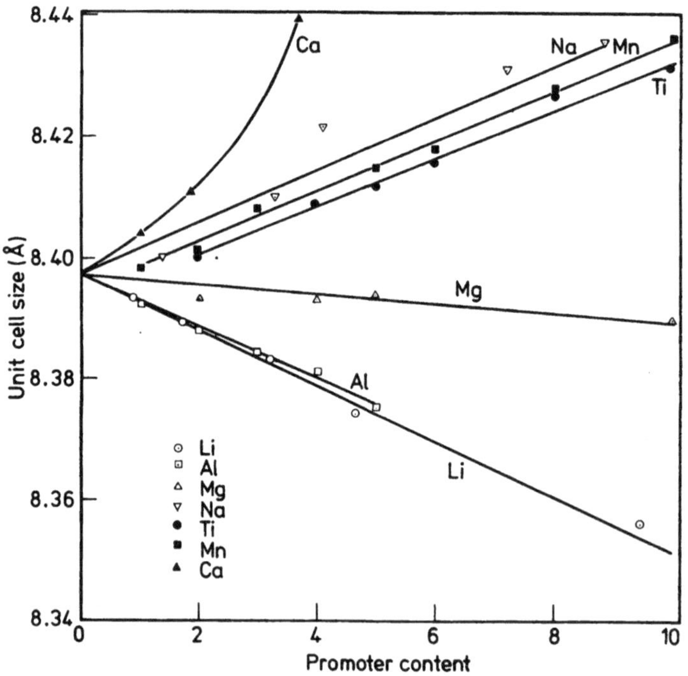

Figure 12. The influence on the unit cell size of magnetite (in Å) of various promoters which go into solid solution with magnetite. Promoter content is in g atom cation per 100 g atom Fe

of the cations [27]. The lower the value of this ratio, (i.e. the more "acidic" the cation) the bigger the area promotional effect. Thus Al_2O_3, TiO_2 and Cr_2O_3 strongly increase the area and MgO has less influence. MnO, Li_2O and CaO have little effect while Na_2O actually results in a decrease in area [27, 31, 35]. Some typical results are depicted in Figure 13. The "fluxing" effect of Na_2O appears to be unrelated to the fact that it is capable of entering into solid solution with magnetite because K_2O which does not go into solid solution also has a strong negative effect on the surface area [36]. The lowering of the area by the strong bases is due to the formation of larger iron crystallites and not simply due to pore plugging by the alkali. It was found that the iron crystallite size, as measured by X-ray line broadening, does increase with increasing alkali content. Silica, which on its own has little if any effect on the area has an indirect effect in the cases where the strong alkalis (Na_2O or K_2O) are present. SiO_2 combines with the alkalis and so lowers the basicity of the catalyst and this results in higher surface areas.

In wüstite-rich fused iron catalysts the influence of the structural promoters MgO and Al_2O_3 on the area of the reduced catalysts is less pronounced than for magnetite-rich catalysts [32]. For the former the area at first increases but then levels off as the promoter content is increased whereas with the latter the area continues to increase.

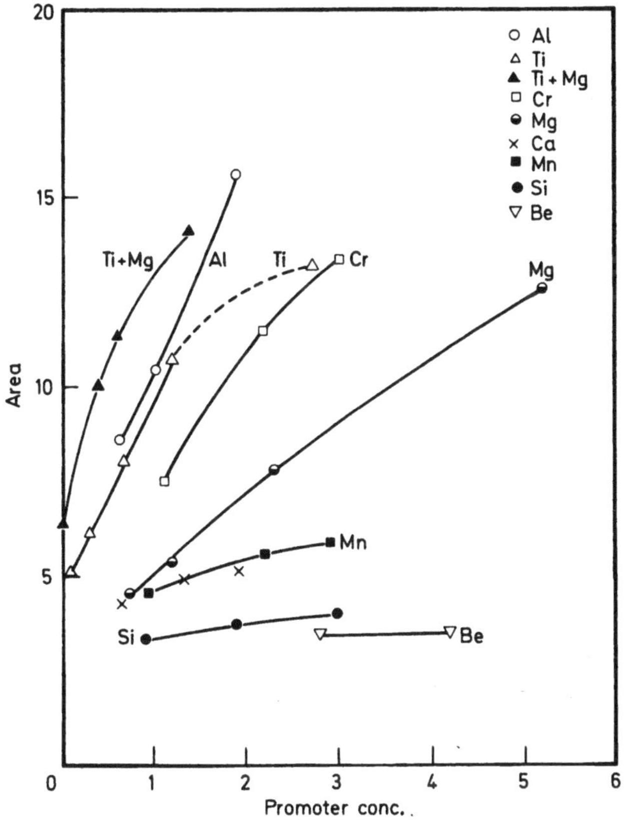

Figure 13. The surface area of the fully reduced catalysts in m^2 (g unreduced sample)$^{-1}$ as a function of the promoter content (g atom promoter cation per 100 g atom Fe)

It is of interest to know what fraction of the total surface area is metal atoms and also to what extent the alkali covers the surface of the catalyst. To this end Emmett [37] and co-workers developed the CO and CO_2 chemisorption techniques to measure the "free Fe" and "K_2O" areas respectively. Similar adsorption studies were carried out on many promoted iron catalysts [27, 35, 36]. Figure 14 illustrates the influence of various promoters on the volume of CO adsorbed which is taken to reflect the metallic iron surface area. Comparing the results with the total areas (Figure 13) it can be seen that in general the higher the BET areas the higher are the metallic areas. Promotion with TiO_2 gives an odd result in that whereas the total area increases progressively with increasing TiO_2 content the CO chemisorption first increases and then decreases progressively [27]. As a percentage of the total area MgO gives the highest metal area while TiO_2 and CaO give low values.

The CO_2 chemisorption studies [36] in general also appear to make sense in that the surface coverage by CO_2 increases as the alkali content increases. Also the chemisorption coverage of catalysts promoted with the stronger alkali, K_2O, is higher than when the weaker alkali Li_2O is used. SiO_2, which forms silicates with the alkalis and so should lower the overall "basicity",

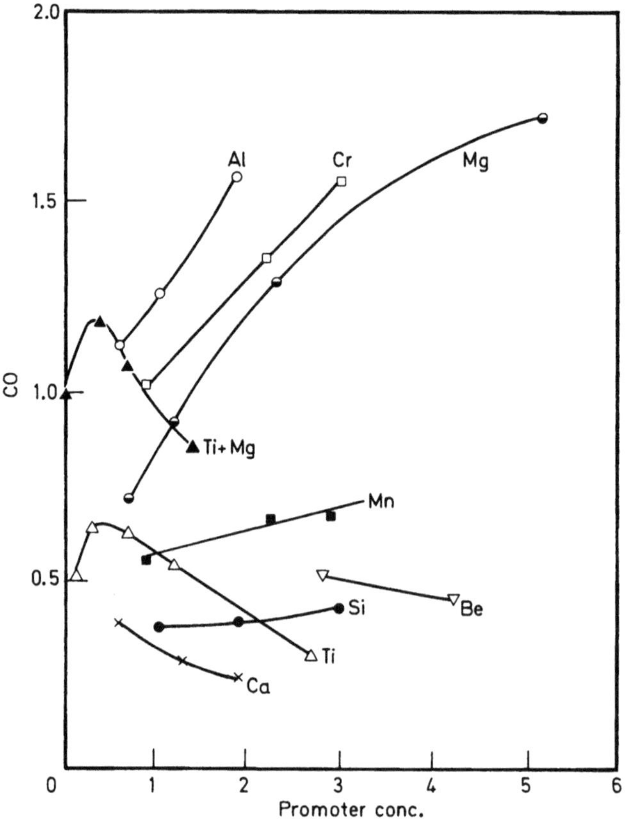

Figure 14. The volume of CO chemisorbed by the fully reduced catalysts (cm^3 (STP) per gram unreduced sample) as a function of promoter content (g atom cation per 100 g atom Fe)

does decrease the CO_2 chemisorption. These findings fit the known Fischer-Tropsch performance of such catalysts. Thus Li_2O is inferior to K_2O as a promoter and the presence of SiO_2 does result in a lower oil selectivity. In spite of this, however, the absolute values of the CO_2 coverages are not reliable measures of the synthesis performance. For instance, CaO promotion also increases the CO_2 chemisorption but it has no effect on the selectivity or activity of the catalyst in the F-T synthesis. It is also of interest to note that chemisorbed N_2 markedly lowers the amount of CO_2 subsequently chemisorbed [38] and also that even pure unpromoted reduced iron chemisorbs some CO_2. Both these findings indicate that CO_2 also chemisorbs on the iron surface and not only on the alkaline additives. If the lowering of CO_2 chemisorption by chemisorbed nitrogen is taken to mean that the latter depresses the surface basicity this does fit the fact that nitriding of iron catalysts leads to a shift in the selectivity away from wax and olefins [2]. Such a shift typically occurs when the basicity of a catalyst is lowered.

4. *Influence of Promoters on Synthesis*

The most important promoters for fused iron catalysts are undoubtedly the stronger alkalis of Group I. K_2O in particular is the most commonly employed and is the only really essential promoter required for industrial iron catalysts. The main function of the structural promoters (e.g. Al_2O_3, Cr_2O_3, MgO etc.) is to increase and stabilize the available active iron surface area to the levels required for high conversion. If the catalysts are employed at high temperatures, the reaction rate appears to be diffusion controlled and increasing the internal surface area of the catalyst produces little, if any, benefit. In such cases the area need not be higher than a certain required minimum.

Usually the observed influence of promoters or impurities (other than K_2O) on the synthesis performance of catalysts can be linked to their chemical interaction with the alkali. Thus silica, a common impurity in the iron oxide sources used to prepare fused catalysts, will react with the alkali to form less basic silicates and thus influence the catalyst's performance [36]. In general the more acidic promoters like TiO_2 and Al_2O_3 will, at the same K_2O content, result in a lower basicity catalyst than one promoted with for example, MgO. The total surface area as such also influences the catalyst's basicity. For a fixed amount of alkali the higher the area the lower the basicity.

The influence of increasing the K_2O content of catalysts is not always consistent and it depends on the type of catalyst and on the conditions under which it is being tested. In Table 11 it can be seen that when fused catalyst is used at low temperatures in a fixed bed reactor the activity decreases and also the alcohol selectivity decreases as the K_2O content is increased. When fused catalysts are employed at higher temperatures in a fluidized bed, the activity at first increases as the K_2O content is raised and then flattens out at high conversion levels (cf. Table 12). The CH_4 selectivity progressively decreases. The acid and alcohol selectivity increase as does the olefinity of the hydrocarbons. Thus the activity can either increase or decrease with K_2O promotion depending on the conditions. It has been observed in other studies that the activity passes through a peak as the K_2O content is increased

Table 11. Synthesis performance of fused magnetite promoted with a fixed amount of Al_2O_3 and varying amounts of K_2O. Fixed bed operation at about 473 K

K_2O content/relative units	Activity/relative units	CH_3OH [a] selectivity
0	64	6.6
0.4	60	4.8
0.8	55	4.9
1.1	50	3.6
2.6	29	2.7

[a] %C atom basis

Table 12. Synthesis performance of promoted fused magnetite catalysts. Amounts and types of alkalis varied. Fluidized bed operation at 603 K

Alkali	Amount of alkali/ relative units[a]	Activity $CO + CO_2$ conversion/ percent	CH_4 selectivity[c]	Water acids selectivity[c]	$\frac{C_3H_6}{C_3H_8}$ ratio	NAC selectivity[b,c]	Carbon deposition[d] rate/
K_2O	0	63	41	0.05	0.2	0.3	2.8
	1.75	90	20	0.2	3	1.7	1.5
	2.1	90	15	0.3	5.7	2.6	1.5
	2.65	92	13	0.5	7.6	3.4	3.0
	3.4	94	12	0.7	11	3.8	4.5
	3.95	93	10	0.8	12	4.0	5.5
Na_2O	0	63	41	0.05	0.2	0.3	
	2.3	78	15	0.15	1.8	2.7	
	4.5	85	13	0.2	2.3	2.0	
	6.8	84	14	0.2	2.0	2.5	
	9.1	87	16	0.2	1.7	2.7	

[a] Alkali levels all expressed on molar basis
[b] NAC = Non Acid Chemicals (alcohols and ketones)
[c] All selectivities, % carbon atom
[d] g C per 100 g Fe per unit time

[2]. The influence of alkali on activity is not unique to iron catalysts. The promotion of a Ni methanation catalyst with sodium was found to increase and then decrease the activity as the Na content was increased [39]. As can be seen in Table 12 the influence of Na_2O is much less pronounced than that of K_2O. Thus while Na_2O does influence the selectivity when compared to that of a catalyst containing no alkali, further Na_2O additions seem to have almost no effect which is contrary to the case of K_2O promotion. The product from Na_2O catalysts is less olefinic than that obtained from K_2O promoted catalysts.

Table 13 compares the influence of SiO_2 on Na_2O and K_2O promoted catalysts. In the former case the addition of SiO_2 has no effect on selectivity whereas for K_2O containing catalyst the addition of SiO_2 markedly effects both the selectivity and the activity of the catalyst. This effect has also been observed by other workers [2].

It has been observed [40] that the manner in which the K_2O was added to the iron catalyst in a fluidized bed (i.e. whether by impregnation or simply by adding the loose powder) had relatively little effect on the catalyst's performance. Investigations at Sasol have shown similar effects. When finely ground potassium silicate was simply added to the pre-reduced powdered iron catalyst the overall performance was not markedly inferior to that of a catalyst in which the potassium silicate was fused together with the iron oxide and then milled and reduced. This is not really surprising when it is borne in mind that microscopic investigation of the latter kind of catalyst has shown that separate potassium silicate phases are present, i.e. the K_2O is in any event not homogeneously distributed in milled fused catalyst. It was also found that the performance of a catalyst prepared by fusing pure Fe_3O_4 with only Na_2O is little different from that of a Na_2O promoted catalyst prepared from iron oxide containing silica. In the former case the Na_2O goes into solid solution and so is automatically distributed throughout the fused catalyst mass while in the latter case separate sodium silicate occlusions are present in the catalyst and so the distribution of the Na_2O is relatively poor. A possible explanation of these observations is that under synthesis conditions the alkali is capable of diffusing over the catalyst surfaces. Given

Table 13. Synthesis performance of promoted fused catalysts. SiO_2 varied with alkali content fixed. Fluidized bed at 593 K

Alkali type	SiO_2/Alkali molar ratio	$CO + CO_2$ conversion/%	CH_4 Selectivity/ % carbon atom	$\frac{C_3H_6}{C_3H_8}$ ratio
Na_2O	0.6	86	13	2.8
	2.1	83	13	2.4
	4.2	81	13	3.1
K_2O	0.8	85	10	6.1
	4.2	78	27	0.9

time it will distribute itself evenly over all the catalyst particles in the reactor and so be able to exert its influence on the majority of the active iron sites. It must not be deduced that maldistribution of the K_2O promoter is of no consequence since it does result in some differences in performance.

C. Sintered, Cemented and Impregnated Iron Catalysts

Catalysts prepared by sintering finely divided iron oxides together with the desired promoters behave very similarly to fused catalysts. This is not surprising since fusion represents the extreme case of sintering. The choice of preparing catalysts by sintering or fusion is dependent on the economics of the preparation steps involved and on the desired strength of the catalyst particles.

Cemented catalysts are formed by binding iron oxide powders with compounds like $Al_2(NO_3)_3$, sodium borate or potassium waterglass. These catalysts were found to be less active than sintered or fused catalysts [2]. This is understandable when the influence of compounds capable of chemically reacting with the key K_2O promoter is borne in mind. Thus silica or alumina binders will result in the lowering of the basicity of the K_2O and thus render it less effective as a promoter. Binders can also lower the activity by physically blocking off or blanketing the active iron sites.

Catalysts prepared by thermal decomposition of metal nitrates were found to be less active than those produced by precipitation [2].

Impregnating high area carriers such as alumina, silica aluminas or silica gels with iron salts such as the nitrate also yields inferior catalysts. The activity as well as the wax selectivity is low compared to precipitated catalysts. Promotion of these catalysts with relatively high amounts of K_2O does little to improve their performance. Once again this can be ascribed to the interaction of the K_2O with the carrier which is inevitably the major component in this kind of catalyst. If impregnated catalysts are to be satisfactory the carrier must not interact chemically with either the iron or the alkali promoter and this proviso eliminates most of the common carriers. Since carbon shows no evidence of adversely effecting either activity or selectivity it could be a satisfactory carrier for iron catalysts in the F-T reaction. A wide-pore active charcoal is preferred to the more common high area, narrow pore material since the former should have less diffusion restrictions.

In general catalysts prepared by impregnating supports with iron sulphates or chlorides are of low activity. The use of nitrates or of oxalates results in more active catalysts.

D. Other Metal Catalysts

Ruthenium is a very active catalyst for the hydrogenation of carbon monoxide. Its activity at low temperatures is higher than that of the common F-T catalysts Ni, Co and Fe. It is a versatile catalyst, in that at the higher temperatures it is an excellent methanation catalyst (it is the most active of the Group VIII metals [41]) while at low temperatures and high pressures it produces large amounts of very high molecular mass waxes [4]. This latter aspect has been reviewed by Schulz [42]. It is most active in the pure metal form, i.e. supports and/or promoters appear to have no beneficial effect. Even under conditions of high wax yields Ru tends to have a high CH_4 selectivity, 10 percent or more. (At similar wax selectivities iron catalysts have CH_4 selectivities below 5 percent.) In a recent study [43], however, CH_4 selectivities below 5 percent have been achieved with 1 percent Ru on γ-Al_2O_3. Unpromoted as well as K_2O-promoted samples were investigated. The degree of conversion and the selectivity to wax increases as the pressure is increased [42]. The waxes are oxygen free and highly paraffinic. Studies at Sasol have shown that under suitable process conditions, e.g. at low conversion levels, ruthenium catalyst, like iron, can produce light hydrocarbons with high olefin and alcohol contents. Ruthenium has a high potential as a catalyst for converting synthesis gas to a variety of hydrocarbons as is evidenced by the activity of many research workers in this field [44, 45, 46, 47].

The activity of the other noble metals of Group VIII has also been investigated. Ru and Os were found to be moderately active while Pt, Pd and Ir were of low activity [4]. The products produced with Rh catalyst contained oxygenated material. Vannice [47, 48] compared the activities of the Group VIII metals supported on Al_2O_3 and reported that the activities decrease in the order Ru, Fe, Co, Rh, Pd, Pt and Ir. Under the conditions used CH_4 was the dominant product for all the catalysts.

The US Bureau of Mines [49] investigated molybdenum as a F-T catalyst with the objective of developing a catalyst resistant to sulfur poisoning. Although reasonable activities were achieved these were much lower than that of iron catalysts. Even though Mo catalysts were active in the presence of H_2S the activity was markedly lower than in its absence. Studies with Mo catalysts were also carried out at Sasol and it was found that in order to achieve reasonable conversions high temperatures (673 K and higher) were required. The CH_4 selectivity was 90 percent. Impregnation with K_2O and the use of a CO rich gas (H_2/CO ratio about 1) lowered the CH_4 selectivity to about 50 percent. However, this catalyst had a lower activity and also its rate of activity decline was higher than that of the unpromoted catalyst. A chromium catalyst was also investigated but its activity was much lower than that of Mo.

Under present day environmental restraints virtually all sulfur must be removed from fuels. Hence there is probably little incentive to develop sulfur resistant catalysts since if sulfur is not removed from the synthesis gas it will have to be removed downstream of the F-T reactors.

6. Catalyst Aging and Poisons

Fischer-Tropsch catalysts can lose activity as a result of factors such as
(a) the conversion of the active phase (e.g. metal) to an inert phase (e.g. oxide)
(b) the loss of active surface area due to crystalline growth (i.e. sintering),

(c) the loss of active area due to the deposition of carbonaceous material (i.e. fouling) and
(d) the chemical poisoning of the surface by for example sulfur.

A. Phase Changes During Synthesis

The ease of reducibility of the oxides of Fe, Co, Ni and Ru increases in the order as given and so it follows that the ease of reoxidation of the metals in a H_2/H_2O atmosphere will decrease in the same order. In keeping with this it is found that under normal F-T synthesis conditions Co, Ni and Ru are not oxidised while in the case of iron, Fe_3O_4 is invariably present [2]. Similarly while Ni and Co can form stable carbides by reacting with pure CO at normal F-T temperatures these phases are not found in used F-T catalysts. With iron catalysts iron carbides are always formed. The metallic iron phase is in fact not stable under normal F-T conditions. When metallic Fe is found in used catalysts this indicates that a layer of inert oxide has blanketed the metal. In a pure H_2 atmosphere the carbides of Fe, Ni and Co all readily reduce to the metal and therefore the stability of these carbides in H_2/CO atmospheres simply depends on the relative rates of the carbiding and reducing reactions.

Figure 15 illustrates the phase changes that occur in an iron catalyst in the case of a fluidized bed reactor operating at about 600 K. The catalyst as charged is 100 percent metallic iron. The metal is very rapidly converted to a mixture of magnetite and iron carbides. Initially an unstable carbide phase, designated pseudo cementite [50], is formed. Its X-ray diffraction pattern is similar to that of cementite but there is a difference in the number

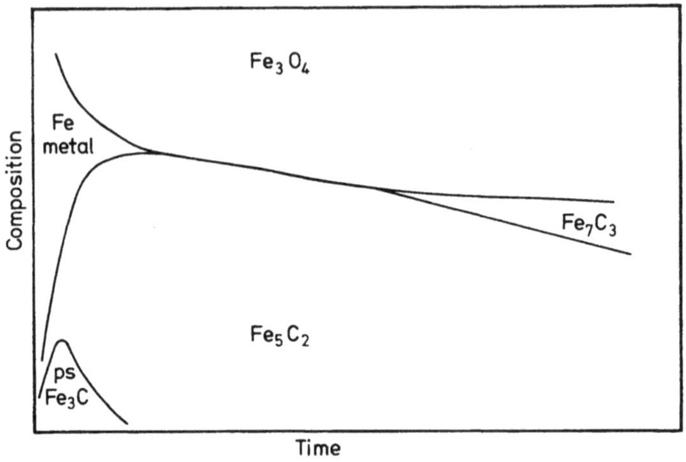

Figure 15. The change in the composition of iron catalysts during the F—T reaction. The phases are those which are present according to X-ray diffraction analysis. At time zero the catalyst is 100% metallic Fe. The units are undefined since the rates of the phase changes depend on alkali content. The figure is only intended to illustrate the trends

of lines present. This phase disappears after a few hours and Hägg carbide (Fe_5C_2) is then the only carbide present. In the older literature Hägg carbide was usually given the formula Fe_2C. Comparison of its X-ray diffraction pattern with that of Mn_5C_2 led to the conclusion that the correct formula is Fe_5C_2 [51, 52]. After several days another carbide phase, called Eckstrom Adcock appears and its concentration slowly increases with time. Eckstrom and Adcock [53] found that when operating at 2.7 MPa pressure and at 633 K the content of this carbide rose to 90 percent but in the Sasol commercial reactors such a high level is not reached. The formula of this carbide has been designated Fe_7C_3 [54]. It is of interest to note that in the Sasol 5 cm ID fixed fluidized bed units no trace of the Fe_7C_3 carbide phase has been observed when the units are operated under the same conditions as used in the commercial Synthol units. Only when the pilot units are operated at considerably higher pressures (>6.0 MPa) does the Eckstrom Adcock carbide appear.

There appears to be no clear correlation between the catalyst's activity and the types of carbide present. There is thus no reason to associate ageing with carbide type or content. Other workers have also reported that there is no difference between the activity or selectivity of Hägg carbide and cementite [57].

Iron carbides are much more resistant to oxidation than metallic Fe. If used catalyst is subjected solely to the total reactor effluent at 613 K, by running a second reactor in tandem, the rate of oxidation of the catalyst in the second reactor is not readily detectable. A typical exhaust gas has a $H_2O:H_2:CO_2:CO$ ratio of 2:5:1.8:0.1. At these H_2O/H_2 and CO_2/CO ratios metallic iron oxidises rapidly. Analyses of the different size fractions of the catalyst from a fluidized bed reactor indicate that as the particle size decreases the carbide/oxide ratio increases, with the very fine particles containing no oxide at all. A probable explanation of this is as follows. As the synthesis gas diffuses into the porous catalyst particles the H_2 and CO is consumed in the F-T reaction and H_2O and CO_2 is produced. Thus the gas becomes progressively less reducing as it penetrates into the particles and beyond a certain depth it must become oxidising. The larger the particle the greater the portion of it which will be in an oxidising atmosphere and hence the higher the oxide content. The gas within the core of the larger particles must have H_2O/H_2 and/or CO_2/CO ratios higher than that of the reactor effluent gas as the latter gas is not oxidising. This is to be expected as (except for the case of 100 percent conversion) the ratio of the reactants to products outside the particles always must be higher than within the catalyst particles. Scheuerman [2] found that for reduced Fe_3O_4 catalysts the activity was the same for samples reduced from 30 to 100 percent. In agreement with this the Bureau of Mines found that reduction beyond 25 percent did not further increase the activity [56]. (6 to 8 mesh fused catalyst particles were used in a fixed bed.) Research at Sasol (using fluidized beds) has confirmed that for fused catalysts reduction beyond a certain level does not result in higher activity. It must be noted that since reduction proceeds from the outside towards the centre of the catalyst particles, partially reduced particles will have unreduced cores. All the above results are in keeping with the observation that during synthesis the cores of the larger catalyst particles re-oxidise and so become inert during F-T synthesis. It thus follows that there is no point in reducing the entire particle in the first place.

It has been observed that iron catalysts containing K_2O carburized more rapidly and completely than those that do not contain K_2O [55]. Studies at Sasol carried out in fluidized bed reactors showed that the higher the K_2O content the higher the level of the carbide in the catalyst at the end of the run. This occurs in spite of the fact that K_2O promotion enhances the conversion (cf. Table 12) and thus should result in a more oxidising gas at the reactor exit. This ability of K_2O to keep the iron in the active carbided state could be one of the main roles of the K_2O promoter.

Anderson and co-workers [2] extensively studied the effect of pre-nitriding iron catalysts. Such catalysts were reported to be more resistant to re-oxidation in synthesis than carbided catalysts.

The overall impression gained from the foregoing observations is that the oxidation of some of the carbide phase is not a major cause of the loss of activity that occurs under normal Fischer-Tropsch synthesis conditions. Under abnormal conditions, such as feeding a very high CO_2 content gas and simultaneously running at high conversion levels (which results in high partial pressures of both CO_2 and of H_2O) the catalyst does oxidise more rapidly and the rate of activity decline is more marked. In such a case the decline is certainly linked with the oxidation of the active carbide phase. Other workers have also observed that changing the feed gas from 3 H_2/1 CO to 3 H_2/1 CO_2 results in the oxidation of the carbide to Fe_3O_4 [58].

B. Sintering

Freshly reduced and carbided precipitated (SiO_2 supported) iron catalysts typically have a BET area of about 200 $m^2\ g^{-1}$. In keeping with this, the lines of the x-ray diffraction pattern are very broad and indistinct which indicates the presence only of very small carbide crystallites. Used catalyst (which has lost about 20 percent of its original activity) has an area of about 50 $m^2\ g^{-1}$ and the x-ray diffraction lines are much sharper. Both measurements indicate that crystal growth has occurred. When precipitated catalysts are prepared with lower contents of the SiO_2 support, the observed rate of activity decline is markedly higher. This can be associated with the expected lower resistance to sintering of such catalysts.

When water vapour is deliberately added to the feed gas (thus increasing the average water vapour pressure throughout the catalyst bed) the rate of activity decline increases. It has also been shown that catalyst near the exit of the reactor (where the water vapour pressure is at its highest) is more deactivated than catalyst higher up in the reactor. These observations fit the commonly observed fact that water vapour enhances the rate of sintering of high area catalysts. The overall conclusion appears to be that for precipitated iron catalysts the loss of area is a contributing factor to the observed loss of activity.

On carbiding fused reduced iron catalysts with pure CO no change in BET area is observed. Similarly it is found that for catalysts which have been used in the F-T synthesis for only a short while the area is still very close to that of the freshly reduced catalyst. At the high temperatures (> 573 K) at which these catalysts are employed high-area carbon is deposited on the catalyst surface. The total measured surface area of a used catalyst is thus higher than that of the fresh catalyst. Due to the known presence of residual insoluble carbonaceous deposits on the catalyst surface (see the next section) the measured areas are of little value as far as trying to decide whether sintering has contributed to the observed loss of activity. Carbon deposition causes particle disintegration and it is feasible that this results in the exposure of more accessible active surface sites. This would partly compensate for the active sites lost for other reasons. (Note that precipitated catalysts are operated at low (< 523 K) temperatures where carbon deposition does not occur and hence this compensating effect is absent.)

C. Fouling

Under normal F-T process conditions products ranging from methane to high molecular mass waxes are always formed. Since the latter is a liquid under synthesis conditions the pores of the catalyst will become wholly or partially filled with wax. In the case of catalysts operating under wax producing conditions the catalyst pores can be filled with wax within minutes. (In the Sasol fixed bed reactors about 50 percent of the hydrocarbon product is in the liquid state inside the reactor.) The liquid wax must slow down the rate of diffusion of the reactants within the catalyst pores and so slow down the rate of the reaction. This is borne out by the fact that when the catalyst is periodically washed in situ with a lighter liquid solvent there is a sharp increase in activity. The practice of periodic solvent treatment was applied in Germany for iron and cobalt catalysts [2]. In the Sasol fixed bed reactors this technique is not worth applying since the pores are filled with wax again within minutes of the treatment and the period of high activity is thus very short lived.

In the case of cobalt catalysts used in low pressure synthesis periodic hydrogen reactivation [2] at high temperatures (e.g. 673 K) was commonly applied. This resulted in the hydrocracking of the wax to volatile compounds and the effect was therefore the same as solvent extraction. It was found that for catalysts operating at higher synthesis pressures, the rate of activity decline was much lower and also that periodic hydrogen treatment was less effective than it was in the case of synthesis at low pressure. It was also found that the wax extracted from the cobalt catalyst used at the higher pressure had a lower average molecular mass than the wax in the low pressure case. All these observations probably can be explained as follows: Under high pressure a large fraction of the hydrocarbon products are in the liquid phase and since all the products are formed within the catalyst pores this liquid must be oozing out continuously from the catalyst pores carrying with it the high molecular mass wax fractions. This prevents the accumulation of these waxes in the catalyst pores. This explains why at high synthesis pressure the catalyst does not loose activity as rapidly, why the average molecular mass of the wax on the catalyst is lower and why hydrogen reactivation is less effective than in the case of low pressure synthesis.

For catalysts operating under high wax selectivity conditions it has been found that if the synthesis temperature is lowered (which results in a shift to even longer-chained waxes) the rate of decline in activity is not adversely affected. Thus long-chained waxes while retarding the diffusion of reactants do not truly foul the catalyst surface.

A distinction should be made between high molecular mass wax (which is soluble in the hydrocarbon products) and insoluble carbonaceous deposits. While waxes simply retard the rate of diffusion of reactants the insoluble deposits permanently decrease the number of active sites and this constitutes true surface fouling. Solvent extraction can remove only the soluble material while high temperature hydrogen reduction is potentially capable of removing both types (unless the carbonaceous deposit has coked to a high degree in which case its hydrogenation rate will be very low).

When iron catalysts operate at higher temperatures (>573 K), aromatic compounds are also formed and their presence makes the formation of coke-like deposits highly probable. A used iron catalyst from a high temperature fluidized bed reactor was successively extracted with heptane, xylene and finally pyridine and the extracted waxes were analysed. The H/C atomic ratios were found to be 1.7, 0.9 and 0.8 respectively. Infra-red analyses confirmed that the extracted waxes were progressively more aromatic in nature. When the pyridine extracted catalyst was treated with pure hydrogen it was found that above 623 K waxes, oils and lighter hydrocarbon gasses were evolved. This indicates that pyridine-insoluble carbonaceous material was present on the catalyst. The evolved hydrocarbons do not originate from the hydrogenation of iron carbides or of some other form of elemental carbon. If a fresh fully reduced catalyst is carbided with pure CO (with the simultaneous deposition of elemental carbon) and the sample is then reduced with hydrogen the only product formed is CH_4.

When an aged iron catalyst from a fluidized bed reactor is treated with hydrogen above 623 K and then again subjected to F-T synthesis it is found that the activity of the catalyst returns close to its original value. This reactivation appears to be due to the removal of the carbonaceous deposits rather than to the reduction of any inert iron oxide. This is because the degree of reactivation is as marked for catalysts containing less than 5 percent iron oxide as for catalysts containing 60 percent iron oxide. It is also of interest to note that as iron catalysts age the selectivity shifts progressively towards lower molecular mass products. On hydrogen reactivation the selectivity again returns to that of a fresh catalyst. Since K_2O promotion is associated with both higher activity and selectivity it seems reasonable to assume that the most active sites are those in the immediate vicinity of the alkali promoter. These sites are likely to be fouled first and so the remaining less active sites which also have a lower selectivity will make a bigger fractional contribution to the product spread and this will result in the observed selectivity shift. There are several patents which deal with the reactivation of fluidized iron and other catalysts by hydrogenation [59, 60, 61]. A recent Auger study [62] indicated that when an iron catalyst is treated with synthesis gas (of H_2/CO ratio 3/1) at 573 K and 6 bar pressure the surface is rapidly covered by carbonaceous material. It was claimed that the deposit poisoned the catalyst.

D. Poisoning by Sulfur and Other Compounds

It has long been known that all sulfur compounds rapidly deactivate iron, nickel and cobalt catalysts. From published data [63] it is expected that FeS will be reduced by H_2 at about 673 K if the H_2S content is below about 40 mg H_2S m_n^{-3}. However, in practice it is found that at such sulfur levels the catalyst deactivates rapidly. Fischer [64] recommended 1 to 2 mg m_n^{-3} as the practical upper limit for S in synthesis gas and this is commonly quoted in the literature. However, in order to assure minimal poisoning the sulfur concentration should be ten times lower than this value. This is illustrated by the results given in Table 14 for a fluidized iron catalyst. In a fixed bed iron catalyst a noticeably steadier conversion is obtained when the S content is lowered from 0,2 to 0.02 mg S m_n^{-3}, The Lurgi Rectisol purification process routinely produces gas containing about 0.03 mg S m_n^{-3}.

In fluidized bed reactors sulfur will be deposited on all catalyst particles but in fixed bed reactors the sulfur is adsorbed predominantly on the catalyst at the reactor entrance. This was demonstrated by carefully unloading a catalyst bed which had been deliberately sulfur poisoned and then measuring the F-T synthesis activity of the different sections. The very top section was found to be inactive while the subsequent sections were progressively more active. Since in a normally deactivated catalyst (low sulfur in feed gas) the middle section of the catalyst bed is more active than the catalyst at the reactor exit (see section 6. B) the above finding indicates that the sulfur must have penetrated at least as far as the centre of the catalyst bed (i.e. about 6 m).

The relative effectiveness of C_2H_5SH and H_2S as poisons in fixed bed iron catalysts was studied and it was found that the organic sulfur resulted in a

Table 14. Influence of the sulfur content of synthesis gas on the rate of activity decline for fluidized iron catalyst at about 593 K

Sulfur content of synthesis gas/mgS m_n^{-3}	Drop in the percent conversion per day
0.1	very low
0.4	0.25
2.8	2.0
28	33

higher rate of deactivation than the inorganic sulfur. This could indicate that H_2S reacts with iron more readily than does C_2H_5SH. Thus H_2S would be more completely adsorbed in the upper catalyst layers while C_2H_5SH will penetrate deeper into the catalyst bed and so poison a greater portion of the catalyst. (Similar reasoning has been applied to explain why thiophene is a more effective poison than for instance COS [2]). In the case of a short catalyst bed (eg as used in a laboratory scale apparatus) the reverse effect would of course be observed, namely that H_2S is a more effective poison than C_2H_5SH.

A fluidized iron catalyst which has been poisoned by sulfur is not readily reactivated. High temperature reduction (up to 723 K) with pure hydrogen has no reactivating effect. Even re-oxidation followed by refusion and reduction is ineffective unless the re-oxidation step is carried out very thoroughly to burn away all traces of sulfur. This has also been observed by other workers. [2] In fixed bed reactors it has been observed that hydrogen reduction at 648 K has an adverse effect and this has been ascribed to the re-distribution of sulfur over the whole bed [2].

In the review by Madon and Shaw [65] several patents are mentioned which claim that K_2O promotion is effective for delaying or preventing poisoning by sulfur. Synthesis tests at Sasol have shown, however, that the use of alkali getters in iron catalyst do little if anything to prevent sulfur poisoning. If a K_2O promoted reduced iron catalyst is uniformly, and progressively poisoned by H_2S it has been found that the amount of CO chemisorbed decreases linearly with the amount of H_2S adsorbed. Since CO is assumed to selectively chemisorb on the exposed reduced iron surface the above result appears to indicate that H_2S did not combine with the surface K_2O in preference to the metallic iron. This supports the finding that alkali promotion does not effectively prevent sulfur poisoning.

Since the strong alkalis of Group I promote the activity of metal catalysts it may be expected that highly electronegative elements will act as poisons. Thus sulfur lowers the activity by removing electrons [66] whereas K_2O increases the activity by donating electrons to the metal. For the same reason iron oxide is an inactive catalyst. Chloride and bromide ions deactivate iron catalysts but, fluoride ions appear to have no adverse effect. P_2O_5 has no effect on the activity even at levels of 10 percent. Up to 1 percent Sb and Bi have no adverse effect but at the 10 percent levels they completely deactivate iron catalysts. Low levels of Pb have little effect but 0.5 wt. percent poisons the catalyst as does Sn down to levels of 0.1 wt. percent.

7. Carbon Deposition on Iron Catalysts

Unter the F-T synthesis conditions normally used for nickel, cobalt and ruthenium based catalysts little deposition of carbon occurs. The temperatures used are, however, usually below 523 K. At such temperatures iron catalysts do form carbides but little or no deposition of elemental carbon is observed.

This occurs only at higher temperatures. When carbon is deposited on iron catalysts the particles swell and also disintegrate. This powdering of the catalyst particles results in the plugging of fixed bed catalyst reactors. For this reason iron catalysts are not used in fixed bed reactors at temperatures above about 550 K. In a fluidised catalyst bed the problem of bed plugging cannot arise but the fines which are produced as a result of catalyst disintegration have a high carbon content and hence have a low particle density. Because of this the fines are readily carried out of the reactors by the effluent gas and will foul the downstream equipment and also the heavy oil products. Due to the swelling of the individual particles the whole fluidized catalyst bed expands.

The deposited carbon originates from the CO in the synthesis gas. After chemisorption on the catalyst the CO can dissociate to atomic carbon and oxygen. The oxygen reacts with either H_2 or CO. The overall reaction is $CO + CO\ (\text{or } H_2) \rightarrow C + CO_2\ (\text{or } H_2O)$.

When only CO is present the reaction is called the Boudouard reaction. If metallic iron is present the atoms of carbon migrate into the iron lattice and interstitial carbides are formed. After the "saturation" of the metal lattice nuclei of elemental carbon are formed and these continue to grow. Such deposits of carbon within the carbide crystals introduce severe stresses which result in the disintegration of the catalyst particles.

Since iron oxide is inactive for the F-T synthesis it is presumably also inactive for the carbon formation reaction. The experiments carried out at 823 K by Manning and Reid [67] lead them to the same conclusion but surprisingly it was also deduced that iron carbide was inactive. When CO was passed over Hagg carbide at 598 K the Boudouard reaction occurred at a high rate [68, 69]. According to x-ray diffraction analysis the only phases present in the catalyst were Hägg carbide and traces of magnetite. Since iron carbide is readily reduced by hydrogen to metallic iron at normal F-T temperatures it is debatable whether the active site on the surface of an iron carbide crystal is in the "reduced" or "carbided" state during F-T synthesis. It was found [69] that the addition of only a small amount of H_2 (or of any compound which resulted in the formation of molecular H_2, eg H_2O, C_2H_5OH or CH_3COOH) resulted in a marked increase in the rate of the Boudouard reaction. Figure 16 illustrates the effect of water vapour. A possible explanation is that the hydrogen reduces the surface carbide and that the metallic iron sites are more active than the carbided iron sites. It was found that whereas molecular nitrogen had no effect on the CO decomposition rate, the presence of ammonia lowered the rate. The effect of NH_3 was temporary as the rate returned to normal on ceasing the NH_3 addition. The presence of hydrocarbons, (paraffins or olefins) had no observable effect at 598 K. (Note that metallic iron is also carbided by light hydrocarbons such as, CH_4 or C_4H_8. The carbide phase formed, however, is cementite Fe_3C, as distinct from Hägg, Fe_5C_2, which is formed when CO or synthesis gas is used.)

A study of the influence of structural promoters on the rate of the Boudouard reaction showed that it was proportional to the iron surface area [68].

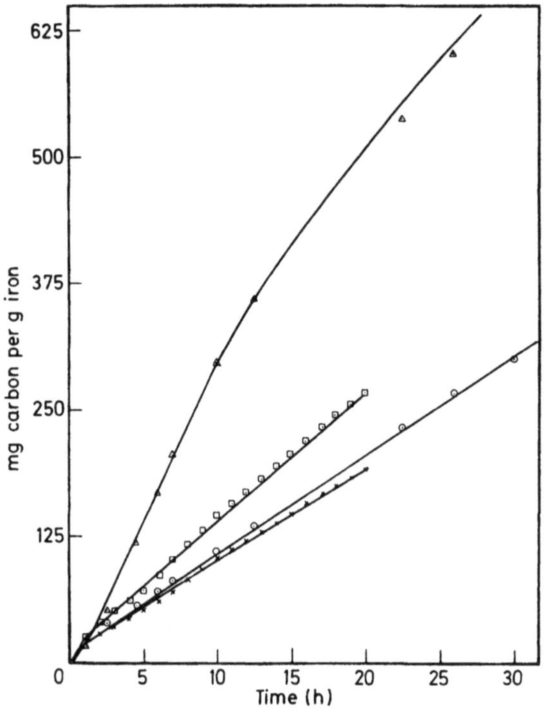

Figure 16. The effect of water vapour on the Boudouard reaction. O, fluidized bed with no water added; △, with 2.9% water vapour added. + Fixed bed with no water added; □, with 2.9% water added

Figure 17 shows that although Al_2O_3 promotion markedly increases the overall reaction the rate per unit metallic iron area was not much influenced. The chemical promoters such as K_2O and Na_2O on the contrary, did increase the intrinsic rate. This latter effect is in agreement which the established observation that as the alkali promotor is increased the rate of carbon deposition during the F-T synthesis also increases [2]. At zero alkali content, however, the rate is also high and so it appears that there is an optimum alkali level for minimum carbon deposition. An example of the influence of the alkali content on the carbon deposition rate during F-T synthesis is given in Table 12. When SiO_2 was also present in alkali promoted catalysts the intrinsic rate of the Boudouard reaction was lowered [68]. This is predictable from the effect is has on the basicity of the alkali (see the discussion in section 5.B.4). Silica has a similar effect on the rate of carbon build-up on alkali promoted catalysts in the F-T synthesis. When a large amount of SiO_2 is added, however, the rate of carbon deposition is markedly increased. This again indicates that to minimise carbon deposition during F-T synthesis an optimum catalyst basicity is required. It has been claimed that chloride lowers the rate of carbon deposition on a fluidized iron catalyst [70]. This result fits the observed link between the basicity of the iron surface and the rate of carbon deposition. Many studies have been carried out in the past on the influence of various additives on the rate of carbon deposition during the F-T synthesis [1, 2] and many contradictory findings have resulted.

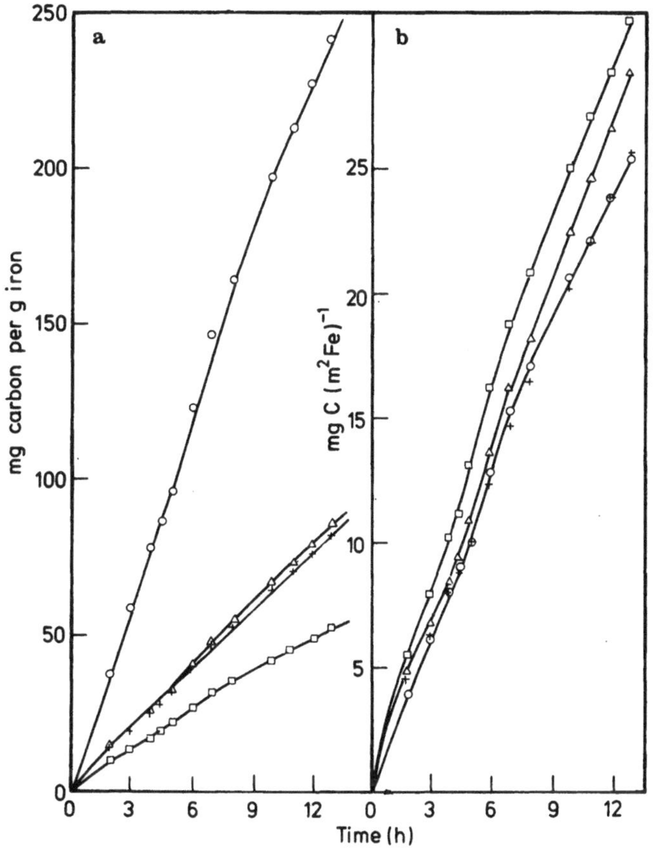

Figure 17. The effect of promoters on the rate of the Boudouard reaction. 17A represents the absolute rates and 17B the intrinsic carbon deposition rates. + unpromoted sample; ○ promoted with about 2.2 g Al_2O_3; △ promoted with about 2.3 g SiO_2 and □ promoted with 1.35 g CaO per 100 g Fe

Thus promotion by Cu was claimed to lower the rate of carbon formation [2], but studies at Sasol did not confirm this. Promotion of fused catalysts with the oxides of Mo, Cu, Zn and Mn had little effect on the carbon deposition rate. Bauklok and Hellbrügge [71] found that promotion with Cr_2O_3 resulted in a decreased rate of carbon monoxide decomposition over iron. The rate of carbon deposition is very sensitive to the temperature. The activation energy of the Boudouard reaction is 113 kJ mol^{-1} [69] which is higher than the apparent value for the F-T reaction. Under typical synthesis conditions for a fluidized iron catalyst the rate of carbon deposition increases by 50 percent for a 10 K rise in average bed temperature. According to the ternary diagrams of White et al. [72], carbon deposition should not occur under the above conditions if all the reactions involving elemental carbon are in thermodynamic equilibrium. It is known, however, that the rates of reaction of carbon with H_2 or H_2O at normal F-T temperatures are slow and so equilibrium is not likely to occur. Curves produced by the US Bureau of Mines [73] give the H_2/CO ratios above which there should be no carbon deposition, but in practice heavy carbon deposition does occur on iron catalysts at these supposedly safe H_2/CO ratios. It is clear that kinetic rather than equilibrium considerations dictate whether and at what rate carbon will deposit.

A large number of runs have been carried out at Sasol using fused iron catalysts in order to establish the relation between the carbon deposition rate and the composition of the gas within the reactor [74]. The composition of the fresh feed gas, the total pressure and also the ratio of fresh feed to recycle gas flow were altered over wide ranges. It was found that there was a satisfactory relation between the carbon deposition rate and the value of $p_{CO}/p_{H_2}^2$ at the reactor entrance (p_{CO} and p_{H_2} are the partial pressures in bar of CO and H_2 respectively). The strong inverse dependence of the rate on the hydrogen partial pressure had also been observed in earlier investigations [21, 75]. Table 15 gives two sets of results for two differently promoted catalysts. The synthesis temperature of the two sets were also not the same. The relation between carbon deposition and $p_{CO}/p_{H_2}^2$ is illustrated in Figure 18. As can be seen from Tables 15 to 17 the value of the simple ratio H_2/CO of the gas in the reactor does not correlate with the rate of carbon deposition. The partial pressure of CO_2 is apparantly also of no consequence as was demonstrated by the fact that a four-fold variation in p_{CO_2} had no clear effect on the rate of carbon deposition [74]. (In this series of tests the entrance p_{CO} and p_{H_2} were maintained constant and only p_{CO_2} was varied). An interesting feature of the factor $p_{CO}/p_{H_2}^2$ is that when the throughput of the reactor is increased by either increasing the total pressure or by decreasing the recycle to fresh gas ratio the rate of carbon deposition decreases. This is illustrated in Tables 16 and 17 respectively. Note that for both sets of experiments the

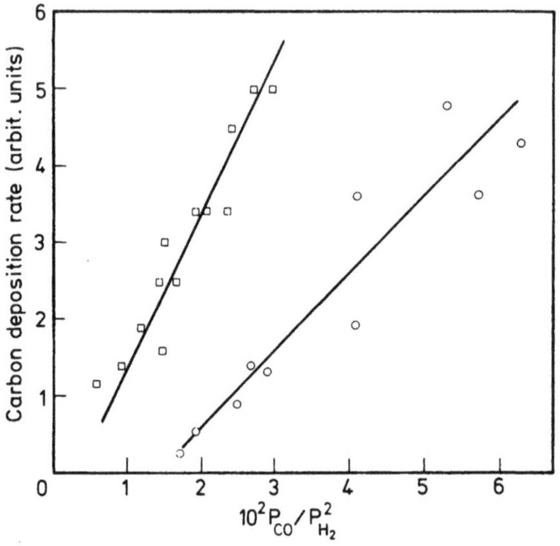

Figure 18. The carbon deposition rate (arbitary units) as a function of the ratio $p_{CO}/p_{H_2}^2$. The two sets of results are for two differently promoted catalysts operated at two different temperatures. The plots □ and ○ are respectively for catalysts B and A in Table 15. Catalyst B has a higher basicity than catalyst A

Table 15. The influence of reactor entry gas composition on carbon deposition rate

Set	Entrance partial pressure/0.1 MPa			p_{CO}/p_{H_2} ratio	$p_{CO}/p_{H_2}^2 \times 100$	Carbon deposition rate[a]
	H_2	CO	CO_2			
A	11.2	2.2	2.2	0.20	1.7	0.24
	9.6	1.8	1.9	0.19	1.9	0.53
	8.3	1.7	1.8	0.20	2.5	0.86
	7,8	1.8	0.7	0.23	2.9	1.3
	7.2	1.4	1.5	0.19	2.7	1.4
	7.5	2.3	3.9	0.31	4.1	1.9
	6.6	1.8	1.0	0.27	4.1	3.6
	6.9	2.7	6.8	0.39	5.7	3.6
	6.3	2.5	2.7	0.40	6.3	4.3
	5.5	1.6	1.2	0.29	5.3	4.8
B	32	6.4	6.4	0.20	0.6	1.2
	24	5.1	4.6	0.21	0.9	1.4
	16.4	4.0	4.1	0.24	1.5	1.6
	17.0	3.6	1.3	0.21	1.2	1.9
	15.6	3.4	3.7	0.22	1.4	2.5
	12.5	2.6	2.5	0.21	1.7	2.5
	13.4	2.7	1.1	0.20	1.5	3.0
	9.2	1.6	0.6	0.17	1.9	3.4
	11.5	2.8	3.4	0.24	2.1	3.4
	8.6	1.8	0.7	0.21	2.4	3.4
	8.5	1.7	1.5	0.20	2.4	4.5
	8.1	1.8	1.7	0.22	2.7	5.0
	8.8	2.3	1.6	0.26	3.0	5.0

[a] g C per 100 g Fe per unit time

Table 16. The influence of pressure on carbon deposition

Set	Total pressure 0.1 MPa	Total feed		$CO + CO_2$ converted/ mol hr^{-1}	Carbon[a] deposition rate
		$p_{CO}/p_{H_2}^2 \times 100$	H_2/CO ratio		
A	9.4	7.7	4.3	49	5.9
	12.9	3.7	5.2	68	3.3
	14.6	3.3	5.3	80	2.1
	18.1	2.3	5.8	94	1.6
	21.5	1.7	5.9	120	1.1
B	20.8	2.3	5.0	106	4.5
	30.8	1.7	4.8	157	2.5
	60.8	0.9	4.7	320	1.4
	75.8	0.6	5.0	377	1.2

[a] g C per 100 g Fe per unit time

Table 17. The influence of recycle/fresh feed ratio on carbon deposition

Recycle/ Fresh Feed ratio	Entrance partial pressures/0.1 MPa		$p_{CO}/p_{H_2}^2 \times 100$	Total feed H_2/CO ratio	$CO + CO_2$ converted/ mol hr^{-1}	Carbon[a] deposition rate
	H_2	CO				
1.2	10.7	1.7	1.5	6.3	117	0.4
2.0	7.9	1.4	2.2	5.6	103	1.1
3.0	5.9	1.0	2.9	5.9	78	1.4
4.0	4.8	0.8	3.5	6.0	63	2.8

[a] g C per 100 g Fe per unit time

linear velocity of the total feed gas was kept constant. In the pressure series the ratio recycle to fresh feed flow was also kept constant. If elemental carbon is considered to be part of the product spectrum then it would be expected that as the amount of F-T synthesis work done increases the amount of carbon deposited would also increase. However, as can be seen from the Tables 16 and 17, this was not the case. As the moles $CO + CO_2$ converted to hydrocarbons increased the rate of carbon deposition actually decreased. This was due to the fact that under the experimental conditions employed the value of $p_{CO}/p_{H_2}^2$ at the reactor entrance decreased when the fresh gas throughputs were increased. Another advantage of using higher pressures is that lower H_2/CO ratio gases can be used as fresh feed without the occurence of excessive carbon deposition on the catalyst.

A tentative derivation of the factor $p_{CO}/p_{H_2}^2$ has been made [74]. It was assumed that the reaction scheme depicted in Figure 19 represents the basic reaction steps in the F-T and in the carbon-forming reactions. According to this scheme a chemisorbed carbon monoxide molecule can either dissociate into a carbon and an oxygen atom or it can react with hydrogen in the non-dissociated state to form an oxygenated surface complex. The dissociated carbon atoms either migrate away and add onto growing carbon deposits or they are hydrogenated to methane. Note that according to this

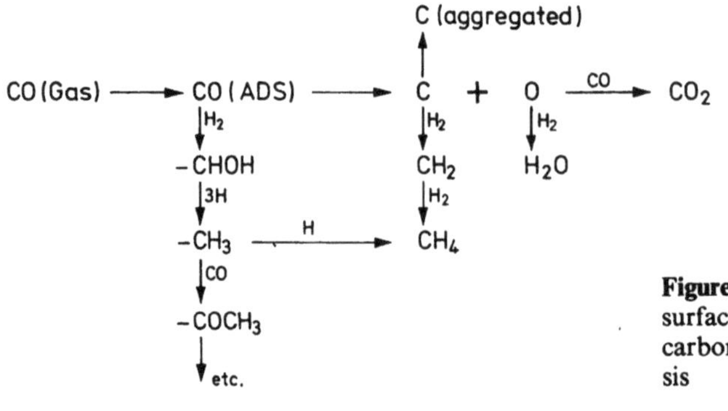

Figure 19. A scheme depicting the surface reaction steps leading to carbon deposition and F-T synthesis

reaction scheme there are two routes along which CH_4 is formed, one via the hydrogenation of single carbon atoms and the other via the hydrogenation of the oxygenated carbon complex. The consequence of this will be discussed in section 9.

8. Product Selectivity

A. Thermodynamic Considerations

In the temperature range commonly used for F-T synthesis it is known that the actual selectivity found in practice is very different from that expected from thermodynamic calculations. Starting with a 1.0 H_2/CO ratio gas Tillmetz [76] calculated that at 0.1 MPa pressure the main products should be methane, carbon dioxide and graphite. The calculated amounts of higher hydrocarbons were found to be negligible. In practice, however, the CH_4 is low, the carbon is negligible and the bulk of the product is higher molecular mass hydrocarbons. Table 18 compares a typical selectivity obtained over an iron catalyst with that calculated by Christoffel et al. [77] for 600 K, 1.6 MPa and a gas with an initial H_2/CO ratio of 2.0. From the table it can be seen that the relative differences are large. As the pressure was increased from 0.1 to 6 MPa the calculated amount of the heavier hydrocarbons increased about a 100 fold but their consentrations were still very small compared to that of CH_4.

Table 19 gives the partial pressures and molar ratios of various compounds at the exit of a fluidized iron catalyst bed operating at about 600 K, 2.0 MPa and a recycle to fresh feed ratio of 2. The table compares the actual pressures of the products with those which would be expected if the given equations were in thermodynamic equilibrium. The actual CO, H_2O and H_2 partial pressures at the bed exit are used in these calculations. Again it is clear that under actual F-T conditions the reactions forming hydrocarbons from

Table 18. Comparison between thermodynamically predicted [77] and typical actual selectivity of products

Compound	Calculated selectivity/ g per m_n^3 (H_2 + CO)		Typical practical selectivity/wt. %
	No carbon deposition	With carbon deposition	
CH_4	170	150	10
C_2H_4	1.7×10^{-7}	9.7×10^{-8}	4
C_2H_6	9.6×10^{-3}	6.1×10^{-3}	4
C_3H_6	1.1×10^{-9}	4.3×10^{-10}	12
C_3H_8	2×10^{-6}	9.3×10^{-7}	2
C_5H_{12}	1.4×10^{-13}	3×10^{-14}	2
C_2H_5OH	3×10^{-9}	1.9×10^{-9}	2

Table 19. Comparison of the calculated partial pressures of various products with those actually found at the reactor exit

Compound or ratio	Actual partial pressure (0.1 MPa) or ratio	Calculated[a] pressure (0.1 MPa) or ratio	Equation
H_2	6.46		
CO	0.11		
CO_2	0.94		
H_2O	1.91		
CH_4	8.68	3.1×10^7	$CO + 3\,H_2 \rightleftharpoons CH_4 + H_2O$
C_2H_4	0.31	54.5	$2\,CO + 4\,H_2 \rightleftharpoons C_2H_4 + 2\,H_2O$
C_2H_6	0.48	9.9×10^7	$2\,CO + 5\,H_2 \rightleftharpoons C_2H_6 + 2\,H_2O$
C_3H_6	0.63	2.5×10^4	$3\,CO + 6\,H_2 \rightleftharpoons C_3H_6 + 3\,H_2O$
C_3H_8	0.09	1.6×10^9	$3\,CO + 7\,H_2 \rightleftharpoons C_3H_8 + 3\,H_2O$
C_2H_6/C_2H_4	1.55	1.8×10^6	$C_2H_4 + H_2 \rightleftharpoons C_2H_6$
C_3H_8/C_3H_6	0.14	6.4×10^4	$C_3H_6 + H_2 \rightleftharpoons C_3H_8$
CH_3OH	0.0006	0.0005	$CO + 2\,H_2 \rightleftharpoons CH_3OH$
C_2H_5OH	0.0170	0.21	$2\,CO + 4\,H_2 \rightleftharpoons C_2H_5OH + H_2O$
C_3H_7OH	0.0038	8.22	$3\,CO + 6\,H_2 \rightleftharpoons C_3H_7OH + 2\,H_2O$
C_2H_6/C_2H_5OH	28	4.8×10^8	$C_2H_5OH + H_2 \rightleftharpoons C_2H_6 + H_2O$
C_2H_4/C_2H_5OH	18	264	$C_2H_5OH \rightleftharpoons C_2H_4 + H_2O$
C_3H_6/C_3H_7OH	200	3,000	$C_3H_7OH \rightleftharpoons C_3H_6 + H_2O$

[a] The actual partial pressure of H_2, CO, CO_2 and H_2O (column 2) are used

CO and H_2 are far from being in equilibrium. The conversion of olefins to paraffins is far from complete and the same applies to the alcohol hydrogenation or dehydration reactions.

As in the case of the higher molecular mass paraffins and olefins the amount of oxygenated hydrocarbons (alcohols, aldehydes, ketones and acids) are also present in much higher concentrations than expected from thermodynamic calculations. It appears, however, that these oxygenates readily interact amongst themselves. Weitkamp [78] found that at the exit of a fluidized iron catalyst bed the following reactions appeared to be close to equilibrium:

$$C_2H_5OH \rightleftharpoons CH_3CHO + H_2$$

$$CH_3CHO + H_2O \rightleftharpoons CH_3COOH + H_2$$

$$(CH_3)_2CO + H_2 \rightleftharpoons (CH_3)_2\,CHOH\,.$$

Table 20 illustrates the relationship between several oxygenated compounds at the exit of a fluidized iron catalyst bed. (The results were taken from the same experiment used in Table 19). Several of the molar ratios presented in the table are not markedly different from the ratios expected if the reactions

Table 20. Comparison of the molar ratios of various oxygenates as obtained and as calculated assuming equillibrium (data taken from the same experiment as for Table 19)

Molar ratio			Equation
Ratio	Observed value	Calculated value	
$(CH_3)_2CO/C_2H_5OH$	0.2	0.07	$C_2H_5OH + C \rightleftharpoons (CH_3)_2CO$[a]
CH_3COOH/C_2H_5OH	0.14	0.11	$C_2H_5OH + H_2O \rightleftharpoons CH_3COOH + 2\,H_2$
CH_3COOH/CH_3CHO	2.7	0.64	$CH_3CHO + H_2O \rightleftharpoons CH_3COOH + H_2$
C_2H_5OH/CH_3CHO	19	5.7	$CH_3CHO + H_2 \rightleftharpoons C_2H_5OH$
$(CH_3)_2CO/(CH_3COOH)^2$	640	1.8×10^5	$2\,CH_3COOH \rightleftharpoons (CH_3)_2CO + CO_2 + H_2O$
$CH_3COOC_2H_5/C_2H_5OH$	very low	0.0065	$C_2H_5OH + CH_3COOH \Rightarrow \rightleftharpoons CH_3COOC_2H_5 + H_2O$
$i{-}C_3H_7OH/(CH_3)_2CO$	0.27	0.29	$(CH_3)_2CO + H_2 \rightleftharpoons i{-}C_3H_7OH$
$i{-}C_3H_7OH/n{-}C_3H_7OH$	0.23	3.7	$n{-}C_3H_7OH \rightleftharpoons i{-}C_3H_7OH$

[a] In this equation carbon is taken as graphite. In practise the C is more likely to be isolated carbon atoms and thus its thermodynamic constants should be somewhat different from that of graphite.

are in equilibrium. This suggests that alcohols, aldehydes, ketones and acids interact rapidly with one another under F-T synthesis conditions. This was confirmed in tests carried out in laboratory scale reactors in the temperature range 473 to 598 K. Whenever ethanol, acetaldehyde or ethyl acetate was injected individually into the feed gas all three compounds were always found in the exhaust gas. At the higher temperatures acetone also was always present.

B. Interrelation Between the Product Carbon Numbers

At Sasol copious information has been generated regarding the product selectivities obtained with various iron catalysts operating under different conditions. In these studies fixed, slurry and fluidized bed reactors have been used. Variously promoted precipitated, sintered or fused catalysts of different particle sizes have been investigated. Synthesis temperatures ranged from 453 to 658 K and pressures from 1 to 7.5 MPa. The relative amounts of CO, H_2 and CO_2 in the fresh feed gases have been varied over wide ranges and the fresh feed to recycle gas ratio has also been varied. The methane selectivities have ranged from about 5 to 80 percent and hard wax selectivities from zero to about 50 percent. A study of all the data generated revealed that there was a clear interrelationship between the products [79]. Figures 20, 21 and 22 demonstrate several examples of this. It thus appears that if the selectivity of any one particular carbon number species is altered then the selectivities of all the other species will also shift by a predictable amount.

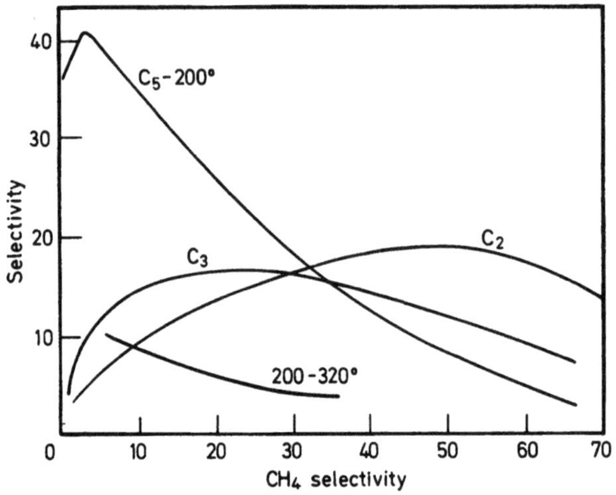

Figure 20. The relation between the selectivity of various product cuts and the CH_4 selectivity (all selectivities, % carbon atom)

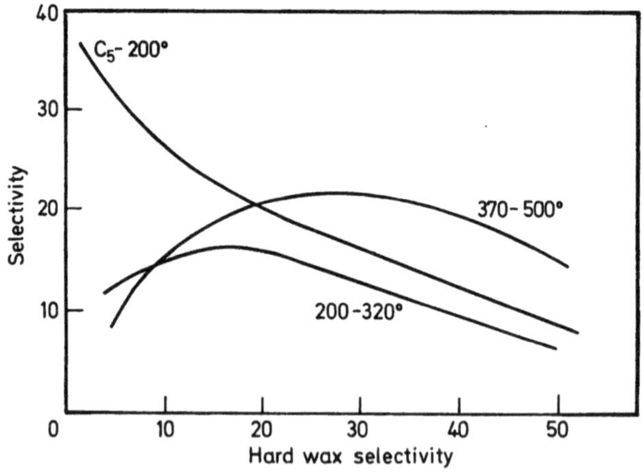

Figure 21. The relation between the selectivity of various product cuts and the hard wax selectivity. Hard wax is the product boiling above 773 K (all selectivities, % carbon atom)

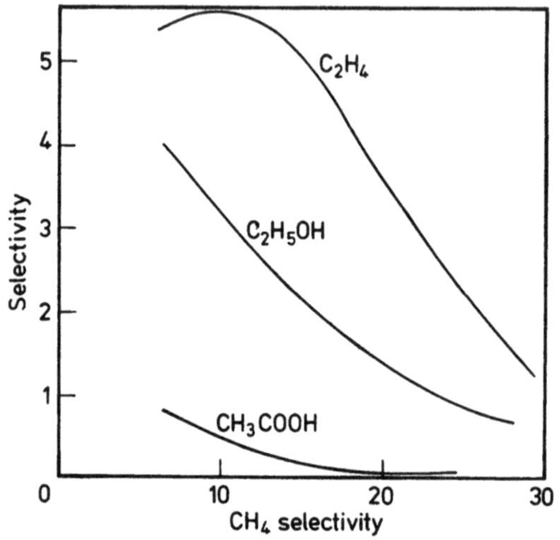

Figure 22. The relation between the selectivity of various C_2 products and the CH_4 selectivity (all selectivities, % carbon atom)

Table 21. Selectivity of various catalysts (data from Storch [1])

Catalyst type		F-T conditions			Selectivity wt %			
		Temperature/K	Pressure/ 0.1 MPa	H_2/CO ratio of fresh feed	Gas C_1-C_4	Gasoline	Diesel	Wax > 593 K
(Co + Ni) precipitated supported catalyst		466	10	2.0	38	45	15	2
Co (precipitated)		401	4	1.87	18	41	23	18
		<403	1	2.0	30	43	19	8
		513	10	2.0	38	35	21	6
Fe (ex carbonyl) cemented		503	20	—	9	16	18	57
Schwartzheide tests with various Fe catalysts	Brabag[a]				23	18	16	43
	Lurgi				33	19	13	35
	Ruhrchemie	473–498	10	1.25	39	22	13	26
	IG Farben				43	25	11	21
	KWI				43	25	14	18
	Rheinpreussen				44	29	14	13

[a] Name of firm or institute

This finding holds irrespective of the means by which the selectivity shift has been brought about, i.e. whether by a change in the synthesis process conditions or by a change in the catalyst's formulation. The selectivities of the compounds at the opposite ends of the product spectrum (namely methane and high molecular mass wax) can attain very high values whereas all the intermediate products can attain only limited maximum values. Thus in the F-T synthesis it does not appear to be possible to produce more than about 18 wt. percent of C_2 and about 16 percent of C_3 hydrocarbons, about 41 percent of gasoline (the C_5 to 473 K cut), about 18 percent of the 473 to 593 K cut and about 22 percent of medium wax (the 643 to 773 K cut).

The data presented in Table 21 shows that the forgoing conclusions are not unique to the catalyst types and synthesis conditions used in the Sasol tests. Note that the light ends (C_1 to C_4) and the heavy ends (>593 K) vary over wide ranges, but the gasoline and diesel (i.e. the intermediate cuts) do not exceed 45 and 23 wt. percent respectively.

Although there are several proposed mechanisms for the F-T process (see section 9) all of them assume that the growth of the hydrocarbon chains is a stepwise process, i.e. one carbon entity at a time adds on to the growing molecule. Such mechanisms readily lend themselves to a simple computing treatment.

Figure 23 represents a reaction sequence in which the probability of chain growth is 0.4 (and hence that of chain termination 0.6). The computation is extended to a predetermined carbon number product and the selectivities of the individual products are obtained by dividing the number of carbon atoms desorbed as those products by the total number of all carbon atoms desorbed. The probability of chain growth can then be changed and new sets of selectivities computed [79]. Figure 24 illustrates the results obtained. In these calculations it was arbitrarily assumed that the probability of chain growth did not vary with the chain length and that there were no chains longer than 120 carbon atoms. In spite of the likelihood that this treatment is oversimplified the predicted maximum selectivities of the various product cuts are not too different from the values found in practice. Thus the calculated maximum selectivities of C_3, the C_5 to C_{11} cut, the C_{12} to C_{18} cut and the C_{24} to C_{35} cut are about 19, 47, 25 and 22 percent respectively while the actual maximum observed values for the C_3, gasoline, diesel and medium wax selectivities are about 16, 42, 18 and 22 percent respectively. The only notable misfit is the C_2 selectivity. Figure 24 indicates that the maximum C_2 selectivity is expected to be about 30 percent while in practice the maximum found is only about 18 percent. If it is assumed that the probability of chain growth of the C_2 entity is double that of all the others, then the calculated selectivity maximums come to 17 percent for C_2, and 21 percent for C_3.

It is commonly observed for iron as well as for cobalt [45, 97] and ruthenium [43, 45] catalysts that the C_2 selectivity is lower than the C_1 and also than the C_3 selectivity. According to Figure 24 this should not occur. Either the actual C_2 selectivities are lower than the values expected from the reaction sequence depicted in Figure 23 and/or the CH_4 selectivities are higher than

	Carbon atoms desorbed	Total carbon desorbed
100 C_1 units $\xrightarrow{0.6}$ 60 CH_4	60	60
↓0.4		
40 C_1 + 40 new C_1 units		
↓		
40 C_2 units $\xrightarrow{0.6}$ 24 C_2H_6	48	108
↓0.4		
16 C_2 + 16 new C_1 units		
↓		
16 C_3 units $\xrightarrow{0.6}$ 9.6 C_3H_8	28.8	136.8
↓0.4		
6.4 C_3 units + 6.4 new C_1 units		
↓		
6.4 C_4 units $\xrightarrow{0.6}$ 3.84 C_4H_{10}	15.36	152.16
↓		
etc.	etc.	etc.

Figure 23. A scheme for calculating the carbon-number product distribution. In the example given the probability of chain growth is 0.4 and thus the probability of chain termination is 0.6

expected. Several examples of this are given in Table 22. For the iron examples the calculated and observed selectivities of C_1, C_3 and C_4 agree reasonably well but but as the CH_4 selectivity increases the discrepancy in the C_2 selectivities increases. In the case of the Co and Ru examples most of the actual C_2 selectivities also appear to be somewhat lower than expected but in addition the CH_4 selectivities are clearly higher than expected. Several factors could contribute to these apparent misfits in the product distribution. Olefins, especially C_2H_4, can be hydrocracked to CH_4 [132]. Ethylene incorporates more readily than longer chained olefins [44, 123, 132]. The probability of chain growth of the C_2 surface entity could be higher than for all other hydrocarbon entities [79]. (Mathematically this will give the same result as a higher incorporation rate for C_2H_4.) In the cases where the CH_4 selectivity is "too high" a mechanistic explanation might be applicable (see section 9). More detailed studies are needed to elucidate the discrepancies in the C_1 and C_2 selectivities.

Investigations of the carbon number spectrum based on the assumption of a stepwise chain mechanism were carried out in considerable detail by several workers. Herington [80] found that the probability of chain growth in F-T synthesis over a cobalt catalyst did not change much with the chain

Table 22. Comparison between actual and calculated selectivities[a, c])

Catalyst	Fe	Fe	Fe	Co	Co	Ru	Ru
Reference	Sasol[b]	Sasol[b]	Sasol[b]	97	43	43	43
C_1	2.7 (2.1)	11.0 (10.2)	50 (52)	12.0 (1.3)	8.4 (2.2)	3.7 (2.0)	14.8 (1.7)
C_2	3.6 (3.6)	10.8 (13.9)	17 (29)	2.1 (2.3)	1.9 (3.7)	1.2 (3.4)	3.5 (3.0)
C_3	4.6 (4.6)	14.3 (14.3)	12 (12)	3.0 (3.0)	4.7 (4.7)	4.4 (4.4)	3.8 (3.8)
C_4	5.2 (5.2)	12.5 (12.8)	6 (5)	3.8 (3.6)	4.0 (5.3)	3.8 (5.0)	2.8 (4.4)

[a] The calculated selectivities were obtained in the manner illustrated in Figures 23 and 24. The C_3's were matched to the values actually observed in the synthesis. The figures in brackets are the calculated ones

[b] The Sasol selectivities include hydrocarbons and oxygenates

[c] All selectivities, % carbon atom

length of the hydrocarbon. Anderson [1, 2, 81] investigated the spectrums of a number of different fixed bed catalysts. The plots of log W_n/n against carbon number gave straight lines over a fairly large carbon-number range (W_n is the mass fraction and n the carbon number). This indicated that the probability of chain growth was fairly constant. Weitkamp [82] evaluated the data from fluidized iron catalyst beds in a similar manner and found that the probability of chain growth (α) was given by

$$\alpha = 0.606 + 0.012\, n$$

In most of the above cases the C_2 data points were out of line with the other carbon-number compounds. This has been found to be the case also over iron catalysts at Sasol. If α is constant its value can be calculated from the molar ratio in which any two carbon-number products are formed in the F-T synthesis [2].

$$\Phi_n/\Phi_x = \alpha^{n-x}$$

where Φ_n and Φ_x are the number of moles of hydrocarbon containing n and x carbon atoms respectively.

More recently the Flory equation [83] as modified by Schulz [84] has been applied to the F-T product spectrum [85]. (The Flory equation deals with the product distribution in a polymerization process). As pointed out by Anderson [86] and Madon [129] the Schulz-Flory treatment is no different from the earlier treatment. The Schulz-Flory equation is is

$$\log \frac{W_n}{n} = n \log \alpha + \log \frac{(1-\alpha)^2}{\alpha}$$

where W_n is the mass fraction, n the carbon number and α the probability of chain growth. Application of the equation leads to the conclusion that the maximum mass selectivity obtainable for C_2 is 30 and for the C_5 to C_{11} cut is 48 percent. This agrees exactly with the prediction from the scheme shown in Figure 23 (see the curves in Figure 24). This is not surprising since the bases for the two schemes are the same.

Studies of the selectivities of the alcohols of different carbon numbers have shown that, as found for the hydrocarbons, there is a regular relationship between the alcohols [2]. The distribution curves for alcohols are similar to those of the $n + 1$ hydrocarbons. Thus under conditions where it is found that the C_3 hydrocarbon is the most abundant hydrocarbon produced then the most abundant alcohol produced is C_2H_5OH.

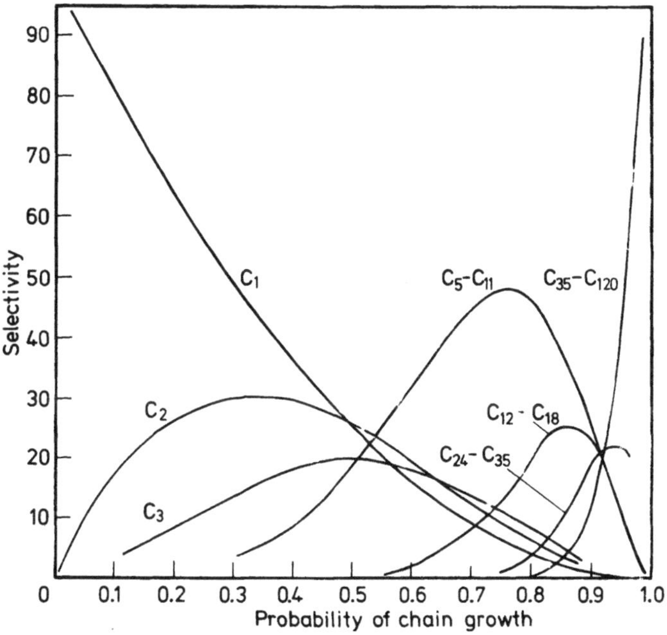

Figure 24. Plots of various selectivities (percent carbon atom basis) as calculated according to the scheme in Figure 23. The highest C number was arbitarily taken to be 120

C. Product Types and Selectivity Control

1. Product Types

As discussed in section 8.B the carbon-number distribution of the products is a function of the probalility of chain growth. The value of the probability can be varied by using different types of catalysts (e.g. iron, cobalt or ruthenium), by adjusting the promoter content of the catalyst (e.g. the amount of K_2O) and by altering the synthesis conditions (e.g. the temperature or the partial pressures of the reactants). There is thus a high degree of flexibility in the F-T process.

A more detailed analysis of the products show that when the catalysts or the process conditions are altered changes occur not only in the overall carbon-number distribution, but also in the types of compounds that are present in the different carbon number cuts. Pichler [87] and co-workers analysed in great detail the C_4 to C_{17} products obtained in the fixed and fluidized iron catalyst bed processes at Sasol as well as the products of a cobalt catalyst. Table 23 gives a few summarized extracts from this study. The cobalt catalyst's products are more paraffinic than those of the iron catalysts. This is in keeping with previous results [1, 2] and with the fact that cobalt is known to be a more active hydrogenating catalyst than iron. Pichlers [87] analyses and also those of Weitkamp [82] showed that as the carbon number increased from 6 to 17, the olefinity decreased. This trend continues at higher carbon numbers (e.g. see Table 4, fixed bed case). The hard wax produced at Sasol has an olefin content of less than 5 percent. (This wax has a carbon number distribution which typically ranges from 30 to 320, the peak being at about 50 [88]).

From Table 23 it can be seen that the hydrocarbons are predominantly straight chained. Of the two iron catalysts the one which operated at the lower

Table 23. Summarised detail of the Types of compounds present[a] in different carbon number products [87]

Carbon No	Compound type	Process		
		Cobalt catalyst at 463 K, 0.1 MPa	Sasol fixed bed Fe at 493 K, 2.7 MPa	Sasol Synthol Fe at 593 K, 2.2 MPa
6	Straight chain paraffin	56	45	18
	Straight chain α olefin	4.0	25	42
	Straight chain internal olefins	21	26	19
	Branched chain paraffins	13	2.4	6
	Branched chain olefins	6.1	1.5	14
	Aromatics		0.002	0.2
8	Straight chain paraffins	54	41	11
	Straight chain α olefins	3	35	33
	Straight chain internal olefins	4	19	11
	Branched chain paraffins	16	1.5	7
	Branched chain olefins	11	3	21
	Aromatics		0.17	5.2
10	Straight chain paraffins	51	42	10
	Straight chain α olefins	3	37	26
	Straight chain internal olefins	3	15	9
	Branched chain paraffins	25	2	10
	Branched chain olefins	1	3	22

[a] amount of compounds in wt.%

temperature yielded the larger fraction of straight chained products. Branching is apparantly not solety a temperature effect as the cobalt catalyst operating at 463 K produced more branched products than did the iron catalyst at 493 K. Pichler [87] found that for an iron catalyst at 593 K the degree of chain branching of the products increased as the carbon-number increased from 4 to 17. However, at the low synthesis temperature (493 K) there was little change in the degree of branching over this carbon number range. A study [89] of Sasol wax (produced at about 498 K) showed that the degree of branching decreased with increasing carbon number. As the synthesis temperatures was increased the degree of branching increased. All branches appeared to be methyl branches, and were statistically distributed along the chain. [90] Weitkamp [82] reported that the amount of alkali promoter had no effect on the degree of branching of the hydrocarbons produced with fluidized iron catalysts.

Small amounts of aliphatic dienes and also cyclic diolefins have been found in the gasoline produced at Sasol. At low temperatures (about 490 K) virtually no aromatics are produced (cf. Table 23), but as the temperature is increased the aromaticity of the product increases (cf. Table 24). Studies at Sasol also found that as the H_2/CO ratio of the gas within the catalyst bed decreases the aromaticity of the oil increases. Weitkamp [91] showed that the concentration of aromatics in successive fractions increased almost linearly with the carbon number. Benzene was present in only small amounts. The quantities of saturated and olefinic C_5 and C_6 ring compounds were also low. It was suggested that aromatics and olefins are derived from common precursors. The thermodynamically unfavoured mono-alkybenzenes predominate and this suggests that aromatics are formed by linkage between the first and sixth carbon atom in the hydrocarbon chains. The formation of aromatics does not appear to be a secondary reaction of desorbed products. It was found at Sasol that the extensive recycling of the light oil product did not increase the aromaticity of the oil.

Table 24. The influence of temperature on the selectivity of fluidized iron catalyst (Sasol)

Temp./ K	Selectivity/% C atom				$\frac{C_3H_6}{C_3H_8}$	350–463 K cut analysis	
	CH_4	Water soluble chemicals					
		Alcohols	Ketones	Acids	Product ratio	Bromine number	Aromatic content/ wt. %
583	10	2.3	0.4	0.3	6	109	4
603	14	2.3	0.8	0.4	10	94	8
623	17	1.6	1.1	0.2	9	92	10
633	20	1.1	1.3	0.2	8	93	13
643	23	0.8	1.2	0.1	6	88	18
653	28	0.5	0.8	0.1	4	85	26

2. Promoters

Promotion of iron catalysts with Group I alkalis has a strong influence on the product selectivity. Other promoters have a lesser influence. This subject has been covered in sections 5.A.3. and 5.A.4. To recap briefly, it is not only the amount of alkali, but also the chemical interaction of the alkali with any other promotors or impurities which determines the basicity of the catalyst. As the basicity increases the selectivity shifts towards the heavier products, more acids are produced and the products are generally more olefinic. (See Tables 8 and 12).

3. Temperature

The influence of temperature on the selectivity is consistent for all F-T catalysts. As the operating temperature is increased, the product selectivity shifts to lighter molecular mass compounds. Nickel and ruthenium catalysts operating at low temperatures (<463 K) produce a high proportion of wax, but at higher temperatures (>573 K) methane is the main product formed. At 433 to 448 K cobalt catalysts produce large amounts of alcohols [92] (40 percent of the non aqueous liquid product), but at 473 K only about 1 percent of oxygenated organic molecules are formed. Over cobalt catalysts the percentage of olefins in the various carbon-number cuts either remained constant or decreased as the temperature was increased [2]. As can be seen in Table 24 Sasol fluidized iron catalyst also showed all the trends mentioned above, namely, as the temperature increased the CH_4 selectivity increased, the olefinity of the products and the alcohols and acids selectivities decreased. Note that the ketone selectivity first increased and then decreased. As mentioned in section 8.A ketones appear to be formed from alcohols, but the reaction occurs only at the higher temperatures. The initial increase in ketone selectivity with increasing temperature is therefore understandable. Table 25 demonstrates the influence of the temperature on the selectivities of a Sasol precipitated iron catalyst operating in a fixed bed under near differential conditions (i.e. high feed gas flow, low percentage conversion). The selectivity shifts away from the heavier products and the olefinity did not change much with increasing temperature.

Table 25. The influence of Temperature on the selectivity of a fixed bed Sasol iron catalyst

Temperature/K	Hard wax[a] selectivity/ % C atom	Olefins in diesel cut/ %
486	47	45
500	34	39
510	24	33
520	17	40

[a] Hard wax is that fraction boiling above 773 K

4. *Gas Composition and Pressure*

Using a cobalt catalyst at 453 K and atmospheric pressure Kölbel [93] found that changing the H_2/CO ratio of the feed gas from 0.56 to 2 resulted in no change in the CH_4 selectivity. Other studies with cobalt catalyst [2] showed, however, that on lowering the H_2/CO ratio of the feed gas the average molecular mass of the product increased as did the olefin and alcohol content. Using a Co:Ni:ThO_2:Mn:Kieselguhr catalyst Ruhrchemie tests [2] showed that a 1.0 H_2/CO ratio gas resulted in a lower CH_4 selectivity and a higher olefin content in the liquid hydrocarbon product than a 2.0 H_2/CO feed gas. Studies at the US Bureau of Mines [94] with a hot gas recycle process using iron lathe turnings as catalyst at about 580 K showed that on increasing the H_2/CO ratio of the fresh feed gas from 1.36 to 2.91 the $C_1 + C_2$ selectivity increased from about 47 to 60 percent and the olefinity of the product decreased. Overall it seems clear that the higher the H_2/CO ratio in the reactor the lighter the hydrocarbon products and the lower their olefinity.

Using supported cobalt catalysts Ruhrchemie [2] and the US Bureau of Mines [95] demonstrated that when the synthesis pressure was increased from 0.1 to 0.7 MPa the selectivity hardly changed. Fischer and Pichler [96], however, found that on increasing the pressure from 0 to 15 MPa the wax selectivity of their cobalt catalyst increased up to 1.5 MPa and then decreased. The operating temperatures at the higher pressures, however, were higher than those used at the lower pressures. It thus seems likely that the lower wax selectivity at the high pressures was rather due to the increased temperature than to the higher pressure. Martin [97] also reported that increasing the operating pressure of a cobalt catalyst, from 0.1 to about 1.5 MPa shifted the hydrocarbon selectivity towards the heavier products. Roelen [2] found over precipitated iron catalysts that up to about 2.0 MPa the wax selectivity increased with increasing pressure. There was little change in the olefinity of the liquid products. Anderson [98] showed that when the pressure was raised from about 0.8 to 2.1 MPa the average molecular mass of the hydrocarbons decreased slightly for precipitated and sintered iron catalysts whereas over fused iron catalysts it increased. The yield of oxygenated molecules increased with pressure. With fused iron catalysts it was found that the degree of unsaturation was not influenced by pressure [56]. Friedman and Schlesinger [99] investigated the influence of pressure up to 10.3 MPa over nitrided fused iron catalysts. They reported that the hydrocarbon selectivity shifted to the heavier products, the oxygenate content increased, but the olefinity of the liquid products showed no clear trend as the pressure was increased. Hall [21] reported that for iron catalysts operating in both fixed and slurry beds the average molecular mass of the products and the oxygenate selectivity increased while the olefinity decreased when the pressure was increased from 2.0 to 4.0 MPa. Pichler [4] reported that when the pressure over a ruthenium catalyst was increased up to about 100 MPa the wax selectivity increased markedly. Most of the studies thus indicate that on increasing pressure the hydrocarbon selectivity shifts towards the heavier products, the oxygenate selectivity increases, but the olefinity of the products is not much altered.

The observed influence of pressure is not likely to be due to the value of the total pressure per se but rather due to one or more of the reactant or product partial pressures. The US Bureau of Mines [94] found that at about 580 K an increase in the partial pressure of either water or CO_2 resulted in a large increase in the methane selectivity of an iron catalyst. Contrary to this other studies at the same institution found that with fused iron catalysts at 513 K the CH_4 production appeared to be inversely proportional to the concentration of CO_2 and H_2O in the reactor. It was deduced that water vapour inhibited CH_4 formation. Kölbel [105] studied the F-T reaction with $CO + H_2O$ as feed at pressures from 0.1 to about 10 MPa. He reported that as the pressure increased the alcohol and acid selectivities increased markedly. Studies with an alumina supported ruthenium catalyst in the temperature range 443 to 573 K showed that the partial pressure of CO was the dominant parameter which controlled the selectivity spectrum [45].

Investigations at Sasol [79] showed that with precipitated iron catalysts at low temperatures the H_2/CO ratio within the reactor was the most important parameter which controlled the selectivity but with fused catalysts at higher temperatures the CO_2 partial pressure appeared to dictate the product selectivity. The influence of the total pressure was ascribed to the partial pressure of CO_2 which increased as the total pressure increased.

Studies of the hydrogenation of CO_2 with cobalt, nickel and iron catalysts showed that methane was the major product formed [100, 101, 102]. At the same temperatures and pressures CO hydrogenation gave large yields of liquid products. In a more recent study with rhodium as catalyst it was also found that CO_2 hydrogenation produced only CH_4 [104]. Küster [103] reported iron catalysts to be ineffective for CO_2 hydrogenation but on promotion with small amounts of cobalt and copper, some liquid hydrocarbon was produced. Promotion of cobalt and iron catalysts with K_2O resulted in an increase in the production of the higher molecular mass hydrocarbons [102, 103].

From the forgoing brief review it appears that the F-T product selectivity over the different catalysts is not always simply related to a single factor such as the H_2/CO ratio within the catalyst bed, but that the actual partial pressures of CO, H_2, CO_2 and H_2O or some more complex ratio of these partial pressures may be involved. Although the true mechanism of the F-T synthesis is not absolutely clear (cf. section 9), it seems likely that the greater the proportion of the catalyst surface sites covered by hydrogen the greater will be the probability of chain termination. Conversely, the greater the proportion covered by the carbon containing entities, the greater will be the probability of chain growth. Thus the ratio between the chemisorbed hydrogen and the chemisorbed carbon entities should determine the carbon-number distribution of the products. If it is assumed that the relative amounts of hydrogen and carbon entities adsorbed is proportional to the partial pressures of H_2 and of $CO + CO_2$ respectively then the factor which controls the selectivity could have the general form $ap^x_{H_2}/(bp^y_{CO} + cp^z_{CO_2})$. For the

Sasol fixed bed iron catalyst the selectivity was observed to correlate with the p_{H_2}/p_{CO} ratio while in the fluidized iron catalyst cases the selectivity correlated with $1/p_{CO_2}$. [79] Both these ratios are in fact different cases of the general form $ap^x_{H_2}/(bp^y_{CO} + cp^z_{CO_2})$. One of the factors which might influence the relative values of the constants a, b, c, x, y and z is the temperature of the catalyst surface. Note that there is about 100 K difference in the operating temperature of the fixed and fluidized iron catalysts. The forgoing rationalization attempts are speculative and further refinements are clearly called for. Tables 26 to 30 and Figures 25 and 26 illustrate a cross section of the results obtained at Sasol. The precipitated iron catalyst was used in a fixed bed reactor at 498 K. The three promoted fused iron catalysts operated in a fluidized bed reactor at 593 to 608 K. Irrespective of the feed gas composition, pressure or recycle ratio, the total feed gas linear velocity at the reactor entrance was always held constant for both the fluidized and the fixed bed reactors. Note that due to the interrelationship between the different carbon-number products (cf. section 8.B) it is only necessary to give the selectivity of a single product, e.g. CH_4 or hard wax, in order to define (in principle) the whole spectrum.

From the data it can be seen that the factors which influence the product selectivity are different for the fixed and fluidized beds. In the case of the fixed bed increasing the pressure has no apparent effect on the carbon-number selectivity (see Table 26 sets A and B). Increasing the CO_2 partial pressure also seems to have no effect (Table 26 set C). However, there is a strong relation between the H_2/CO ratio and the selectivity (set D). As the H_2/CO ratio decreases the hard wax selectivity increases, i.e. the whole spectrum shifts to higher molecular mass products. For the fused catalysts, however, the CH_4 selectivity is clearly influenced by the total pressure. As the pressure increases the CH_4 selectivity decreases. (See Table 27, sets A and B). In each of these two sets of results the H_2/CO ratio increases slightly, but in spite of this the CH_4 selectivity decreases. Hence the H_2/CO ratio does not appear to play a major role in the control of selectivity. Set C in Table 27 illustrates that the partial pressure of CO_2 appears to play an important role. As the CO_2 pressure increases the methane selectivity decreases. Again this occurs in spite of an increasing H_2/CO ratio. The influence of the total pressure could hence in reality be due to the influence of the CO_2 pressure. Figures 25a, b and c show the relation between the CH_4 selectivity and the reciprocal of the CO_2 pressure for three catalysts. The relationship is fairly satisfactory for the low and medium basicity cases, but not for the high basicity catalyst. A better correlating factor appears to be $p^{0.5}_{H_2}/(p_{CO} + p_{CO_2})$. This is depicted in Figures 26a, b and c. The points are still somewhat scattered, but the overall trend is clearer. The factor $p^{0.5}_{H_2}/(p_{CO} + p_{CO_2})$ is of course simply another case of the general factor $ap^x_{H_2}/(bp^y_{CO} + cp^z_{CO_2})$.

In the experiments depicted in Table 27d different amounts of water vapour were deliberately added to the total feed gas. In spite of the big changes in the value of p_{H_2O} the CH_4 selectivity hardly altered. From this it is deduced

Table 26. Reaction over fixed bed precipitated iron catalyst[c], 498 K

	Reactor entry pressure/0.1 MPa				H_2/CO ratio	Selectivity/% C atom			Diesel olefinicity/%	Operating mode
	Total	p_{CO_2}	p_{H_2}	p_{CO}		Hard wax	NAC[a]	acids[b]		
A	7.8	0.08	4.34	2.28	1.9	33	1	0.006	39	Near differential (high fresh feed flow, no recycle, low % conversion)
	14.7	0.15	8.19	4.31	1.9	31	—	0.012	37	
	18.1	0.18	10.08	5.30	1.9	34	1.3	0.014	40	
	21.5	0.21	11.97	6.30	1.9	32	1.2	0.018	35	
B	11.4	0.66	5.91	2.35	2.5	18	3.1	0.21	15	Recycle/fresh feed ratio 1.26; H_2 + CO conversion 60% (on fresh feed basis)
	27.6	1.39	14.07	5.87	2.4	18	3.8	0.55	17	
C	27.7	1.24	14.3	6.40	2.2	22	4.0	0.55	22	Recycle/fresh feed ratio 1.1; H_2 + CO conversion 50%
	27.5	4.05	12.8	6.10	2.1	23	4.0	0.62	25	
D	28	3.78	16.6	2.79	7.3	13	3.4	0.1	7	Recycle/fresh feed ratio about 2 high conversion levels
	28	1.52	13.3	3.63	3.7	22	2.6	0.2	14	
	28	2.00	12.1	5.58	2.2	32	2.1	0.3	27	
	28	1.38	9.96	5.37	1.9	37	2.1	0.3	31	
	28	1.52	10.8	10.8	1.0	48	9.5	0.2	52	

[a] Water soluble non-acidic products
[b] Water soluble acidic products
[c] The same catalyst was used in all cases. In series A and B the fresh feed gas composition was the same throughout. The linear gas velocities were the same in all cases

Table 27. Reaction over fluidized bed promoted fused iron catalyst[e], 593 K

[a]	Reactor entry pressure/0.1 MPa					H_2/CO ratio	Selectivity/% C atom			$\frac{C_2H_4}{C_2H_6}$ ratio	Activity[b]
	Total	p_{CO_2}	p_{CO}	p_{H_2}	p_{H_2O}		CH_4	NAC[c]	acids[d]		
A	9.4	1.1	0.90	3.86	0.04	4.3	26	1.4	0.20	0.24	48
	12.9	1.45	1.17	5.17	0.04	5.2	19	1.7	0.29	0.26	68
	14.5	1.52	1.31	5.66	0.04	5.3	17	1.9	0.34	0.25	80
	18.0	1.86	1.58	7.31	0.04	5.8	15	2.3	0.47	0.45	94
	21.5	2.00	1.93	9.66	0.04	5.9	11	3.5	0.81	0.67	120
B	10.8	0.85	1.02	4.27	0.04	4.2	18.3	2.0	0.2	0.5	54
	20.8	1.50	1.95	8.53	0.04	4.4	13.8	3.5	0.6	1.2	106
	40.8	2.82	3.58	14.22	0.04	4.0	10.5	4.1	1.1	1.2	207
	60.8	3.79	5.26	24.6	0.04	4.7	6.8	6.8	1.7	1.3	315
	75.8	4.55	6.70	30.7	0.04	4.6	5.5	7.1	1.9	0.9	375
C		1.03	1.65	7.24	0.04	4.4	19	1.5	0.2	0.28	98
	18.0	2.20	1.31	7.45	0.04	5.7	15	2.3	0.47	0.45	96
		3.10	1.31	7.17	0.04	5.5	9	2.6	0.63	0.34	98
		5.10	1.03	8.07	0.04	7.8	8	3.3	1.3	1.0	97
D		0.67	1.79	9.84	0.04	5.5	12	3.5	0.6	1.3	107
		0.92	1.80	9.62	0.72	5.3	12	3.2	0.7	1.6	105
	20.8	1.20	1.83	10.0	1.40	5.5	12	3.2	1.0	2.0	98
		1.57	1.84	10.3	2.79	5.6	11	5.9	3.2	2.0	83

[a] Sets A and C are for medium basicity and B and D for high basicity catalysts
[b] Moles $CO + CO_2$ convented per hour
[c] Water soluble non-acidic products
[d] Water soluble acidic products
[e] The linear gas velocities were the same in all cases

Table 28. Reaction over fluidized bed low basicity fused iron catalyst, 593 K

CH_4[a]) Selectivity	Partial pressure at entrance/0.1 MPa			$\frac{p_{H_2}^{0.5}}{p_{CO}+p_{CO_2}}$	$\frac{p_{H_2}}{p_{CO}}$	$\frac{1}{p_{CO_2}}$
	CO_2	CO	H_2			
28	3.30	0.56	3.14	0.46	5.6	0.30
36	2.61	2.16	7.08	0.56	3.3	0.38
40	0.99	1.88	10.12	1.11	5.4	1.01
41	1.80	3.35	18.10	0.83	5.4	0.56
42	1.03	2.42	10.47	0.94	4.3	0.97
47	0.68	1.47	8.19	1.33	5.6	1.47
48	0.81	0.82	7.1	1.63	8.6	1.23
49	0.73	1.62	10.9	1.40	6.7	1.37
55	0.57	1.16	9.84	1.81	8.5	1.75
56	0.39	0.80	8.7	2.48	10.9	2.56
57	0.60	1.52	10.84	1.55	7.1	1.67
57	1.31	0.29	5.10	1.41	17.6	0.76
71	0.07	0.82	12.3	3.94	15.0	14.3
73	0.02	0.68	12.8	5.11	18.8	50

[a] percent methane, C atom basis

that p_{H_2O} need not be taken into consideration in so far as the control of the carbon number distribution is concerned. As H_2O chemisorbs more strongly than H_2, CO or CO_2 it will lower the number of available reaction sites and so lower the activity (cf. Table 32) but it will not necessarily change the ratio of chemisorbed H_2 to chemisorbed CO + CO_2 and thus the selectivity need not be influenced by the partial pressure of water.

From Figure 26 it is clear that as the basicity of the iron catalyst increases the slopes of the curves decrease. Thus when the alkali promoter level is high the gas composition has less influence on the product spectrum. It can be seen that by suitable manipulation of the gas composition and the catalyst basicity the CH_4 selectivity in a fluidized iron catalyst bed can be altered at will between 80 and 5 percent (C atom basis). It has been mentioned that other investigators found that the hydrogenation of CO_2 with various catalysts resulted in the formation of mainly methane. Amongst the results presented in Table 30 there are several cases (marked) in which the fresh feed consisted of CO_2 and H_2 only. As can be seen methane is certainly not the main product when CO_2 is hydrogenated over K_2O promoted iron catalysts at high temperature.

It is realized that correlating the overall selectivity obtained in an integral reactor with a factor related to the composition of the gas at the reactor entrance may be an oversimplification. Thus, for example, as the synthesis gas enters a catalyst particle and migrates towards the centre it reacts on route and so the concentration of reactants and products changes within the particle. The same overall picture is relevant over the length of the catalyst bed. Under normal F-T synthesis conditions the catalyst pores are either

Table 29. Reaction over fluidized bed medium basicity fused iron catalyst, 593 K

CH_4[b]) Selectivity	Partial pressure[a]) at entrance/0.1 MPa			Partial pressure at exit/0.1 MPa				pressure ratios at entrance		
	H_2	CO	CO_2	H_2	CO	CO_2	H_2O	$\frac{p_{H_2}^{0.5}}{p_{CO} + p_{CO_2}}$	$\frac{p_{H_2}}{p_{CO}}$	$\frac{1}{p_{CO_2}}$
8	7.52	1.72	4.28	5.0	0.2	5.3	2.4	0.46	4.4	0.23
9	6.97	1.66	2.76	4.2	0.1	3.4	2.1	0.60	4.2	0.36
10	7.38	1.59	2.34	4.3	0.1	2.5	2.1	0.69	4.6	0.43
11	9.66	1.93	2.0	7.0	0.2	2.3	2.3	0.79	5.0	0.50
12	7.24	2.48	1.86	4.1	0.1	2.6	1.8	0.62	2.9	0.54
13	6.90	2.76	2.83	3.8	0.1	4.2	1.8	0.47	2.5	0.35
15	7.31	1.58	1.86	4.6	0.2	2.3	2.0	0.79	4.6	0.54
17	5.66	1.31	1.52	3.9	0.1	2.1	1.8	0.84	4.3	0.66
17	6.90	1.72	0.76	4.6	0.1	1.2	1.7	1.06	4.0	1.32
17	6.07	2.00	1.59	3.1	0.1	2.1	1.7	0.69	3.0	0.63
18	6.00	1.79	1.24	3.3	0.1	1.8	1.4	0.81	3.4	0.81
19	7.10	1.72	0.90	5.0	0.1	1.4	1.7	1.02	4.1	1.11
19	5.17	1.17	1.45	3.1	0.1	1.8	1.4	0.87	4.4	0.69
22	9.93	1.66	0.69	8.4	0.1	0.9	2.1	1.34	6.0	1.45
22	7.38	1.66	0.62	5.4	0.1	1.0	1.7	1.19	4.4	1.61
24	9.66	1.38	0.34	7.7	0.1	0.5	1.5	1.81	7.0	2.94
27	10.14	1.31	0.28	8.3	0.1	0.4	1.5	2.00	7.7	3.57

[a] After the water gas shift has equilibriated
[b] percent methane, C atom basis

completely filled with liquid wax or the surface is covered with a film of wax. The solubility and the rates of diffusion of the different reactants and products in the liquid wax will be different. Hence the actual concentrations at the catalyst sites could be markedly different from those in the gas phase. In spite of all these complications there nevertheless should be a correlation between the partial pressures in the gas phase and the selectivity of the process.

With iron catalysts it is generally found that when the selectivity shifts away from the light products the selectivity of the oxygenated products increases. Simultaneously the olefinity of the hydrocarbons increases (see Tables 26 and 27). For both fluidized and fixed bed iron catalysts, the selectivity of the oxygenates increases when the total pressure is increased (see Tables 26 and 27). In Table 27d it can be seen that as the water vapour pressure increases, the acid selectivity and the olefinity of the C_2 cut increase. This effect might be due to the lowered conversion level (cf. Table 32) as well as to the increased water vapour pressure. It has been frequently demonstrated [2] that as the space velocity of the fresh feed gas is increased (resulting in a lower conversion level), the oxygenate content and the olefinity of the products increase. Since it has been found that the alcohols and acids appear to be in equilibrium with one another (cf. Table 20), the relative amounts of alcohols depend

Table 30. Reaction over fluidized bed high basicity fused iron catalyst, 608 K

CH_4 Selectivity[b]	Partial pressure at entrance/0.1 MPa[a]				Partial pressure at exit/0.1 MPa				Pressure ratios at entrance		
	CO_2	CO	H_2	H_2O	CO_2	CO	H_2	H_2O	$\frac{p_{H_2}^{0.5}}{p_{CO} + p_{CO_2}}$	$\frac{p_{H_2}}{p_{CO}}$	$\frac{1}{p_{CO_2}}$
22.1	0.06	1.45	11.55	0.02	0.21	0.05	10.53	1.51	2.25	8.0	16.7
20.8	0.49	0.98	4.90	0.08	0.69	0.11	3.84	0.87	1.51	5.0	2.04
18.3	0.85	1.02	4.27	0.15	1.11	0.12	3.04	0.98	1.10	4.2	1.18
17.7	0.23	1.64	11.59	0.06	0.29	0.05	9.65	1.82	1.82	7.1	4.3
15.0[c]	2.07	1.04	13.22	0.94	1.91	0.18	11.42	2.95	1.17	12.7	0.48
14.5	0.51	1.84	9.31	0.10	0.72	0.08	6.81	1.97	1.30	5.1	1.96
13.8	1.50	1.95	8.53	0.24	1.89	0.18	6.00	2.16	0.85	4.4	0.67
13.7	0.56	1.90	9.64	0.10	0.79	0.14	7.32	1.94	1.26	5.1	1.78
13.6	4.88	3.38	28.9	1.29	4.86	0.48	24.0	6.25	0.65	8.6	0.20
13.5[c]	2.34	1.12	13.45	1.00	2.30	0.20	11.9	2.85	1.06	12.0	0.43
13.4	0.10	2.96	11.75	0.01	1.08	0.24	10.14	2.08	1.12	4.0	10.0
11.7	0.72	1.92	9.74	0.10	0.98	0.16	7.39	2.07	1.18	5.1	1.39
11.4	2.77	0.64	11.48	1.59	2.27	0.15	9.07	4.07	0.99	17.9	0.36
10.9	1.17	3.66	17.66	0.21	1.63	0.19	13.0	3.92	0.87	4.8	0.85
10.8	1.62	2.41	11.29	0.25	1.92	0.29	8.03	2.82	0.83	4.7	0.62
10.5	2.82	3.58	14.22	0.35	3.30	0.40	8.60	4.40	0.59	4.0	0.35
9.8	1.24	2.11	6.98	0.14	1.76	0.14	4.22	1.98	0.79	3.3	0.81
8.0	1.65	2.31	6.76	0.20	2.42	0.17	4.09	1.87	0.66	2.9	0.61
7.1[c]	6.72	1.71	9.14	1.34	7.58	0.53	7.62	2.71	0.36	5.3	0.15
6.9	1.53	2.24	6.90	0.16	2.19	0.16	4.11	2.08	0.70	3.1	0.65
6.8	3.79	5.26	24.6	0.61	4.50	0.31	16.8	6.72	0.55	4.7	0.26
6.5[c])	9.03	1.83	7.66	1.24	9.88	0.83	6.72	2.05	0.25	4.2	0.11
6.0	17.74	5.03	17.77	2.09	19.96	2.06	13.45	5.48	0.19	3.5	0.06
5.5	4.55	6.70	30.73	0.71	5.35	0.56	21.01	8.31	0.49	4.6	0.22

[a] After the water gas shift has equilibriated
[b] percent methane, C atom basis
[c] Fresh feed gas consisted only of CO_2 and H_2. The CO results from the water gas shift reaction and is introduced into the reactor with the recycle gas.

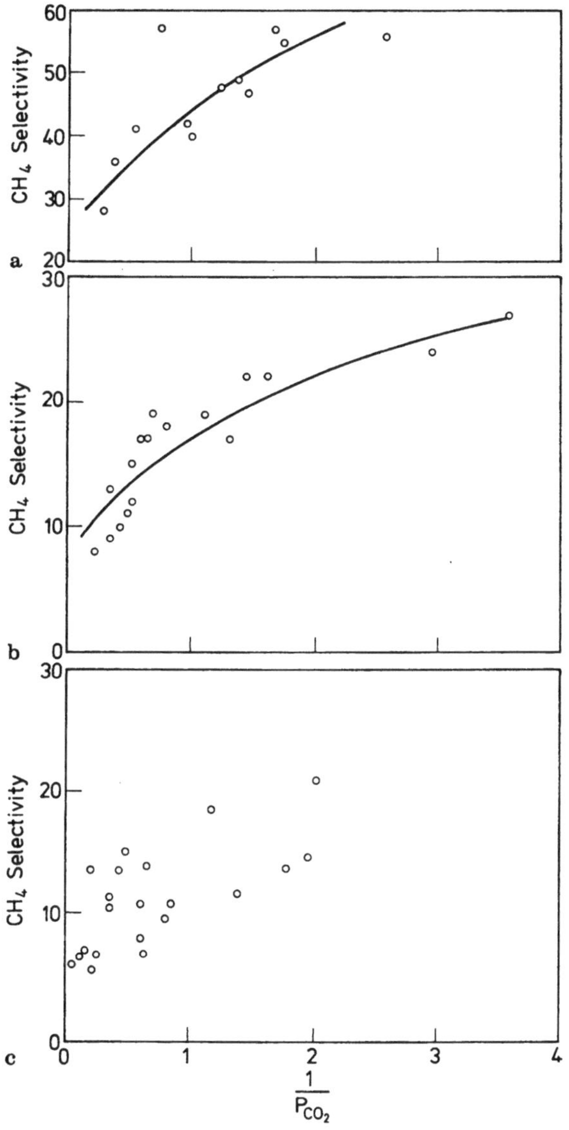

Figure 25. Methane selectivity as a function of $1/p_{CO_2}$ at the reactor entrance of fluidized iron catalysts. Figure 25a is for a low basicity catalyst (cf. Table 28); Figure 25b is for a medium basicity catalyst (Table 29) and 25c is for a high basicity catalyst (Table 30). The unit of pressure is 0.1 MPa (bar)

on the H_2O and H_2 partial pressures. In fluidized iron catalyst beds it has been observed that as the H_2/H_2O ratio of the exhaust gas increases the acid selectivity decreases and the alcohol selectivity increases. When the overall system is strongly hydrogenating both the acid and the alcohol selectivity will naturally decrease.

5. Secondary Reactions

The probability that a product, e.g. wax, is formed in one part of the catalyst bed and then subsequently cracked in another part has been investigated

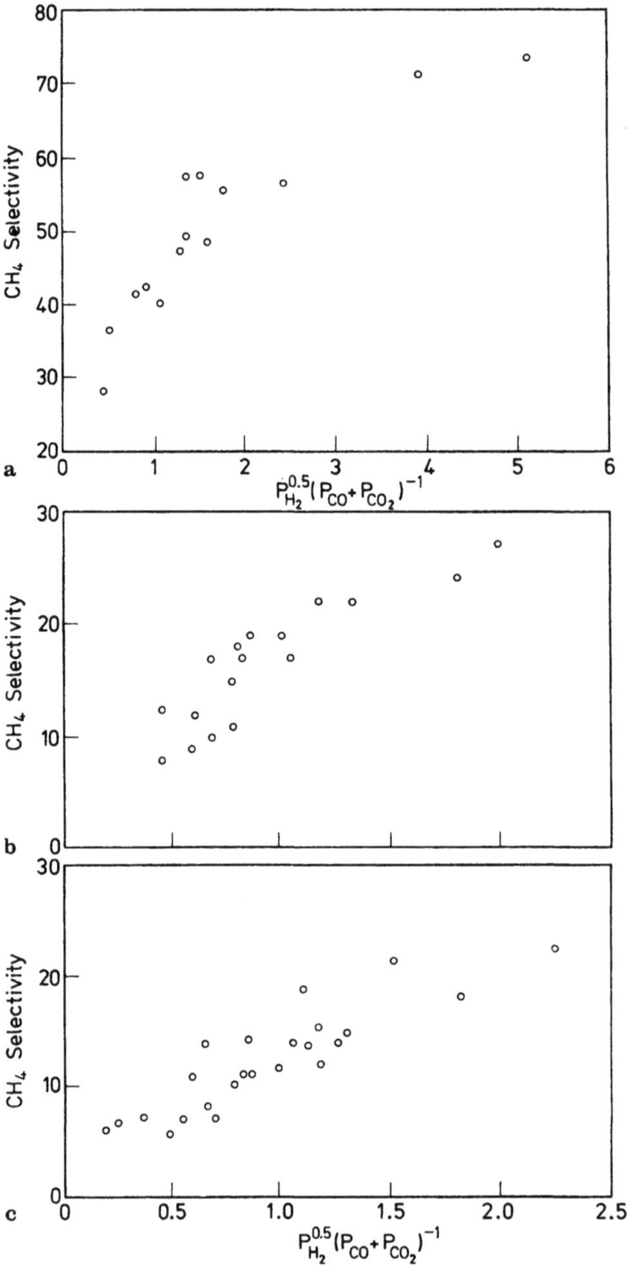

Figure 26. Methane selectivity as a function of $p_{H_2}^{0.5}(p_{CO} + p_{CO_2})$ at the reactor entrance of fluidized iron catalysts. Figures 26a, b and c are for low, medium and high basicity catalysts respectively. See Tables 28, 29 and 30. The unit of pressure is 0.1 MPa (bar)

by various workers [2]. Japanese investigators found that with Ni—Co—SiO_2 catalysts hydrocracking of the hydrocarbon products occurred at 458 K, but only when the conversion levels were high. Taylor [106] found that ethane cracking was slow over cobalt catalysts under normal F-T synthesis conditions. Craxford [107] studied the hydrocracking of waxes over cobalt under different conditions and concluded that the rate of hydrocracking was slow when carbon monoxide was present. This is in keeping with the finding that hydrocracking was only observed at high conversion levels [108]. In a more recent study with ruthenium catalyst Dalla Betta [109] also concluded that chemisorbed CO inhibited the hydrocracking of products. Kölbel [110] found that for Co and Ni catalysts operating below 473 K hydrocracking accounted for less than 5 percent of the CH_4 produced while for an Fe catalyst the figure was below 3 percent at temperatures up to 573 K. The US Bureau of Mines reported that with iron catalysts the oil was cracked above 573 K [94]. Studies at Sasol were carried out in which the light oil product formed in a fluidized iron catalyst bed at 608 K was recycled back to the reactor over a prolonged period in a 2/1 ratio (i.e. 2 volumes of oil recycled per volume of oil produced). No net loss of oil was observed. The boiling point distribution of the oil, its olefinity and aromatic content were normal. Thus there was no indication that the oil had been chemically altered even under conditions of high synthesis gas conversion (about 93 percent). Over the same iron catalyst hexane was completely cracked to CH_4 at 598 K in a pure hydrogen atmosphere.

Whereas longer chained olefins are not hydrogenated under F-T synthesis conditions over iron catalysts at about 608 K, ethylene is extensively hydrogenated when it is added to the reactor feed gas. This is in keeping with the known fact that rate of olefin hydrogenation decreases rapidly with increasing chain length [5]. Note that under normal F-T synthesis conditions over fluidized iron catalysts the olefin/paraffin ratio of the C_2 and C_3 hydrocarbons are typically 1 and 7 respectively. Schulz [132] added ^{14}C-labelled C_2, C_3, C_4 and C_{16} olefins to the synthesis gas over both iron and cobalt catalysts and found that the major portion of the olefins were hydrogenated to the paraffins. Some incorporation of the olefins with growing chains also occurred.

Under normal F-T conditions with fluidized iron catalysts various low molecular mass alcohols, acids and ketones have been added to the feed gas. In all cases sizable fractions of the oxygenates are destroyed, being either hydrogenated or incorporated in the end products. Hence the extent to which olefins and oxygenated molecules emerge from the reactor must be dependent on the availability of vacant catalyst sites. This in turn is dependent on the degree of CO conversion.

6. *Maximization of Selective Products*

As discussed in section 8.B there are strong indications that the F-T reaction involves a stepwise growth mechanism and that the product distribution is dependent on the probability of chain growth. In practice a wide distribution of products is obtained. If the selectivity of a particular narrow range of products were to be markedly improved this would require two sharp changes in the probability of chain growth, one on either side of the desired carbon number range. Thus, for example, if predominantly C_3 to C_5 hydrocarbons were required the probability of chain growth of the C_1 and C_2 surface entities would have to be high while that of C_5 and higher entities would have to be low. This obviously would require a highly tailored catalyst. Alternatively a completely different breed of catalyst with a quite different mechanism of chain growth would be required. This represents a very challenging field of research for the future.

Large deviations from the calculated distribution of products have been reported for Ru on Al_2O_3 catalysts. Everson [45] found that at 1.2 MPa, 510 K and a 3 H_2/1 CO ratio gas the C_5 to C_{12} mass selectivity was about 60 percent. Madon [43, 129] reported a C_5 to C_{25} mass selectivity up to about 99 percent at 3.1 MPa, 514 K and a 2 H_2/1 CO ratio gas. Calculations (cf. section 8.B) predict a maximum of about 50 percent for C_5 to C_{12} and about 74 percent for C_5 to C_{25}. Further investigations of Ru—Al_2O_3 catalysts should be pursued, preferably on a larger scale and at higher space velocities.

Several remarkably high selectivities for low molecular mass olefins have been claimed in patents. Japanese workers reported a C_2H_4/CH_4 ratio of 15/1 over copper impregnated Al_2O_3 catalyst. [111] Using a co-precipitated Mn—Fe catalyst (Mn/Fe ratio 9) Kölbel [112] claimed a C_2, C_3 and C_4 selectivity of about 38, 15 and 26 wt. percent respectively. The product was higly olefinic and only traces of CH_4 were formed. The experiments were carried out with a feed gas having a H_2/CO ratio of about 0.7, at 553 K and 1.08 MPa and a conversion of 76 percent was achieved. Using complex mixtures of metals and oxides as catalysts (e.g. 40 ZnO, 12 Cr_2O_3, 3 Cu, 2 MnO, 0.15 Co and 0.02 K_2O on 100 Al_2O_3) Peters [113] claimed to obtain high yields of C_2H_4. The yields of CH_4 produced are not mentioned. Workers at Ruhrchemie [114, 115, 116] claimed to have obtained high yields of C_2 to C_4 olefins on variously promoted iron catalysts. For example, with a 100 Fe:25 Ti:10 ZnO:4 K_2O homogenized oxide mixture as catalyst the following selectivity was achieved; 10.1 CH_4, 33.4 C_2H_4, 21.3 C_3H_6 and 19.9 C_4H_8 wt. percent. It is difficult to accurately assess the commercial viability of any of the above processes without more information, e.g. about the lengths of time during which the claimed high selectivities are achieved.

The influence of cobalt, potassium and tungsten carbide on the olefinity of the products produced with a precipitated iron catalyst has been investigated [117]. Although an improvement in the olefinity of the C_2 cut was achieved the overall carbon number distribution remained typical of iron catalysts at high temperatures. Bartish [118] found that when a Fe—Cu catalyst was supported on magnesium mordenite the selectivity of CH_4 was halved. However, the accompanying changes in the C_2 to C_4 selectivities were not markedly different from the values which would have been expected from the shift in CH_4 selectivity. In a Gulf patent [133] it is claimed that the addition of a small amount of Ru to Co produced a catalyst which showed a marked decrease in CH_4 production. However, this was accompanied by an increase in the average chain length of the hydrocarbon product and so the results are not abnormal.

Selective poisoning of F-T catalysts has been utilized by many investigators in order to alter the selectivity. This subject has been reviewed by Madon and Shaw [65]. It was found that sulfur-poisoned cobalt catalysts resulted in an increase in the olefinity of the products and in the oil selectivity [119, 120]. The same effect was observed for a fluidized fused iron catalyst [121]. In all cases, however, the poisoning was accompanied by a loss of activity. Dalla Betta [122] also reported that H_2S inhibited the formation of CH_4 relative to the other products over Ni and Ru catalysts. Davis and Wilson [70] found that the halogens Cl and Br lowered the CH_4 selectivity of fluidized iron catalysts and increased the olefinity of the C_2 to C_{11} products. Again, however, the catalyst's activity was considerably decreased.

Another method of altering the overall selectivity is to recycle selected product cuts so that they can be incorporated in further chain growth. Thermodynamically C_2H_4 should have a greater tendency to incorporate than higher carbon-number olefins [2]. Smith's [123] results with synthesis gas plus C_2H_4 over a cobalt catalyst indicated that C_2H_4 was readily incorporated into the F-T products. The increased oil produced, however, contained a high concentration of oxygenates and this indicates that the system had become markedly less hydrogenating. This effect would also have contributed to the increase in the oil selectivity. It was demonstrated that propene and butene could also be incorporated into the F-T products on a cobalt catalyst [124]. In a more recent study Dwyer and Somorjai [131] showed that over iron single crystal surfaces the addition of small amounts of C_2H_4 or C_3H_6 to the synthesis gas resulted in a shift in the selectivity to the heavier products. Kölbel [125] deduced that the 543 to 603 K oil cut was capable of being incorporated in a slurry bed process with an iron catalyst at 513 K. At Sasol an olefin rich light oil cut (in the gasoline range) was recycled to both a fluidized bed iron catalyst operating at a 608 K and to a slurry bed iron catalyst at about 508 K. No evidence of incorporation of the gasoline was found. The contact times of the recycled oils, however, were not as long as those in the case of Kölbel's experiments.

Research geared to a marked improvement in the yields of narrow preselected product cuts in the F-T synthesis remains a worthwhile objective. However, at present compromises must be made. Alternatively secondary product work-up is needed to increase the yield of the desired cut. The exceptions are of course the one-carbon compounds such as CH_4 or CH_3OH which can be produced in very high yields by operating at conditions where the probability of chain growth is very low. In order to obtain high yields of C_2 to C_4 hydrocarbons it will be necessary to accept high yields of CH_4 also. From the product interrelation curves (e.g. Figure 20), it can be deduced that a C_2 to C_4 selectivity of about 40 percent can be achieved but that the CH_4 selectivity will then range from 20 to 40 percent while the oil (C_{5+}) will range from about 38 to 16 percent. The above type of distribution will of course be acceptable in areas where both the CH_4 and the light oil can be marketed. To maximise gasoline a fluidized iron catalyst system is recommended which gives a product spectrum as shown in Table 2. If the +473 K material is catalytically cracked, all C_3 and C_4 olefins are oligomerized and the low molecular mass alcohols are converted to gasoline then an overall gasoline selectivity of about 75 percent can be achieved. The highest yield of diesel can be achieved by using the fixed bed iron catalyst under conditions which yield a product spectrum similar to that given in Table 2. By hydrocracking the waxes and oligomerizing the C_3 to C_6 olefins an overall yield of about 18 percent C_1 to C_4 gas, 20 percent gasoline and 58 percent diesel can be obtained. (Wax cracking yields about 70 percent diesel while olefin oligomerazion can yield 75 percent diesel).

Since in practice both diesel and gasoline is needed in high yields a more versatile process would be one which employs both the fixed and the fluidized bed reactors. Another big advantage of such a combination is that the quality (cetane number) of the total diesel pool would be easier to maintain. This is due to the fact that wax hydrocracking produces a very high quality diesel while olefin oligomerization produces a poorer quality diesel. (Cetane number about 70 and 35 respectively.)

The Mobil process which first converts synthesis gas to methanol and then converts the methanol to gasoline represents an alternative approach to the maximisation of the production of gasoline [126]. The ZSM-5 catalyst yields about 76 percent C_5 gasoline of high quality. The balance is C_1 to C_4 hydrocarbons. With alkylation the total gasoline can be about 88 percent. No diesel is made however. Mobil has also investigated the direct conversion of synthesis gas to gasoline with "dual" function catalysts. Either the ZSM-5 type zeolite was loaded with a metal active for the hydrocarbon synthesis or a physical mixture of the zeolite and a F-T catalyst was employed [127, 128]. However, the gasoline selectivities were much lower and the gas yields much higher than in the methanol route process. Due to the role of the zeolite the diesel production was negligible in all cases. More recently Mobil [130] found that with a 1 to 4 mixture of promoted fused iron and ZSM-5 catalyst a gasoline selectivity of 60 percent was achieved. This can be compared with the 75 percent which can be achieved by F-T synthesis followed by olefin oligomerization and wax cracking. The advantage of the Mobil approach is that these secondary reactions are effectively carried out in situ by the zeolite. Also the Mobil straight run gasoline, having a higher octane value, needs less downstream processing. A few disadvantages are as follows: (i) the lack of versatility (e.g. no diesel is produced); (ii) coke build up on the zeolite necessitates periodic regeneration; (iii) if the metal component is oxidised during the regeneration (as will happen with iron) the catalyst will have to be re-reduced each time.

9. Mechanism

Due to the wide diversity of the products in the F-T synthesis it is not surprising that several different mechanisms have been proposed [2, 6]. Each scheme accounts satisfactorily for the product spectrum, but the evidence used to support the individual theories is usually indirect in nature and can sometimes be interpreted in more than one way. There are three main categories. In the first it is assumed that upon chemisorption the carbon monoxide molecule dissociates completely into carbon and oxygen atoms. The carbon atoms are individually hydrogenated to CH_x entities and these then link up to form the products (e.g. see Figure 27). In the second category it is assumed

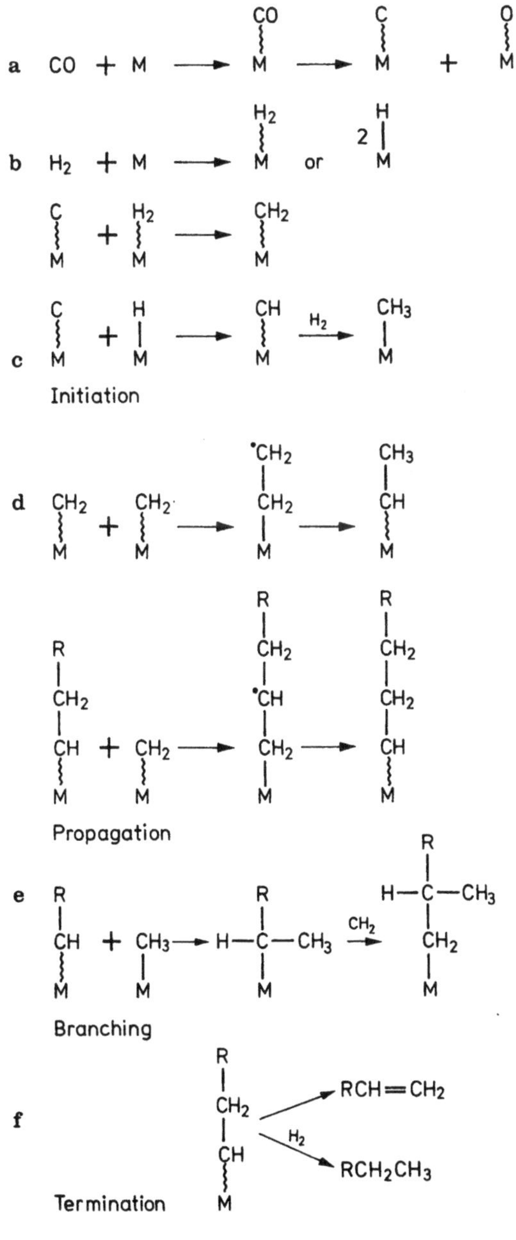

Figure 27. Reaction scheme A. The CO dissociates completely on adsorption. The symbol } denotes a bond of indefinite order

that the chemisorbed CO does not dissociate but is directly hydrogenated to an oxygen containing species. Chain growth occurs by H_2O elimination between neighbouring species (e.g. Figure 28). The key feature of the third category is the CO insertion reaction. Here the CO molecule also does not dissociate and it inserts into either M-H or M-C bonds, where M represents a catalyst surface site (e.g. Figure 29). The very first theory (Fischer [134]) belonged to the CO dissociation category. This was later discarded in favour of the non-dissociation concepts. However, in recent years new and strong evidence in favour of the original dissociation theory has been produced.

Fischer [134, 135] postulated that CO carbided the catalyst surface and that the carbide was then hydrogenated to methylene groups which polymerized to longer — chained hydrocarbons. Craxford and Rideal [136] refined the carbide theory. From studies of the para to ortho H_2 conversion under various conditions they concluded that under normal F-T conditions only molecular hydrogen was involved in the synthesis. Atomic hydrogen was only present when large amounts of CH_4 was produced (i.e. under strongly

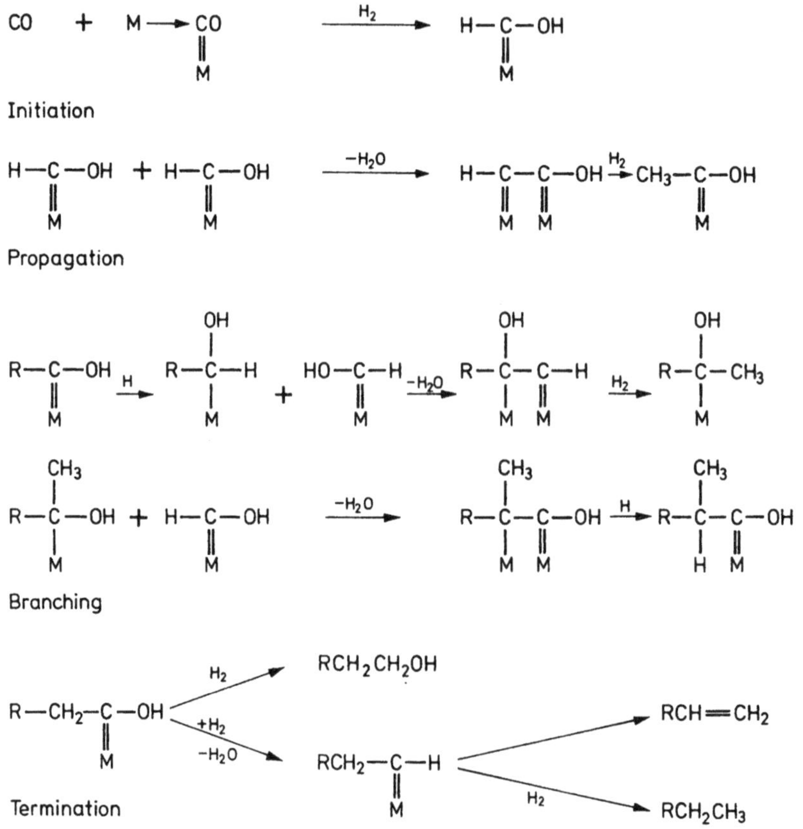

Figure 28. Reaction scheme B. The CO does not dissociate. It is hydrogenated to surface HCOH entities which link up

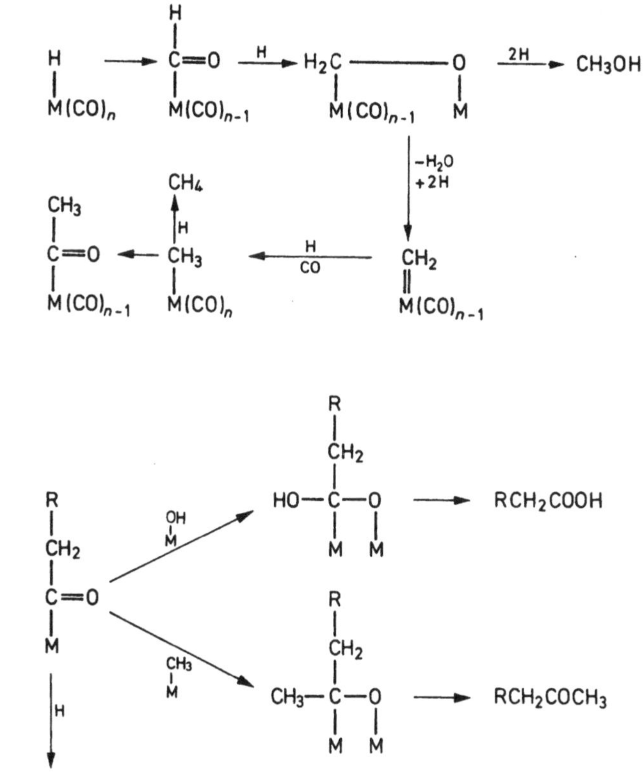

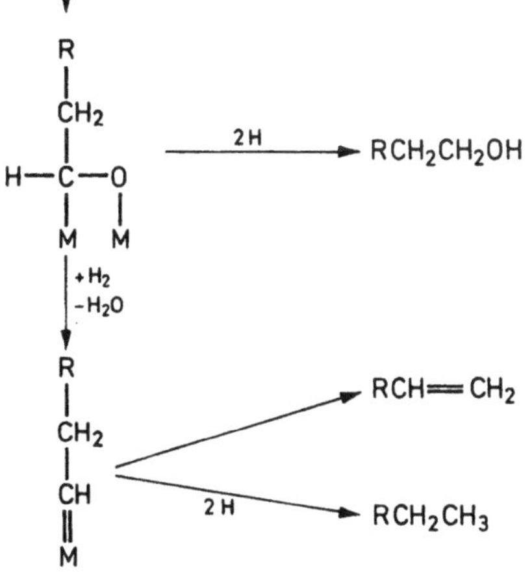

Figure 29. Reaction scheme C. The CO does not dissociate. Stable surface carbonyl compounds are formed. CO insertion occurs. For the sake of simplicity the postulated $(CO)_x$ group has been omitted from many of the surface species

hydrogenating conditions). It was postilated that adsorbed CO reacted with H_2 to form single C atoms. These were then converted to CH_2 entities which linked up to form long chains. The chains were cracked by hydrogen to yield a wide range of hydrocarbons. Anderson [2] evaluated the carbide theory in depth and concluded that it was not satisfactory. One of the key objections was that the theory does not predict the formation of oxygen containing products, e.g. alcohols. Emmett [137] calculated that only CH_4

and short chain paraffins could be produced by the hydrogenation of bulk phase carbide. (The thermodynamic properties of single surface carbon atoms may of course be markedly different from that of the bulk carbide phase). Emmett [138] investigated the carbide theory by depositing ^{14}C as a tracer on cobalt and iron catalysts. It was found that hydrocarbons did not exchange with surface carbide but that carbon monoxide did. From the analysis of the F-T products it was found that appreciable amounts of radio active CH_4 was formed but the radio activities of the longer chained products were lower. It was concluded that less than 20 percent of the F-T synthesis proceeded by carbide reduction. The results, however, do not unambiguously disprove the carbide mechanism.

A key fact to bear in mind when considering possible reaction mechanisms is that very high selectivities of oxygenated molecules can readily be obtained on normal F-T catalysts by altering the reaction conditions. With the standard $Co—ThO_2—MgO—SiO_2$ catalyst the yield of oxygenated product is about 1 percent at synthesis temperatures in the range 453 to 473 K. At lower temperatures (433 to 448 K), the alcohol content of the liquid organic product was found to be 40 percent [92]. K_2O promoted fused iron catalysts operating at about 463 K yielded an oil which contained over 30 percent alcohols, but at about 493 K only a small fraction of oxygenated molecules was obtained [2]. Bashkirov, Kagan and co-workers carried out many tests with promoted iron catalysts at high pressures [139]. In one case it was found that the alcohol content of the product boiling below 433 K was 86 percent [140]. In the light of such findings it is understandable that oxygenated surface species were postulated to exist on the catalyst. Elvins and Nash [141] were the first to suggest this. Storch, Golumbic and Anderson [1] presented a detailed reaction scheme in which HCOH entities are the basic building blocks. Figure 28 illustrates their proposal. Emmett [142, 143] and co-workers studied the incorporation of a series of radio actively labelled alcohols in the F-T reaction. They concluded that ethanol and n-propanol initiated chain growth since only one alcohol was incorporated per chain. Methanol incorporation increased linearly with the chain length. Their results also indicated that the hydroxyl group defined the point of attachment of the next carbon atom added to the growing chain. Although the interpretation of the results were not unambiguous they did appear to support the concept that oxygenated surface complexes were involved in the chain growth mechanism.

Infra-red techniques are commonly employed to obtain information concerning the bonds that exists on catalyst surfaces. Kölbel [144] reported that under F-T reaction conditions IR spectra indicated the presence of carboxyl, carbonate and carbonyl surface species. Blyholder has studied the adsorbed species on various metallic F-T catalysts for many years. He postulated that alkoxide surface structures were the active intermediates in F-T synthesis [145]. When CO and C_2H_4 were co-adsorbed on iron evidence was found for the insertion of CO into the metal alkyl bond [146]. Blyholder [147] also studied the interaction of CO, H_2 and C_2H_4 on silica-supported Ni and Co catalysts. The results indicated the presence of a carboxylate surface structure which was stable up to 473 K. Leisegang [148] inferred the existence of an intermediate complex between adsorbed CO and H_2 on a cobalt catalyst. The complex was stable up to 468 K. Dalla Betta [149] found no evidence of a HCOH species on a ruthenium catalyst. However, they did find that under synthesis conditions (up to 523 K) the surface was almost entirely covered by chemisorbed CO. The possibility of the existence of a HCOH type species at low surface coverage was, however, not excluded. Formate bands were observed on the Al_2O_3 support but not on the Ru metal.

Kölbel [169] reported that the heat of adsorption of a mixture of H_2 and CO was greater than the sum for the individual gases. This indicated the formation of an oxygenated surface complex. Kölbel [179] also studied the desorption of CO and H_2 from a reduced iron catalyst. From the fact that the desorbed gas always had a H_2/CO ratio of one it was concluded that a HCOH complex existed on the catalyst surface.

Pichler observed that the optimum F-T synthesis conditions for different catalysts was close to those at which the rate of formation of volatile carbonyls was high. He deduced that carbonyl intermediates were involved in the F-T reaction. An elegant mechanism was proposed which accounted for the formation of all the normally observed F-T products [150]. An abreviated illustration of the scheme is given in Figure 29. It seems unlikely that the multicarbonyl complex $RM(CO)_n$ can be stable at normal F-T temperatures [6]. However, the multicarbonyl need not be an essential feature, and CO insertion from a neighbouring surface site is an acceptable alternative. The CO insertion mechanism for chain growth enjoys a wide acceptance [85, 151]. The fact that it has been established that CO insertion reactions do occur in homogeneous catalysis naturally lends credence to the mechanism. An unusual but interesting version of the CO insertion concept has been proposed by Deluzarche [152]. It is suggested that CO inserts into a surface OH group to form M—O—COH which is then hydrogenated to M—O—CH_3. Further CO insertion gives M—O—CO—CH_3 and the process is repeated.

In recent years, there has been a strong resurgence of opinion supporting the original carbide theory [153 to 157]. Chemisorbed CO is assumed to first dissociate completely into C and O atoms and the carbon atoms are only then hydrogenated. Many studies indicate that at normal F-T temperatures CO adsorption on Ni, Co and Fe is dissociative [158 to 161]. Nevertheless, relatively large amounts of undissociated chemisorbed CO is present on the catalyst surface under synthesis conditions [157, 159, 162]. The presence of carbon on F-T catalysts during synthesis has of course been known for many years. Relatively large amounts are deposited on iron catalysts at temperatures above about 573 K. Its presence does not necessarily imply its direct involvement in the F-T mechanism. However, it has been demonstrated that not only CH_4, but also C_2 to C_4 hydrocarbons are produced when deposited carbon is treated with H_2 in the absence of molecular CO [157, 159]. From the results of other investigators it appears that the hydrogenation of surface carbon yields mainly CH_4 [155, 163, 164].

Sachtler et al. [156] studied the incorporation of labelled surface carbon

(^{13}C) into hydrocarbon products during the F-T reaction over Ni at 443 K. Some preliminary studies with Co and Ru were also carried out. It was convincingly demonstrated that not only was $^{13}CH_4$ formed during synthesis, but that many of the C_2 to C_4 product molecules contained more than one (^{13}C) atom. The results clearly support the concept that CO is first dissociated to C and O and the C is then hydrogenated to CH_x entities which link up to form hydrocarbons. The exchange between ^{12}CO (gas) and ^{13}CO (chemisorbed) as well as the oxygen scrambling between ^{12}CO (chemisorbed) and ^{13}C (surface) was also investigated. (In both cases H_2 was absent). They concluded that if oxygen scrambling occurred to any marked degree then appreciable quantities of ^{13}CO should be present in the effluent gas. It was found that during the F-T reaction the ^{13}CO was <3 percent. The fact, however, that some ^{13}CO was present does mean that oxygen scrambling had occurred. Also, part of the ^{13}CO formed on the surface could have participated in the F-T reaction rather than desorb into the gas phase.

Figure 27 illustrates a possible reaction scheme in which CO first dissociates before being hydrogenated. Under normal F-T conditions the products are predominantly straight chained. If it is assumed that hydrogen is present on the catalyst surface in a "molecular" form (albeit possibly highly activated), then it follows that the main partially hydrogenated species will be CH_2. The linking of CH_2 units can result only in straight chains. If on the other hand a relatively large amount of atomic hydrogen is on the surface then one could expect CH, CH_2 and CH_3 units to be present. It is conceivable that the presence of CH units could result in the formation of highly branched products. The proposed sequence of events is as follows. Carbon monoxide dissociates completely (step a). Hydrogen is chemisorbed as a molecule and only a limited amount of atomic hydrogen is formed (step b). The carbon atoms are hydrogenated to CH_2 units but some CH and CH_3 units are also present (step c). Propagation occurs by the CH_2 units linking (e.g. by insertion) with neighbouring surface units (step d). Branching can occur by the addition of CH_3 units to the chain as depicted in step (e). Since there are presumed to be few CH_3 units present the degree of branching will be low. (Note that the Anderson scheme as presented in Figure 28 also requires the presence of atomic hydrogen for chain branching). Chain termination occurs by desorption, which yields the olefin, or by hydrogenation which yields the paraffin. (step f).

As previously mentioned, the obvious weakness of the proposed reaction scheme given in Figure 27 is its failure to account for the very considerable production of oxygenated molecules over iron catalysts and also over cobalt catalyst at the lower temperatures. To accommodate this, compromises have been made. Matsumoto and Bennett [163] suggest a scheme in which methanation occurs via the CO dissociation route while chain growth occurs by CO insertion. Ponec [154] also tentatively supported such a process. Sachtler [156] suggests that the CO insertion reaction occurs as a "side reaction" on oxidised metal patches. With iron catalysts relatively high yields of methanol can be achieved by operating at low temperatures or with partially poisoned catalysts. For instance, with a sulfur poisoned catalyst operating at 603 K the CH_3OH selectivity is about 10 percent whereas it is normally below 0.1 percent. Since methanol is a one-carbon compound it cannot be formed by the insertion of CO into an alkyl radical. The presence of methanol in the products therefore supports the concept that elementary CHOH type complexes must be present on the catalyst surface and not only CH_x entities. The question arises whether the alternative schemes are really different or whether they are simply variations of the same mechanism. An attempt to present a unifying general scheme is presented in Figure 30. Carbon monoxide chemisorbs (step a) and then becomes activated (step b). In this state both the C and the O atoms are loosely associated with one another as well as with the catalyst sites. The active complex can dissociate into separate C and O atoms (step c). The C atoms are either hydrogenated to CH_2 or they migrate away to form aggregates of carbon. Alternatively the active C—O complex can be hydrogenated (step d) to form an activated CH_2O complex (A). This complex can be hydrogenated either to CH_3OH or to CH_2 and H_2O. Thus the CH_2 species can be formed via two closely related routes. Chain growth occurs by the reaction (e.g. insertion) of the activated C—O with hydrocarbon entities as depicted, followed by hydrogenation. Termination occurs by either desorption or hydrogenation of the oxygen-associated surface complex (B). The strength of the linkage between the C and O atoms in this complex determines to what extent hydrocarbons or oxygenated compounds are formed. The strength of the C—O linkage will in turn be dependent on the type of catalyst, the presence of promoters and the temperature etc. Referring to the termination steps illustrated in Figure 30 desorption of the complex B can yield aldehydes while hydrogenation yields alcohols. Chain termination by a surface oxygen atom results in acids. A speculative alternative route to acid formation is via CO_2 insertion as depicted in Figure 30. Note that the apparent presence of carboxylate surface structures on cobalt has been reported by Blyholder [147].

The generally observed increase in alcohol selectivity with increasing H_2 partial pressure (e.g. as when the total pressure is raised) is in keeping with the proposed mechanism i.e., that alcohols are formed by the hydrogenation of the surface complex B. It was mentioned in section 8.C.4 that at high temperatures over iron catalysts aldehydes, alcohols and acids appear to be in thermodynamic equilibrium. There could therefore be some doubt if all three types of oxygenates are primary products. At lower temperatures, however, these compounds are not in thermodynamic equilibrium, e.g. the molar ratio acetic acid/ethanol can be much higher than the equilibrium value. This indicates that the acids must be primary and not secondary products.

According to Pichler's reaction scheme (Figure 29) the formation of ketones could be a primary reaction. Ketones, however, are formed only at higher temperatures, and this could mean that a secondary reaction is involved. Table 20 suggests that acetone may be formed by the reaction of ethanol with surface carbon. This reaction is in fact the basis of a patent for catalyst regeneration [167]. In general, however, there is no consistent

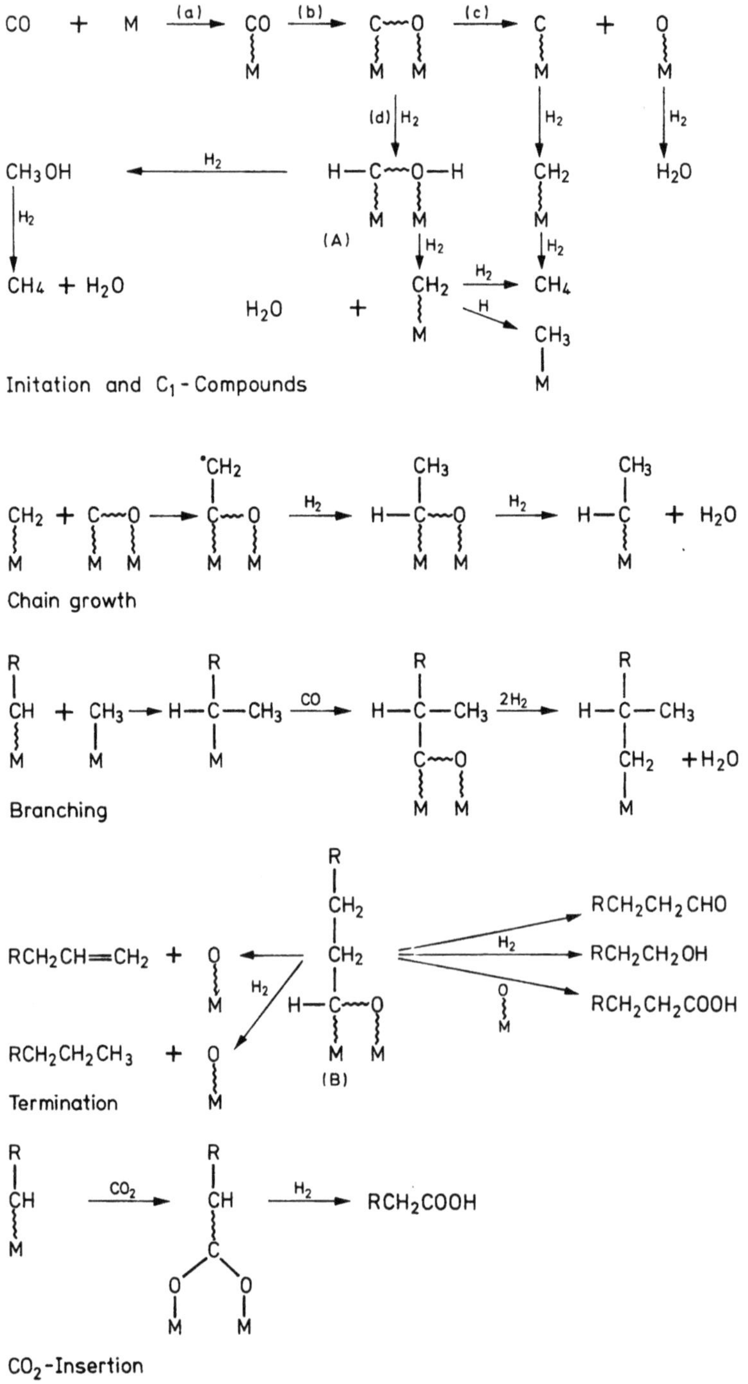

Figure 30. A general reaction scheme which has features of both schemes A and C (Figures 27 and 29). The symbol ⌇ denotes a bond of indefinite order

correlation between the ethanol and ketone selectivies over iron catalysts at high temperatures. When acetic acid is introduced into reactors operating under normal F-T conditions a large amount of acetone is formed. By analogy with the formation of ketones from the thermal decomposition of metal carboxylates this result is not surprising. Again, however, no clear correlation is observed in practice between acetic acid and acetone production in the F-T synthesis. Since acetone and isopropanol are in thermodynamic equilibrium (cf. Table 20) it is possible that ketones originate from isomerised alcohols but then the reverse can be equally true.

The incorporation of olefins in the F-T reaction, or the hydroformylation reaction (oxo synthesis), can also be described using the proposed general scheme. This is illustrated in Figure 31. The apparant similarities between the oxo and the F-T reactions have been discussed by Olive. [165] The synthesis of methanol, over e.g. supported CuO, also fits into the same scheme. In this case the C—O bond is strong relative to the C—M or O—M bonds and so hydrogenation of the surface complex A in Figure 30 leads to CH_3OH. Union carbide [166] claims a process in which synthesis gas is converted to ethylene glycol and glycerine over Rhodium catalysts at high pressures. Figure 32 illustrates that this process can also be described by the proposed general scheme. The C—O bond remains strong and is never broken.

An alternative scheme to account for chain branching has been proposed by Le Roux and Dry [89]. This is illustrated in Figure 33. The n-alkyl complex is assumed to desorb momentarily to a free radical which isomerizes in the normal way and then readsorbs. The higher the temperature the more readily the alkyl complex will desorb. This agrees with the observation that in the F-T synthesis the products formed at higher temperatures are more isomerized.

As mentioned in section 8.B, it is often found that the CH_4 selectivity is disproportionately high compared to the other carbon number products. According to the reaction scheme in Figure 30 there are apparently two routes along which CH_4 can be formed, one via the hydrogenation of the C atoms and the other via the hydrogenation of the HCOH complex (A) and of the resultant methanol. The additional source of CH_4 would cause the observed "kink" in the carbon number distribution curve.

In the reaction schemes presented in Figures 27 to 32 (except for Figure 29), it is assumed that hydrogen is present predominantly in the molecular state. While this is not an essential feature of any of the schemes there is some indirect evidence which supports the concept. Craxford [136] found that para to ortho H_2 conversion did not occur during the F-T reaction. Chemisorption studies at low temperatures on reduced iron catalysts [168] indicated that H_2 chemisorbed dissociatively. However, when CO was first pre-adsorbed onto the surface, no H_2 was chemisorbed. When H_2 was first pre-adsorbed, CO chemisorption did occur. It is concluded that H_2 cannot compete with much success with CO for vacant adsorption sites. If this still holds under F-T conditions, then it follows that the concentration of atomic hydrogen on the catalyst surface will be low.

The promotion of iron catalysts with K_2O results in a marked improvement

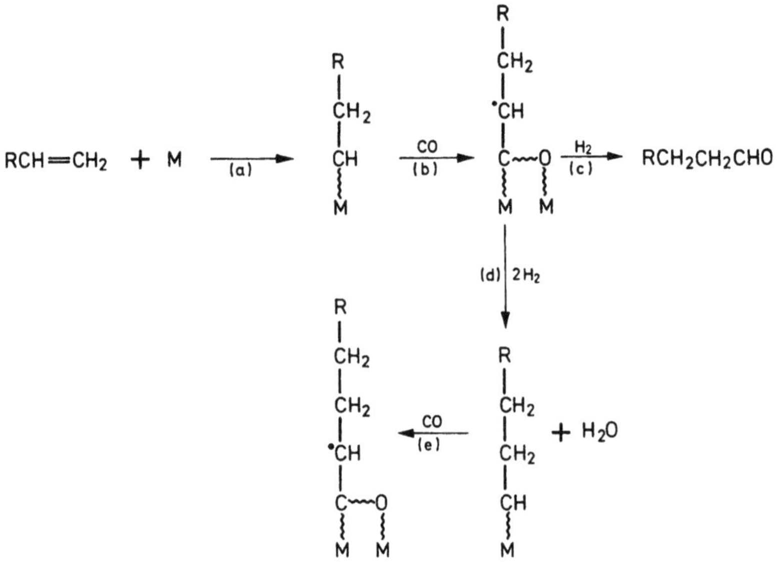

Figure 31. The general reaction scheme applied to olefin incorporation during F—T synthesis and also applied to the hydroformulation reaction (Oxo synthesis). Steps a, b and c represent the hydroformulation reaction. Steps a to e represent olefin incorporation during F—T and further chain growth. The symbol ⌇ denotes a bond of indefinite order

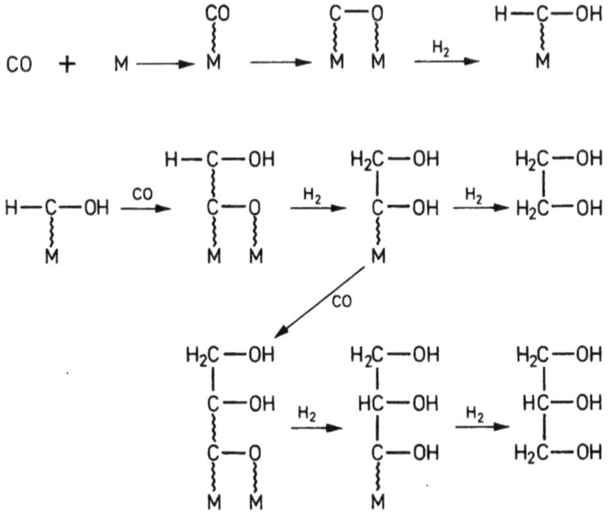

Figure 32. The general reaction scheme applied to the formation of polyhydric alcohols. The symbol ⌇ denotes a bond of indefinite order

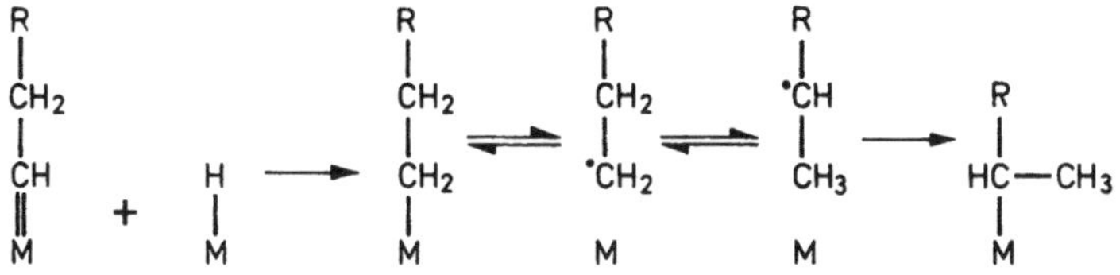

Figure 33. An alternative scheme for chain branching

in the catalysts' performance in the F-T as well as in the ammonia synthesis. The influence of K_2O promotion has consequently been investigated for many years. In 1931 it was shown that K_2O enhanced the electron emission of iron. [170] Russian [171] workers found that for reduced fused iron catalysts the addition of K_2O resulted in a decrease in the work function of the catalyst. When CO chemisorbed on metals it was found that the electrical resistance and the work function of the metals increased [172 to 175]. In the case of H_2 chemisorption, the resistance and work function first decreased and then increased with increasing coverage [174, 176]. From the foregoing results it appears that when CO chemisorbs on a metal F-T catalyst it withdraws electrons from the metal surface atoms. At low coverage hydrogen, on the contrary, donates electrons to the catalyst surface. Since K_2O acts as an electron donor it could be expected that K_2O promotion would enhance the chemisorption of CO and discourage that of H_2 [168, 177]. Measurements of the heats of chemisorption did in fact show that K_2O promotion resulted in a higher heat of adsorption of CO on reduced iron [168, 178] whereas it lowered the heat of adsorption of hydrogen [168]. A recent investigation also demonstrated that the binding energy of CO on K_2O promoted iron was greater than on pure iron [180]. The increase in the strength of the CO chemisorption on the catalyst surface can be equated to the transformation of the molecularly chemisorbed CO to an activated state in which the bonds between the C and O is weakened and those between the catalyst and the C and O atoms are strengthened (as depicted in step (b) in Figure 30). In the extreme it results in the complete dissociation of the CO molecule (step c in Figure 30). If it is assumed that in the activated state the CO entity is more readily attacked by hydrogen (e.g. initiation step d in Figure 30) and/or that it partakes in the insertion reaction more readily (chain growth steps) then this would explain the observed effect of K_2O promotion, namely, an increase in the rate of reaction and a shift in the selectivity to higher carbon number molecules. By enhancing CO and depressing H_2 chemisorption K_2O promotion must automatically result in a lower H_2 surface coverage. This should result in the surface being less hydrogenating and so more acids and olefins should be produced (as is found in practice).

10. Kinetics

The temperature at which a catalyst is used can be taken as a rough indicator of the catalyst's activity, i.e. the lower the temperature, the higher the activity. Applying this measure to F-T catalysts investigated in the past the activity declines in the order Ni, Co and Fe. Although these catalysts were prepared in similar manners the actual available active surface areas were not known. Vannice [48] has measured and reviewed the specific activities of the Group VIII metals. With Al_2O_3 as the carrier he reported that the activities declined in the order Ru, Fe, Ni, Co, Rh, Pd, Pt and Ir [181]. The turnover number of Ru was two orders of magnitude higher than that of Ir. When SiO_2 was used as the carrier a wider range of activities was observed and the activity order was different [182]. The activity declined in the order Co, Fe, Ru, Ni, Rh, Ir and Pd. In order to avoid the influence of the carrier it would be of interest to repeat these studies on the pure metals.

Vannice [182] demonstrated that a "volcano plot" was obtained when plotting turnover numbers against the heat of adsorption of CO. A chemisorbed CO presumably becomes activated by a strong interaction of the C and/or O atom with the catalyst surface. This should lead to the weakening of the C—O bond and hence to subsequent reaction with H_2. The observed increase in activity with increasing heat of adsorption is thus understandable. At high heats of adsorption the C—O bond might be expected to be completely severed and this could result in a high activity. However, it was found that as the heat of CO adsorption increased beyond about 126 kJ mol^{-1} the rate of activity declined. Apparently the C—O bond does not dissociate on noble metals even though the heat of adsorption is high. Alternatively it can be argued that since H_2 chemisorption is needed for reaction the CO chemisorption must not be too strong as then H_2 will not be able to compete for adsorption sites.

For practical catalysts operating under normal F-T conditions the activity is dependent on the particle size of the catalyst. It has been found that when either fused or precipitated iron catalysts are operated in fixed bed reactors under near differential conditions (i.e. a high fresh feed flow and a low conversion) at about 493 K, the activity increases linearly with the amount of external surface area of the catalyst particles. Even in the case of the very fine iron catalysts (less than 100 μm) used in fluidized beds the effect of particle size is in evidence. The smallest particles have the highest content of deposited carbon indicating that per unit mass of iron the smaller particles have done more F-T synthesis work than the larger particles.

These results indicate that either or both outside film or inside pore diffusional resistances are present. The average pore size of the precipitated catalyst is about 5 nm while that of the reduced fused catalyst is about 100 nm. Because of this big difference and the fact that both types of catalyst show a strong dependence on particle size even when the particles are very small, it seems possible that the outside film diffusion resistance is the main factor involved. Because of the wide spectrum of products normally produced some liquid wax is always present on the catalyst. As discussed in section 6.C periodic solvent extraction of the wax on a working catalyst results in an immediate increase in activity, but this is short-lived as a result of the redeposition of the wax film on the catalyst surface. Anderson [56] also reported a strong dependency of the activity of fused nitrided iron catalysts on particle size at about 513 K. It was deduced that only the outside ca. 0.1 mm thick "skin" of the catalyst particles was actively involved in the F-T synthesis. Thus it appears that once the reactants have diffused through the outer wax film it is very rapidly consumed within the outside "skin" of the catalyst particles.

When the process conditions are such that the wax selectivity is low, e.g. when operating at high temperatures, promotion with K_2O enhances the activity of iron catalysts (see Table 12). When conditions are geared at high wax yields, however, increasing the K_2O content increases the wax selectivity and decreases the activity (see Table 8). The promoting influence of K_2O on activity and wax selectivity has been discussed in sections 5.A.3, 5.B.4 and 9. The observed reversal of the influence of K_2O on activity beyond a certain K_2O content is possibly due to the diffusion resistance in the wax film on the catalyst. At the higher K_2O levels the wax formed on the catalyst becomes progressively longer chained and thus more viscous; i.e. the rate of diffusion through this wax decreases. The rate of supply of reactants and hence the reaction rate decreases. When wax together with the total feed gas is pumped into a reactor it appears to have little if any effect on the activity of the catalyst. This is understandable since the catalyst is already saturated with wax. When iron catalysts are nitrided prior to synthesis it is found that the acrivity is higher and also the wax selectivity is lower than for the same catalyst pretreated in the normal way [2]. Studies at Sasol have confirmed these findings. The activity increased by a factor of 1.3 while the wax selectivity decreased by a factor of 7. Here again the increase in activity might be directly due to the lower content and chain length of the wax on the catalyst resulting in a higher rate of diffusion of reactants to the catalyst sites.

The influence of temperature on the rate of reaction, as reflected by the overall apparent activation energy (E), may be dependent on whether or not wax is being produced. Precipitated iron catalysts which produce high yields of wax have an E value of 55 to 62 kJ mol^{-1} in the temperature range 473 to 523 K [79]. Fused iron catalyst which produce much less wax has an E value of about 71 kJ mol^{-1} in the same temperature range [184]. The studies of Vannice [48, 182] on different Group VIII metals were carried out under very low wax selectivity conditions. For Fe the value of E was about 113 kJ mol^{-1}. Anderson [56] found that the activation energy of nitrided iron catalysts increased as the H_2/CO ratio of the feed gas decreased; from 82 kJ mol^{-1} for a 2/1 ratio gas, to 100 kJ mol^{-1} for a 0.7/1 ratio gas. Since the lower ratio gas should produce more wax this finding is contrary to the forgoing speculation.

It should be borne in mind that when liquid wax is present on the catalyst the reactants first have to dissolve in the liquid and then diffuse into the pores where they react. The reaction creates a diffusion gradient, i.e. the higher the reaction rate the higher the diffusion rate. Hence the kinetics of diffusion are directly related with the kinetics of the surface reaction.

The published kinetic equations for the F-T and the related methane synthesis do not present a uniform picture. The subject has been reviewed by several authors [2, 6, 48, 56, 185, 194]. Table 31 illustrates a few examples of kinetic equations reported for each of the four main metal catalysts. As

Table 31. Some kinetic equations for CO hydrogenation over different metals

Metal	Kinetic equation	Ref. No
Ni	$r = kp_{CO}p_{H_2}^{0.5}$	[185]
	$r = kp_{CO}/(1 + kp_{CO})^2$	[186]
	$r = kp_{H_2}^{0.9}p_{CO}^{-0.2}$	[187]
	$r = kp_{H_2}^{0.85}p_{H_2O}^{-0.9}$	[188]
Co	$r = k$ (i.e. independent of reactant pressures)	[189]
	$r = kp_{H_2}^2/p_{CO}$	[190]
Fe	$r = kp_{H_2}^{0.6}p_{CO}^{0.4} - fr^{0.5}p_{H_2O}^{0.5}$	[56]
	$r = kp_{H_2}/(1 + ap_{H_2O}/p_{CO})$	[2]
Ru	$r = kp_{H_2}^2$	[191]
	$r = kp_{H_2}^{1.33}p_{CO}^{-0.13}$	[192]
	$r = kp_{H_2}^2p_{CO}/(1 + mp_{CO} + np_{H_2})^3$	[193]
	$r = kp_{H_2}^{1.5}p_{CO}^{-0.6}$	[157]

can be seen the exponents of the partial pressure of hydrogen vary from zero to two, and for carbon monoxide from one to minus one. Vannice [48] concluded that when the H_2/CO ratio was between 1 and 3 the exponent of H_2 was about 1 and that of CO between zero and —0.5. It is noteworthy that in the case of iron catalysts the partial pressure of water usually appears in the equations. In view of the fact that iron oxidises much more readily than the other F-T metals the sensitivity to H_2O is understandable. Nevertheless H_2O would be expected to chemisorb strongly on the other metals and so compete with CO and H_2 for catalyst sites. Hence H_2O should also play a role in the kinetics, as has in fact been reported in one case for a nickel catalyst [188]. The kinetic expressions vary from simple to complex. Comparison of different equations are complicated by the fact that the various catalysts were prepared in different ways, e.g. the support materials differed, which inevitably results in differences in metallic area and in the pore structures of the final catalyst particles. The types of product also vary from predominantly wax to pure CH_4. These differences should influence the rate of diffusion of reactants to and within the catalyst particles to varying degrees and this could change the form of the derived equations. Many laboratory studies have been carried out in high flow low conversion single-pass reactors. While this approach simplifies interpretations it ignores the possibility that the partial pressures of one or more of the products could have a strong influence on the kinetics. Thus it has been shown at Sasol that in a differential reactor the F-T rate over an iron catalyst is proportional only to the partial pressure of hydrogen [184] whereas in an integral reactor [79] the rate is proportional to $p_{H_2}p_{CO}/(p_{CO} + ap_{H_2O})$. Thus information generated

Table 32. Influence of entry partial pressures on activity of Fe catalysts (integral reactors)

Variable	Fixed bed precipitated catalysts 496 K						Fluidized bed fused catalyst 603 K					
	Partial pressure/0.1 MPa				Total pressure/ 0.1 MPa	Activity[a])	Partial pressure/0.1 MPa				Total pressure/ 0.1 MPa	Activity[a])
	H_2	CO	CO_2	H_2O			H_2	CO	CO_2	H_2O		
p_{H_2}	12.2	5.7	0.2	0	20	97	7.2	1.7	1.0	0	18	89
	15.2	5.6	0.2	0	23	109	10.1	1.6	0.8	0		100
p_{CO}	15.2	5.6	0.2	0	23	109	11.6	1.5	0.1	0	13.8	41
							11.7	3.0	0.1	0	15.8	66
	15.7	8.0	0.2	0	26	134	11.9	4.6	0.1	0	18.8	99
p_{CO_2}	14.3	6.4	1.2	0	27	77	7.2	1.7	1.0	0		98
							7.5	1.3	2.2	0	18.0	96
	12.8	6.1	4.1	0	27	72	7.2	1.3	3.1	0		98
							8.1	1.0	5.1	0		97
p_{H_2O}	8.4	4.3	0.1	0	15	69	9.8	1.8	0.7	0		107
							9.6	1.8	0.9	0.7	20.8	105
	8.4	4.1	0.1	0.35	15	51	10.0	1.8	1.2	1.4		98
							10.3	1.8	1.6	2.8		83
Total pressure	5.9	2.4	0.7	0	11.4	44	4.3	1.0	0.9	0	10.8	54
							8.5	2.0	1.5	0	20.8	106
	14.1	5.9	1.4	0	27.6	114	14.2	3.6	2.8	0	40.8	207
							30.7	6.7	4.6	0	75.8	375

[a] Moles (H_2O + CO_2) produced per hour

in differential reactors could be of no value for designing commercial integral reactors. For laboratory studies the high speed "stirred tank" type of units [195, 196] are recommended as with these the reactant and product concentrations at any point along the length of an integral reactor can be accurately simulated.

At Sasol the composition and pressure of the gas fed to both fixed bed and fluidized bed reactors operating at high conversion levels have been varied over wide ranges. Table 32 illustrates a few examples. The partial pressures at the reactor inlets and the overall conversion activities are given. As can be seen for both fixed and fluidized beds the integral activity increases as the partial pressure of H_2 or of CO is increased. Increasing the CO_2 pressure has little if any effect while increasing the H_2O pressure, decreases the activity. This is in keeping with the the findings of Anderson [56] and Tramm [197]. The influence of H_2O is reversible, i.e. under normal conditions additional water does not permanetly deactivate the catalyst. It has been previously mentioned that the addition of hydrocarbon products to the reactor feeds has no discernible influence on the F-T reaction rate. When the total pressure is increased the overall rate of reaction increases linearly. (Note when the pressure is increased all other factors such as the gas linear velocity, recycle ratio, fresh feed composition etc. was maintained constant). This has also been observed by Hall [21] for fixed, slurry and fluidized beds. The equation $r = mp_{CO}p_{H_2}/(p_{CO} + ap_{H_2O})$ reflects all the influences demonstrated in Table 32. The observation that the activity increases as the H_2/CO ratio of the gas fed to a fixed bed reactor increases from 1.1 to 1.9 also follows from the equation. When the partial pressure of H_2O is very low, as it would be in high flow differential reactors, the equation simplifies to $r = mp_{H_2}$ which is the observed kinetic equation for such reactors [184].

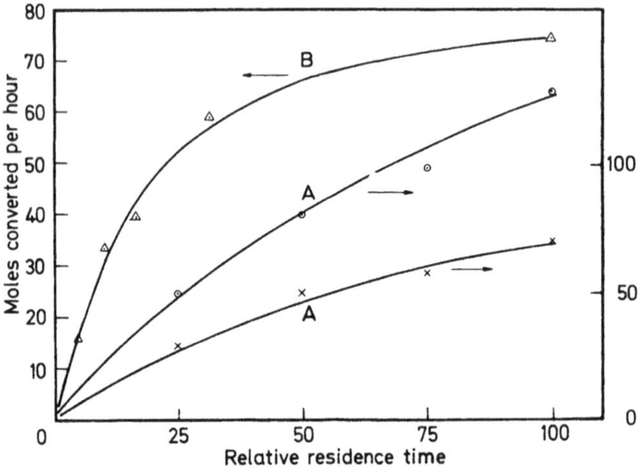

Figure 34. Typical reaction profiles in fixed bed, (A), and fluidized bed, (B), iron catalysts. The symbols show the actual measured amounts of moles converted. The solid lines are for the profiles as calculated with the kinetic equation $r = mp_{CO}p_{H_2}/(p_{CO} + ap_{H_2O})$

In the high temperature fluidized bed reactors the water gas shift reaction is very rapid and is found to be in thermodynamic equilibrium all along the catalyst bed. This simplifies the computation of the partial pressures of the key reactants CO, H_2 and H_2O along the catalyst bed. In the lower temperature fixed bed reactor the shift reaction does not reach equilibrium and so a separate kinetic model for this reaction is also required. Figure 34 illustrates a few examples of the fits obtained between the observed and calculated reaction profiles. Bennett [198] has also applied the equation to data generated in a gradientless reactor and found good fits over several fused iron catalysts. These studies should be extended to different metal catalysts prepared in similar ways. The mode of operations should also be varied from wax to gas production.

In fixed bed reactors a high linear velocity is always recommended as this decreases the temperature rise in the bed near the entrance. When operating in the wax mode the "flattening" of the overall temperature profile benefits the wax selectivity. At high linear velocities the partial pressure of H_2O is also lowered and this benefits the overall kinetics in spite of the shorter gas residence times. The fresh feed conversion need not decrease since the higher linear velocity can be achieved by increasing the recycle gas flow only.

References

1. Storch, H. H.; Golumbic, N.; Anderson, R. B.: The Fischer-Tropsch and related synthesis. New York: John Wiley 1951
2. Anderson, R. B.: Catalysis Vol. IV (Emmett, P. H., ed.) New York: Reinhold 1956
3. Anderson, R. B.: Advan. Catal. Vol. 5 (Frankenburg, Komarewsky, Rideal, ed.) New York: Academic Press Inc., 1953
4. Pichler, H.: ibid. Vol. 4, 1952
5. Bond, G. C.: Catalysis by Metals, New York: Academic Press, 1962
6. Frohning, C. D.: Fischer-Tropsch-Synthese, Chemierohstoffe aus Kohle (Falbe J., ed.) Stuttgart: Thieme 1977
7. Sabatier, P.; Senderens, J. B.: C.R. Acad. Sci., Paris *134*, 514 (1902)
8. BASF: German Pat. 293, 787 (1913)
9. Fischer, F.; Tropsch, H.: Brennst.-Chem. *4*, 276 (1923)
10. Fischer, F.; Hartner-Seberich, R.; Roelen, O.: ibid. *11*, 489 (1930)
11. Ullmanns Encyklopädie d. techn. Chem. *9*, 715 München-Berlin: Urban and Schwarzenberg 1957
12. Forney, A. J.; Haynes, W. P.; Elliott, J. J.; Zarochak, M. F.: Us Bur. Mines Rep. 5841 (1961); 5871 (1961); 6126 (1962); 6609 (1965); ACS Div. Fuel *20* (3), 171 (1975)
13. Haynes, W. P.; Elliott, J. J.; Youngblood, A. J.; Forney, A. J.: Div. Fuel Chem. Sym. Am. Chem. Soc. *14*, 26 (1970)
14. Fischer, F.: Brennst.-Chem. *11*, 500 (1930)
15. Kölbel, H.; Ackermann, P.; Engelhardt, F.: Erdöl u. Kohle *9*, 153, 225, 303 (1956)
16. Farley, R.; Ray, D. J.: J. Inst. Petrol. *50*, 27 (1964)
17. Benson, H. E.: Ind. Eng. Chem. *46*, 2278 (1954)
18. Keith, P. C.: Oil and Gas J. *45*, 102 (1946)
19. Jones, D. H.; Shingles, T.; Kelly, F. T.; Dry, M. E.: Ger. Offen. 2812108 (1977)
20. McGrath.: U.S. Pat. 2,543,327 (1951); 2,598,647 (1947); 2,640,844 (1947)
21. Hall, C. C.; Gall, D.; Smith, S. L.: J. Inst. Petrol. *38*, 845 (1952)
22. Hoogendoorn, J. C.: Sym. Clean Fuels from Coal I.G.T., Chicago, Sept. 1973
23. Hoogendoorn, J. C.: Amer. Gas Assoc. 9th Synth. Pipeline Gas Symp., Nov. 1977

24. Mitchell, A. S.: Chem. Ind. *23*, 924 (1966)
25. Schuit, G. C. A.; van Reijen, L. L.: Advan. Catal., Vol. 10 (Eley, Frankenburg, Komarewsky, ed.) Academic Press N.Y. 1958
26. Gruber, H. L.: Anal. Chem. *34*, 1828 (1962)
27. Dry, M. E.; du Plessis, J. A. K.; Leuteritz, G. M.: J. Catal. *6*, 194 (1966)
28. Fischer, F.; Tropsch, H.: Ges. Abhandl. Kenntnis Kohle *10*, 333 (1930)
29. Hammer, H.; Bittner, D.: Erdöl u. Kohle, Erdgas Petrochemie *31*, 369 (1978)
30. Fischer, F.; Koch, H.: Brennst.-Chem. *13*, 61 (1932)
31. Clarke, J. J.; Dry, M. E.; van Zyl, W. J.: J. S. Afr. Chem. Inst. *15*, 11 (1962)
32. Dry, M. E.; Leuteritz, G. M.; van Zyl, W. J.: J. S. Afr. Chem. Inst. *16*, 15 (1963)
33. Hall, W. K.; Tarn, W. H.; Anderson, R. B.: J. Amer. Chem. Soc. *72*, 5436 (1950)
34. Dry, M. E.; Ferreira, L. C.: J. Catal. *7*, 352 (1967)
35. Dry, M. E.: Brennst.-Chem. *50*, 193 (1969)
36. Dry, M. E.; Oosthuizen, G. J.: J. Catal. *11*, 18 (1968)
37. Brunauer, S.; Emmett, P. H.: J. Amer. Chem. Soc. *62*, 1732 (1940)
38. Clarke, J.; Dry, M. E.; Montano, J. J.; van Zyl, W. J.: Trans. Faraday Soc. *57*, 2239 (1961)
39. Huang, C. P.; Richardson, J. T.: J. Catal. *51*, 1 (1978)
40. Seelig, H. S.; Week, H. I.; Voss, D. J.; Zisson, J.: Div. of Pet. Chem. ACS Meet. Atlantic City, Sept. 1952
41. Vannice, M. A.: J. Catal. *37*, 449 (1975)
42. Schulz, H.: Chemierohstoffe aus Kohle (Falbe, J., ed.) Thieme Verlag, Stuttgart 1977
43. Madon, R. J.; Bucker, E. R.; Taylor, W. F.: Final ERDA Rep. Contract No E (46—1) 8008. July 1977
44. King, D. L.: J. Catal. *51*, 386 (1978)
45. Everson, R. C.; Woodburn, E. T.; Kirk, A. R. M.: J. Catal. *53*, 186 (1978)
46. Dalla Betta, R. A.; Shelef, M.: J. Catal. *48*, 111 (1977)
47. Vannice, M. A.: J. Catal. *50*, 228 (1977)
48. Vannice, M. A.: Catal. Rev. *14* (2), 153 (1976)
49. Schulz, J. F.; Karn, F. S.; Anderson, R. B.: U.S. Bur. Mines Rep. 6974
50. Malan, O. G.; Louw, J. D.; Ferreira, L. C.: Brennst.-Chem. *42*, 209 (1961)
51. Senateur, J. P.; Fruchart, R.; Michel, A.: C.R. Acad. Sci., Paris *255*, 1615 (1962)
52. Senateur, J. P.; Fruchart, R.: C.R. Acad. Sci., Paris *256*, 3114 (1963)
53. Eckstrom, H. C.; Adcock, W. A.: J. Amer. Chem. Soc. *72*, 1042 (1950)
54. Herbstein, F. H.; Snyman, J. A.: Inorg. Chem. *3* (6), 894 (1964)
55. Pichler, H.; Merkel, H.: U.S. Bur. Mines Tech. Paper 718 (1949)
56. Anderson, R. B.; Karn, F. S.; Schulz, J. F.: U.S. Bur. Mines. Bull. 614 (1964)
57. Schulz, J. F.; Hall, W. K.; Dubs, T. A.; Anderson, R. B.: J. Amer. Chem. Soc. *78*, 282 (1956)
58. Frye, C. G.; Pickering, H. L.; Eckstrom, H. C.: J. Phys. Chem. *62*, 1508 (1958)
59. Voorhies, A.: U.S. Patent 2,533,072 (1950)
60. Griffin, L. I.: U.S. Patent 2,487,867 (1949)
61. Rubin, L. C.; West, C.; McGrath, H. G.: U.S. Pat. 2,620,346 (1952)
62. Dwyer, D. J.; Somorjai, G. A.: J. Catal. *52*, 291 (1978)
63. McKinley, J. B.: Catalysis Vol. 5 p. 451 (ed. Emmett P. H.) Reinhold N.Y. 1957
64. Fischer, F.: Brennst.-Chem. *16*, 1 (1935)
65. Madon, R. J.; Shaw, H.: Catal. Rev. *15* (1), 69 (1977)
66. Enikeev, E.; Krylova, A. V.: Kinet. Katal. *3*, 116 (1962)
67. Manning, M. P.; Reid, R. C.: Ind. Eng. Chem. Process. Des. Develop. *16* (3), 358 (1977)
68. Dry, M. E.; Shingles, T.; Botha, C. S. van H.: J. Catal. *17*, 341 (1970)
69. Dry, M. E.; Shingles, T.; Boshoff, L. J.; Botha, C. S. van H.: J. Catal. *17*, 347 (1970)
70. Davis, H. G.; Wilson, T. P.: U.S. Pat. 2,717,259 (1950)
71. Bauklok, W.; Hellbrügge, J.: Brennst.-Chem. *23*, 87 (1942)
72. White, G. A.; Roszkowski, T. R.; Stanbridge, D. W.: Hydrocarbon Process. July 1975, p. 130
73. Greyson, M.: Catalysis Vol. IV, p. 480 (Ed. Emmett, P. H.) Reinhold N.Y. 1956
74. Dry, M. E.: Hydrocarbon Process. *59*, 92 (1980)

75. Arnold, J. H.; Keith, P. C.: Amer. Chem. Soc. Adv. Chem. Ser. *5*, 120 (1951)
76. Tillmetz, K. D.: Chem.-Ing.-Tech. *48* (11), 1065 (1976)
77. Christoffel, von E.; Surjo, I.; Baerns, M.: Chemiker-Z. *102*, 19 (1978)
78. Weitkamp, A. W.; Frye, C. G.: Ind. Eng. Chem. *45*, 363 (1953)
79. Dry, M. E.: Ind. Eng. Chem. Prod. Res. Develop. *15* (4), 282 (1976)
80. Herington, E. F. G.: Chem. Ind. (London) 347 (1946)
81. Friedel, R. A.; Anderson, R. B.: J. Amer. Chem. Soc. *72*, 1212, 2307 (1950)
82. Weitkamp, A. W.; Seelig, H. S.; Bowman, N. J.; Cady, W. E.: Ind. Eng. Chem. *45* (2), 343 (1953)
83. Flory, P. J.: J. Amer. Chem. Soc. *58*, 1877 (1950). Principles of Polymer Chemistry, Cornell Univ. Press, Ithaca, N.Y. 1967
84. Schultz, G. V.: Z. Phys. Chem. B *43*, 25 (1935)
85. Henrici-Olivé, G.; Olivé, S.: Angew. Chem. Int. Ed. *15*, 136 (1976)
86. Anderson, R. B.: J. Catal. *55*, 114 (1978)
87. Pichler, H.; Schulz, H.; Kühne, D.: Brennst.-Chem. *49* (11), 344 (1968)
88. Brink, A.; Dressler, F.: Brit. Polym. J. *1*, 37 (1969)
89. Le Roux, J. H.; Dry, L. J.: J. Appl. Chem. Biotechnol. *24*, 81 (1974)
90. Le Roux, J. H.: J. Appl. Chem. (London) *19*, 230 (1969)
91. Cady, W. E.; Launer, P. J.; Weitkamp, A. W.: Ind. Eng. Chem. *45*, 350 (1953)
92. Gall, D.; Gibson, E. J.; Hall, C. C.: J. Appl. Chem. (London) *2*, 371 (1952)
93. Kölbel, H.; Engelhardt, F.: Erdöl u. Kohle *2*, 52 (1949)
94. Forney, A. J.; Haynes, W. P.; Elliott, J. J.; Zarochak: A.C.S., Div. Fuel *20*, 3 (1975)
95. Anderson, R. B.; Krieg, A.; Friedel, R. A.; Mason, L. S.: Ind. Eng. Chem. *41*, 2189 (1949)
96. Fischer, F.; Pichler, H.: Brennst.-Chem. *20*, 41 (1939)
97. Martin, F.: Chem. Fabrik *12*, 233 (1939)
98. Anderson, R. B.; Seligman, B.; Shultz, J. F.; Kelly, R.; Elliott, M. A.: Ind. Eng. Chem. *44*, 391 (1952)
99. Friedman, S.; Schlesinger, M. D.: U.S. Bur. Mines Bull., 614 (1964)
100. Koch, H.; Küster, H.: Brennst.-Chem. *14*, 245 (1933)
101. Fischer, F.; Pichler, H.: Brennst.-Chem. *14*, 306 (1933)
102. Russel, W. W.; Miller, G. H.: J. Amer. Chem. Soc. *72*, 2446 (1950)
103. Kuster, H.: Brennst.-Chem. *17*, 203, 221 (1936)
104. Sexton, B. A.; Somorjai, G. A.: J. Catal. *46*, 167 (1977)
105. Kölbel, H.; Engelhardt, F.: Erdöl u. Kohle *3*, 529 (1950)
106. Taylor, E. H.; Taylor, H. S.: J. Amer. Chem. Soc. *61*, 503 (1939)
107. Craxford, S. R.: Fuel (London) *26*, 119 (1947)
108. Craxford, S. R.: Trans. Faraday Soc. *42*, 576 (1946)
109. Dalla Betta, R. A.; Piken, A. G.; Shelef, M.: J. Catal. *35*, 54 (1974)
110. Kölbel, H.; Ludwig, H.; Hammer, H.: J. Catal. *1*, 156 (1962)
111. Tsutsumi, S.; Hayashi, K.: Jap. Pat. AG 34 3214
112. Kölbel, H.; Tillmetz, K. D.: Offenlegungsschrift 2507647
113. Peters, K.: U.S. Pat. 2,960,518 (1960)
114. Büssemeier, B.; Frohning, C. D.; Cornils, B.: Hydrocarbon Proc. Nov. 105 (1976)
115. Büssemeier, B.; Frohning, C. D.; Horn, G.; Kluy, W.: South African Pat. Appl. 764763
116. Rottig, W.: South African Pat. Appl. 762412
117. Kitzelmann, D.; Vielstich, W.; Dittrich, T.: Chem.-Ing.-Tech. *49*, 463 (1977)
118. Bartish, C. M.: U.S. Pat. Appl. 644,313 (1975)
119. Myddleton, W. W.: Brit. Pat. 509,325 (1939)
120. Herington, E. F. G.; Woodward, L. A.: Trans. Faraday Soc. *35*, 958 (1939)
121. Layng, E. T.: U.S. Pat. 2,446,426 (1948)
122. Dalla Betta, R. A.; Piken, A. G.; Shelef, M.: J. Catal. *40*, 173 (1975)
123. Smith, D. F.; Hawk, C. O.; Golden, P. L.: J. Amer. Chem. Soc. *52*, 3221 (1930)
124. Kölbel, H.; Ruschenburg, E.: Brennst.-Chem. *35*, 161 (1954)
125. Kölbel, H.; Ackermann, P.: Brennst.-Chem. *31*, 10 (1950)
126. Chang, C. D.; Kuo, J. C. W.; Lang, W. H.; Jacob, S. M.; Wise, J. J.; Silvestri, A. J.: Ind. Eng. Chem. Process. Des. Develop. *17* (3), 255 (1978)

127. Brenan, J. A.: ERDA Publication F E 2276-3 (1976)
128. Mobil Oil Corp. Ger. Pat. 2518097 Nov. 1975
129. Madon, R. J.: J. Catal. *57*, 183 (1979)
130. Caesar, P. D.; Brennan, J. A.; Garwood, W. E.; Ciric, J.: J. Catal. *56*, 274 (1979)
131. Dwyer, D. J.; Somorjai, G. A.: J. Catal. *56*, 249 (1979)
132. Schulz, H.; Rao, B. R.; Elstner, M.: Erdol Kohle *23*, 651 (1970)
133. Kobylinski, T. P.: U.S. Pat. 4,088,691 (1978)
134. Fischer, F.; Tropsch, H.: Brennst.-Chem. *7*, 97 (1926)
135. Fischer, F.; Koch, H.: Brennst.-Chem. *13*, 428 (1932)
136. Craxford, S. R.; Rideal, E. K.: J. Chem. Soc. 1604 (1939)
137. Browning, L. C.; Emmett, P. H.: J. Amer. Chem. Soc. *74*, 1680 (1952)
138. Kummer, J. T.; de Witt, T. W.; Emmett, P. H.: ibid *70*, 3632 (1948)
139. Loktev, S. M.: Higher Aliphatic Alcohols, Chapter 11, Moscow, 1970
140. Kagan, Y. B.; Bashkirov, A. N.; Morozov, L. A.; Kryulov, Y.; Orlova, N. A.: Neftekhimiya *6* (2), 262 (1966)
141. Elvins, O. C.; Nash, A. W.: Nature *118*, 154 (1926)
142. Kummer, J. T.; Emmett, P. H.: J. Amer. Chem. Soc. *75*, 5177 (1953)
143. Kummer, J. T.; Podgurski, H. H.; Spencer, W. B.; Emmett, P. H.: J. Amer. Chem. Soc. *73*, 564 (1951)
144. Kölbel, H.; Ralek, M.; Jiru, P.: Erdöl/Kohle/Erdgas/Petrochem., Brennst.-Chem. *9*, 580 (1970)
145. Blyholder, G.: J. Phys. Chem. *70*, 893 (1966)
146. Blyholder, G.; Goodsel, A. J.: J. Catal. *23*, 374 (1971)
147. Blyholder, G.; Shihabi, D.; Wyatt, W. V.; Bartlett, R.: J. Catal. *43*, 122 (1976)
148. Heal, M. J.; Leisegang, E. Ç.; Torrington, R. G.: J. Catal. *51*, 314 (1978)
149. Dalla Betta, R. A.; Shelef, M.: J. Catal. *48*, 111 (1977)
150. Pichler, H.; Schulz H.: Chem.-Ing.-Techn. *42*, 1162 (1970)
151. Sternberg, H. W.; Wender, I.: Intern. Conf. Coordination Chem. 1959 (Chem. Soc. London)
152. Deluzarche, A.; Hindermann, J. P.; Kieffer, R.; Muth, A.; Papadopoulos, M.; Tanielian, C.: Tetrahedron Lett. *9*, 797 (1977)
153. Joyner, R. W.: J. Catal. *50*, 176 (1977)
154. Ponec, V.: Catal. Rev. *18* (1), 151 (1978)
155. Araki, M.; Ponec, V.: J. Catal. *44*, 439 (1976)
156. Biloen, P.; Helle, J. N.; Sachtler, W. M. H.: J. Catal. *58*, 95 (1979)
157. Ekerdt, J. G.; Bell, A. T.: J. Catal. *58*, 170 (1979)
158. Kishi, K.; Roberts, M. J.: J. Chem. Soc. Faraday Trans *71*, 1715 (1975)
159. Rabo, J. A.; Risch, A. P.; Poutsma, M. L.: J. Catal. *53*, 295 (1978)
160. Broden, G.; Gafner, G.; Bonzel, H. P.: Appl. Phys. *13*, 333 (1977)
161. Low, G. G.; Bell, A. T.: J. Catal. *57*, 397 (1979)
162. Sexton, B. A.; Somorjai, G. A.: J. Catal. *46*, 167 (1977)
163. Matsumoto, H.; Bennett, C. O.: J. Catal. *53*, 331 (1978)
164. Wentrcek, P. R.; Wood, B. J.; Wise, H.: J. Catal. *43*, 363 (1976)
165. Henrici-Olivé, G.; Olivé, S.: J. Molec. Catal. *3*, 443 (1977/78)
166. Pruett, R. E.; Walker, W. E.: U.S. Pat. 3, 833,634 (1974)
167. Kratzer, M. B.: U.S. Pat. 2,632,015 (1953)
168. Dry, M. E.; Shingles, T.; Boshoff, L. J.; Oosthuizen, G. J.: J. Catal. *15*, 190 (1969)
169. Kölbel, H.; Roberg, H.: Ber. Bunsenges. Phys. Chem. *81*, 634 (1977)
170. Brewer, A. K.: J. Amer. Chem. Soc. *53*, 74 (1931)
171. Krylova, A. V.; Kuznetsov, L. D.; Konyukhova, I. N.: Kinet. Katal. *5* (5), 948 (1964)
172. Suhrmann, R.: Z. Elektrochem. *63*, 748 (1959)
173. Sugita, T.; Ebsisawa, S.; Kawasaki, K.: Surface Sci. *11*, 159 (1968)
174. Kölbel. H.: Actes Congr. Intern. Catalyse 2, Paris *2*, 2075 (1960)
175. Wedler, G.; Papp, H.; Schroll, G.: Surface Sci. *44*, 463 (1974)
176. Ponec, V.; Knor, Z. J.: J. Catal. *10*, 73 (1968)
177. Kölbel, H.; Müller, W. K. H.; Schöttle, E.; Hammer, H.: Angew. Chem. *2*, 554 (1963)
178. Kölbel, H.; Haubold, H.: Z. Elektrochem. *65* (5), 421 (1961)

179. Kölbel, H.; Patzschke, G.; Hammer, H.: Brennst.-Chem. *47*, 4 (1966)
180. Broden, G.; Gafner, G.; Bronzel, H. P.: Surface Sci. *84*, 295 (1979)
181. Vannice, M. A.: J. Catal. *37*, 449 (1975)
182. Vannice, M. A.: J. Catal. *50*, 228 (1977)
183. Vannice, M. A.: J. Catal. *44*, 152 (1976)
184. Dry, M. E.; Shingles, T.; Boshoff, L. J.: J. Catal. *25*, 99 (1972)
185. Lee, A. L.; Feldkirchner, H. L.; Tajbl, D. J.: Amer. Chem. Soc. Div. Fuel Chem. prepr. *14*, 126 (1970)
186. Van Herwijnen, T.; Van Doesburg, H.; De Jong, W. A.: J. Catal. *28*, 391 (1973)
187. Luyten, L.; Jurgens, J. C.: Bull. Soc. Chim. Belges *54*, 303 (1945)
188. Saletore, D. A.; Thomson, W. J.: Ind. Eng. Chem. Process Des. Develop. *16* (1). 70 (1977)
189. Anderson, R. B.; Hall, W. K.; Krieg, A.; Seligman, B.: J. Amer. Chem. Soc. *71*, 183 (1949)
190. Brötz, W.: Z. Elektrochem. *5*, 301 (1949)
191. McKee, D. W.: J. Catal. *8*, 240 (1967)
192. Karn, F. S.; Schulz, J. F.; Anderson, R. B.: J. Phys. Chem. *64*, 446 (1960)
193. Phung Quach, Thi Q.; Rouleau, D.: J. Appl. Chem. Biotechnol. *28*, 378 (1978)
194. Mills, G. A.; Steffgen, F. W.: Catal. Rev. *8* (2), 159 (1973)
195. Tajbl, D. G.; Simons, J. B.; Carberry, J. J.; I. & E. C. Fundamentals *5* (2), 171 (1966)
196. Berty, J. M.: Chem. Eng. Progr. *70* (5), 78 (1974)
197. Tramm, H.: 5th World Petroleum Congr. Sect. III, paper 27 (1964)
198. Atwood, H. E.; Bennett, C. O.: Ind. Eng. Chem. Process Des. Develop. *18* (1), 163 (1978)

Chapter 5

Catalytic Reforming of Hydrocarbons

J. H. Sinfelt

Corporate Research Laboratories
Exxon Research and Engineering Co.
Linden, N.J. 07036, USA

Catalytic reforming is one of the most important industrial applications of catalysis. The major reactions in the process involve the production of aromatic hydrocarbons from saturated hydrocarbons via dehydrogenation of cyclohexanes, dehydroisomerization of alkylcyclopentanes, and dehydrocyclization of alkanes. Isomerization, hydrogenolysis, and hydrocracking reactions are also involved in the process. The objective in reforming is to achieve high rates of production of aromatic hydrocarbons as selectively as possible, since they have excellent anti-knock ratings as components of automotive fuels. The process is commonly conducted at temperatures in the vicinity of 770 K and at pressures of 10—35 atm. The catalysts employed in reforming are bifunctional in nature, in that they possess both metallic and acidic components. The metallic component, e. g., platinum, possesses high activity for hydrogenation and dehydrogenation reactions. The acidic component, which consists of the carrier on which the metallic component is dispersed, is active for reactions which are commonly associated with acid catalysis, e. g., isomerization and hydrocracking. The carrier is commonly alumina which contains a small amount of chlorine as a promoter. Platinum on alumina reforming catalysts were used throughout the industry during the 1950s and 1960s, but during the past decade they have been largely replaced by catalysts containing platinum and a second metallic element such as rhenium or iridium. These newer catalysts exhibit improved reforming performance relative to the original platinum catalysts and have generated much interest with regard to their chemical nature and structure.

Contents

1. Introduction

Over the past three decades catalytic reforming has evolved very rapidly to the point where it is now one of the most important industrial applications of catalysis. The process was originally developed to produce gasoline components of high anti-knock quality, in response to the fuel requirements of high compression ratio automobile engines. The objective of the process is to convert saturated hydrocarbons (alkanes and cycloalkanes) in petroleum naphtha fractions to aromatic hydrocarbons as selectively as possible, since the latter have excellent anti-knock ratings [1, 2]. Naphtha fractions are composed of hydrocarbons having boiling points within the approximate range of 320–470 K. Reaction temperatures of 700–800 K and pressures of 10–35 atm. are employed in the process. The catalysts employed commonly contain platinum [3–5] or a combination of platinum and a second metallic element such as rhenium [6] or iridium [7, 8].

A typical catalytic reforming unit consists of a number of fixed bed reactors (frequently four) in series. The naphtha feedstock is vaporized and heated to the desired reaction temperature prior to being admitted to the first reactor. As the components in the naphtha undergo reaction during passage through the catalyst bed, the temperature of the vapor stream decreases continuously due to the endothermicity of the reaction. The major reaction occurring in the first catalyst bed is the dehydrogenation of cycloalkanes to aromatics. The effluent from the first reactor is reheated to the desired reaction temperature and admitted to the second reactor. As reaction occurs, the temperature of the vapor stream decreases again, although to a smaller extent than in the first reactor. The process of reheating followed by reaction is repeated until the hydrocarbon stream has passed through all the reactors. The effluent from the final reactor is cooled and separated into liquid and gaseous products. The liquid product, commonly known as reformate, consists essentially of C_5 through about C_{10} hydrocarbons. Aromatic hydrocarbons in the liquid product are typically present in amounts in the range of 60–70 wt. percent, the actual amount depending on the reaction conditions. The anti-knock quality of the liquid reformate increases with increasing aromatics content. The gaseous product consists of hydrogen and C_1—C_4 hydrocarbons, with a hydrogen concentration which is commonly within the range of 60–90 mole percent. A stream of this gas is recycled to the inlet of the first reactor, where it is combined with the naphtha feed in a ratio of about 5–10 moles of recycle gas per mole of naphtha. Consequently, there is a high partial pressure of hydrogen in the system. The high hydrogen partial pressure is crucial for the maintenance of catalytic activity, since it serves to retard fouling of the catalyst surface by hydrocarbon residues. Such residues are formed via extensive dehydrogenation of chemisorbed hydrocarbons to highly unsaturated species which in turn readily undergo condensation or polymerization reactions. The hydrogen serves to inhibit the formation of highly unsaturated species on the surface, and also to remove hydrocarbon residues via hydrogenolysis reactions [9–11]. Since increasing hydrogen pressure also has the effect of decreasing the yield of aromatic hydrocarbons, the choice of hydrogen pressure for a reformer is a matter of balancing product yields against deactivation rates.

While catalytic reforming is a complex process involving reactions of a large number of hydrocarbons, a reasonable understanding of the chemistry of the process and of the functioning of reforming catalysts has evolved. The understanding has resulted from studies of the reactions of individual hydrocarbons and from investigations of reforming catalysts by a variety of chemical and physical methods. The present article begins with a general review of reforming reactions, including mechanistic features and kinetics. A discussion of the classical platinum-alumina catalyst is then given. Finally, recent advances in reforming catalysts are considered, in particular the development of bimetallic (or, more generally, polymetallic) catalyst systems.

2. Catalytic Reforming Reactions

The major reactions in catalytic reforming are: (a) dehydrogenation of cyclohexanes to aromatics, (b) isomerization of *n*-alkanes to branched alkanes, (c) dehydroisomerization of alkylcyclopentanes to aromatics, (d) dehydrocyclization of alkanes to aromatics, and (e) hydrocracking or hydrogenolysis of alkanes and cycloalkanes to low molecular weight alkanes [1, 12]. The catalysts on which these reactions occur are bifunctional in nature, i.e., they possess two different types of catalytic activity. In addition to having ability to catalyze hydrogenation and dehydrogenation reactions which are characteristic of metallic catalysts, they also are effective for catalyzing hydrocarbon rearrangements which are typical of acid catalysts. These two distinct types of activity are associated with two different components of the catalyst. The hydrogenation and dehydrogenation activities are associated with a metal component (e.g., platinum), while the activity for acid catalysis is due to the presence of acidic sites on the surface of the carrier on which the metal component is dispersed. The carrier is commonly alumina, which is covered by the metal component to an extent of 1 percent or lower in the catalysts typically employed in reforming. Early studies of reactions (a) through (e) on bifunctional catalysts were published by Haensel and Donaldson [5], Ciapetta and Hunter [13, 14], Heinemann et al. [15], and Hettinger et al. [16].

The major part of the anti-knock improvement obtained in the catalytic reforming of petroleum naphthas is due to the formation of aromatics.

Lesser contributions result from the isomerization of straight-chain paraffins to branched paraffins and hydrocracking of hydrocarbons to lower molecular weight hydrocarbons in the gasoline boiling range.

Of all the reactions taking place in catalytic reforming, the dehydrogenation of cyclohexanes occurs by far the most readily. Isomerization reactions also occur readily, but not nearly so fast as the dehydrogenation of cyclohexanes. The limiting reactions in most catalytic reforming operations are hydrocracking and dehydrocyclization, which generally occur at much lower rates.

The thermodynamics of the more important reactions in catalytic reforming can be discussed conveniently by referring to the equilibria involved in the various interconversions among certain of the C_6 hydrocarbons. Some thermodynamic equilibrium constants at 773 K, a typical temperature in catalytic reforming, and heats of reaction are given in Table 1 [1, 17]. The equilibrium constant K in Table 1 is defined in terms of the partial pressures of the reactants and products expressed in atmospheres.

The dehydrogenation of cyclohexane and the dehydrocyclization of n-hexane to yield benzene are seen to be strongly endothermic, so that increasing temperature has a marked effect on improving the extent of conversion to aromatics. Hydrogen partial pressure obviously has a marked effect on the extent of formation of benzene, and from the viewpoint of equilibria alone, it is advantageous to operate at as high a temperature and as low a hydrogen partial pressure as possible to maximize the yield of the aromatic. However, other considerations, such as catalyst deactivation due to formation of carbonaceous residues on the surface, place a practical upper limit on temperature and a lower limit on hydrogen partial pressure in catalytic reforming operations. The thermodynamic considerations in the formation of aromatics from higher molecular weight paraffins and cycloparaffins are qualitatively the same as for the formation of benzene from n-hexane and cyclohexane. Quantitatively, the equilibrium tends to shift more in favor of the aromatic product as the molecular weight increases.

The dehydrogenation of paraffins to olefins, while it does not take place to a large extent at typical reforming conditions (equilibrium conversion of n-hexane to 1-hexene is about 0.3 percent at 783 K and 17 atm hydrogen partial pressure), is nevertheless of considerable importance, since olefins appear to be intermediates in some of the reactions. This matter will be dis-

Table 1. Thermodynamic data on reactions of C_6 hydrocarbons [1, 17]

Reaction	K (at 773 K)	ΔH_R/kJ mole^{-1}
Cyclohexane → Benzene + 3 H_2	6×10^5	221
Methylcyclopentane → Cyclohexane	0.086	−16.0
n-Hexane → Benzene + 4 H_2	0.78×10^5	266
n-Hexane → 2-Methylpentane	1.1	−5.9
n-Hexane → 3-Methylpentane	0.76	−4.6
n-Hexane → 1-Hexene + H_2	0.037	130

cussed in more detail in a subsequent section. The formation of olefins from paraffins, similar to the formation of aromatics, is favored by the combination of high temperature and low hydrogen partial pressure. The thermodynamics of olefin formation can play an important role in determining the rates of those reactions which proceed via olefin intermediates, since thermodynamics sets an upper limit on the attainable concentration of olefin in the system.

The equilibria in the case of isomerization reactions are much less temperature sensitive than in the case of dehydrogenation reactions, since the heats of reaction are relatively small. Furthermore, it is seen that the equilibrium between methylcyclopentane and cyclohexane favors the former, indicating that the five-membered ring structure is more stable than the six-membered ring. In the case of the equilibria between n-hexane and the methylpentanes, it can be seen that the 2-methylpentane is the favored isomer over 3-methylpentane, which is reasonable from the simple statistical consideration that the substituent methyl group can occupy either of two equivalent positions in the former molecule as compared to one in the latter. Thermodynamic calculations would indicate the presence of appreciable quantities of dimethylbutanes at equilibrium (about 30–35 percent of the total hexane isomers at 773 K), but such quantities of these isomers are not observed in reforming. While equilibrium tends to be established rather readily in the case of n-hexane and the methylpentanes, this does not appear to be true for the dimethylbutanes. This suggests that a strong kinetic barrier resists the rearrangement of singly branched to doubly branched isomers [12], which apparently does not exist in the case of the rearrangement of the normal structure to the singly branched structures.

Hydrocracking and hydrogenolysis reactions involve rupture of carbon-carbon bonds followed by hydrogenation. Such reactions are very exothermic and are highly favorable from a thermodynamic point of view. Thermodynamic considerations are of little concern in a discussion of the extent to which these reactions occur in catalytic reforming. Practically speaking, the reactions are limited solely by kinetic factors.

A. Mechanistic Considerations

In discussing the mechanism of hydrocarbon transformations on bifunctional catalysts, it is useful to refer to a reaction scheme originally proposed by Mills et al. [18] to describe the reforming of C_6 hydrocarbons. The scheme is shown in Figure 1. The vertical reaction paths in the figure take place on the hydrogenation-dehydrogenation centers of the catalyst and the horizontal reaction paths on the acidic centers. According to the mechanism, the conversion of methylcyclopentane to benzene, for example, first involves dehydrogenation to methylcyclopentenes on hydrogenation-dehydrogenation centers of the catalyst, followed by isomerization to cyclohexene on acidic centers. The cyclohexene then returns to hydrogenation-dehydrogenation centers where it can either be hydrogenated to cyclohexane or further dehydrogenated to benzene, the relative amounts of these products depending

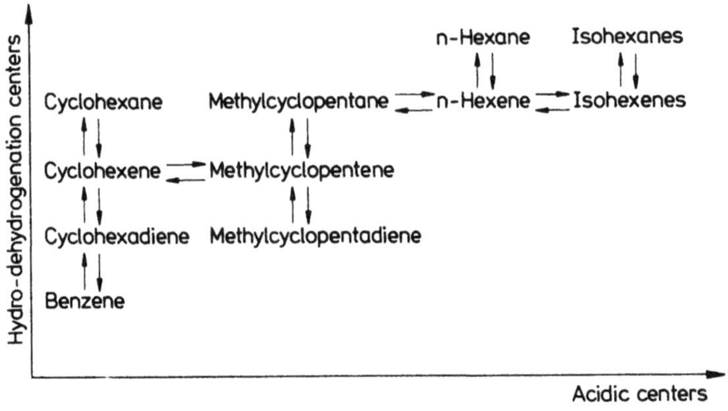

Figure 1. Reaction paths in catalytic reforming of C_6 hydrocarbons [18]

on reaction conditions. The mode of transport of olefin intermediates between metal and acidic sites must be considered in this type of reaction scheme. In the case of isomerization of alkanes or dehydroisomerization of alkylcyclopentanes, a reaction sequence involving transport of olefin intermediates from metal to acidic sites via the gas phase gives a good account of much experimental data [19–21].

Investigations of the isomerization of alkanes in recent years have provided evidence that the reaction can occur on certain metals, notably platinum, in the absence of a separate acidic component in the catalyst [22–24]. While it has been shown that a purely metal-catalyzed isomerization process can occur, the findings do not challenge the commonly accepted mode of action of bifunctional reforming catalysts in which separate metal and acidic sites participate in the reaction. The available data at conditions commonly employed with commercial reforming catalysts indicate that a purely metal catalyzed process does not contribute appreciably to the overall isomerization reaction on a bifunctional catalyst. Nevertheless, the metal catalyzed isomerization reaction is of interest from the point of view of understanding the nature of hydrocarbon transformations on metal surfaces. It has been suggested that carbonium-ion like intermediates are involved in alkane isomerization reactions on platinum [25], and a specific mechanism has been proposed by Anderson and Avery [26].

In the reaction scheme in Figure 1, the dehydrocyclization of n-hexane is indicated to proceed via formation of n-hexene on dehydrogenation centers, followed by cyclization of the n-hexene to methylcyclopentane on acidic centers. The methylcyclopentane then is converted to benzene in the manner already described. Alternatively, it seems possible that hexadiene may be an intermediate in the reaction sequence. Such a sequence would involve formation of hexadiene on platinum sites, followed by cyclization on acidic centers to the cyclic olefin, methylcyclopentene [1]. However, there is also evidence that dehydrocyclization may proceed by another route which involves only the metal component of the catalyst. Thus, it has been observed that unsupported platinum powders catalyze the dehydrocyclization of n-heptane [23]. Also, Dautzenberg and Platteeuw [27] report that dehydrocyclization of n-hexane to benzene occurs over a catalyst in which platinum is supported on a non-acidic alumina. Since bifunctional catalysis with participation of acidic sites is then presumably eliminated, the activity is attributed to the platinum itself. These workers also suggest that dehydrocyclization may occur via the gas phase cyclization of hexatrienes formed on the platinum. However, the rate of dehydrocyclization, which is in general dependent on hydrogen partial pressure, is a particularly strong inverse function of this variable for the path involving hexatriene intermediates. Consequently, this path is probably not an important contributor to the overall dehydrocyclization reaction at the high hydrogen partial pressures typical of catalytic reforming.

Reactions involving rupture of carbon-carbon bonds with accompanying hydrogenation (i.e., hydrocracking and hydrogenolysis) occur to a significant extent in reforming. Part of the overall reaction requires only the metal component of the catalyst. This part, which we designate as hydrogenolysis here, has been studied rather extensively in recent years [28]. The sequence of steps which is generally postulated for the reaction includes (a) chemisorption of hydrocarbon reactant with dissociation of carbon-hydrogen bonds and (b) rupture of carbon-carbon bonds in the unsaturated hydrocarbon residue formed in the chemisorption step. The species formed in step (b) are then hydrogenated to form the products of the reaction. Another part of the overall reaction involves acidic sites on the carrier, and is designated as hydrocracking. Saturated hydrocarbons presumably undergo hydrocracking by two different routes, one involving only the acidic sites of the carrier [29, 30] and the other involving a cooperative action [31] of these sites with metal sites. In the latter route olefins are regarded as intermediates in a reaction sequence similar to that already described for isomerization, except that the olefin undergoes a fragmentation (cracking) reaction rather than a skeletal rearrangement.

B. Kinetics

In this section we consider results of kinetic investigations of reactions of selected hydrocarbons on reforming catalysts. Although a substantial part of the work has been conducted at temperatures lower than those commonly employed in actual reforming operations, the kinetic information has revealed a great deal about the nature of the reactions which occur.

1. Dehydrogenation of Cyclohexanes

The dehydrogenation of cyclohexane and its homologs to aromatics commonly occurs at temperatures above about 570 K on a number of Group VIII metal catalysts. At temperatures much lower than are required for dehydrogenation to aromatics, however, the carbon-hydrogen bonds in cyclohexanes are readily activated on various Group VIII metals. This is demonstrated by the high activities of the metals for the exchange reaction of cyclohexane with deuterium to yield deuterocyclohexanes, $C_6H_{12-x}D_x$ [32]. By extending studies of the exchange reaction to temperatures sufficiently high for dehydrogenation to aromatics to occur, one can obtain useful information for interpreting the kinetics of the latter reaction. In studies on a nickel catalyst, the exchange of cyclohexane with deuterium was the sole reaction observed at temperatures lower than about 520 K [33]. When the temperature was increased above 520 K, dehydrogenation to benzene occurred simultaneously with the exchange reaction. The rate of dehydrogenation increased more rapidly than the rate of exchange with increasing temperature. At temperatures slightly higher than 573 K, the dehydrogenation to benzene occurred with negligible simultaneous formation of deuterocyclohexanes. At these temperatures, the chemisorption of cyclohexane is therefore effectively irreversible. A similar conclusion was reached earlier in a study conducted in our laboratory on the kinetics of dehydrogenation of methylcyclohexane to toluene on a platinum catalyst [34].

The rate data on methylcyclohexane dehydrogenation on platinum in the temperature range of 588 to 645 K at low conversion levels are consistent with a kinetic analysis based on a simple sequence of irreversible steps:

$$\mathrm{M} \xrightarrow{k_1} \mathrm{M_a} \rightarrow \mathrm{T_a} \xrightarrow{k_2} \mathrm{T}$$

Here M and T represent methylcyclohexane and toluene in the gas phase, M_a represents the initial surface species formed in the chemisorption of methylcyclohexane, and T_a is adsorbed toluene. The transformation of M_a to T_a probably consists of a series of steps involving partially dehydrogenated chemisorbed intermediates. However, at steady-state conditions the rates of all the steps in the sequence are equal, and the kinetic analysis is not complicated by this factor. Assuming that coverage of the active sites by components other than toluene is very small, and equating the rates of the initial and final steps of the reaction sequence, we obtain the equation

$$k_1 p_M(1 - \theta_T) = k_2\theta_T \tag{1}$$

where θ_T represents the fraction of the active sites covered by toluene and p_M represents the methylcyclohexane partial pressure. The left-hand member of equation 1 represents the rate of the initial methylcyclohexane chemisorption step, and the right-hand member represents the rate of the final toluene desorption step. Solving equation 1 for θ_T, we can then write for the rate of toluene desorption, and hence the rate of reaction, the expression

$$r = k_2\theta_T = \frac{k_2\left(\dfrac{k_1}{k_2}\right) p_M}{1 + \left(\dfrac{k_1}{k_2}\right) p_M}. \tag{2}$$

This rate law describes the experimental data very well. From the temperature dependence of k_2, an activation energy of 138 kJ mole^{-1} is obtained for the desorption of toluene from the surface [34]. From the temperature dependence of k_1/k_2, a difference of 125 kJ mole^{-1} between the activation energies of the initial and final steps is obtained, so that the activation energy for chemisorption of methylcyclohexane is 13 kJ mole^{-1}. A low activation energy for chemisorption of methylcyclohexane appears very reasonable, in view of the observation that exchange of cyclohexane with deuterium over platinum takes place readily even at room temperature [32].

The activation energy of 138 kJ mole^{-1} for the desorption step is in fair agreement with a calculation using absolute rate theory. According to absolute rate theory, the rate of desorption is given by

$$r = C_a \frac{kT}{h} \exp(-E/RT) \tag{3}$$

where C_a is the number of molecules adsorbed per unit surface, kT/h is a frequency factor, and E is the activation energy [35]. Substituting an experimentally determined value of 2.6×10^{14} molecules sec^{-1} cm^{-2} at 588 K for r at conditions where the reaction is zero order with respect to methylcyclohexane pressure, and assuming C_a is equal to 10^{15} molecules cm^{-2} of platinum surface at full coverage, a value of 155 kJ mole^{-1} is calculated for the activation energy E. This agrees fairly well with the value of 138 kJ mole^{-1} derived from the temperature dependence of k_2.

For methylcyclohexane dehydrogenation on platinum at low conversion levels, a kinetic analysis involving a sequence of irreversible reaction steps is thus a plausible one. Attempts to describe the kinetics in terms of a reaction scheme invoking adsorption equilibrium between methylcyclohexane and an adsorbed intermediate meet with several difficulties, as has been discussed previously [1, 34]. Since recent findings in the investigation of cyclohexane dehydrogenation on nickel in the presence of deuterium [33] lead to the same conclusion, this may well be a characteristic feature of the kinetics of dehydrogenation of cycloalkanes on metals.

2. *Isomerization Reactions*

The kinetics of isomerization of alkanes will be discussed by referring to a specific example, the isomerization of n-pentane over a platinum on alumina catalyst containing 0.3 wt. percent platinum [21]. The rate measurements were made in a flow system at 645 K at low conversion levels (4–18 percent). The n-pentane was passed over the catalyst in the presence of hydrogen at total pressures ranging from 7.7 to 27.7 atm and at hydrogen to n-pentane ratios varying from 1.4 to 18. Over this range of conditions the rate was found to be independent of total pressure and to increase with increasing n-pentane to hydrogen ratio. The rate data were correlated by an expression of the form

$$r = k[(nC_5)/(H_2)]^n \tag{4}$$

where r and k are the rate and rate constant, respectively, and (nC_5) and (H_2) are the concentrations of n-pentane and hydrogen in the gas phase. The exponent n was equal to 0.5 over the range of conditions studied.

The kinetic data were interpreted in terms of a mechanism involving n-pentene intermediates:

$$nC_5 \underset{}{\overset{Pt}{\rightleftharpoons}} nC_5^{=} + H_2$$

$$nC_5^{=} \xrightarrow[Site]{Acid} iC_5^{=}$$

$$H_2 + iC_5^{=} \overset{Pt}{\rightleftharpoons} iC_5$$

According to this mechanism the n-pentane dehydrogenates to n-pentenes on platinum sites. The n-pentenes are then adsorbed on acid sites, where they are isomerized to isopentenes. The latter are then hydrogenated to isopentane on platinum sites, thus completing the reaction. The isomerization step on the acid sites has commonly been assumed to involve carbonium ion type intermediates [36].

If it is assumed that the isomerization of the n-pentenes on the acid sites is the rate limiting step, the rate of reaction is given by the equation

$$r = k'[nC_5^{=}]_a \tag{5}$$

where k' is a rate constant and $[nC_5^{=}]_a$ represents the concentration of n-pentenes on the acid sites. Assuming that equilibrium is effectively established between n-pentenes, n-pentane, and hydrogen in the initial dehydrogenation step, the concentration $(nC_5^{=})$ of n-pentenes in the gas phase is

$$(nC_5^{=}) = K_1(nC_5)/(H_2) \tag{6}$$

where K_1 is an equilibrium constant and (nC_5) and (H_2) are the concentrations, respectively, of n-pentane and hydrogen in the gas phase. We assume also that equilibrium is established between n-pentenes in the gas phase and on the acid sites, and that the equilibrium can be represented by the equation

$$[nC_5^{=}]_a = K_2(nC_5^{=})^n \tag{7}$$

where the exponent n is expected to have a value between 0 and 1. The rate expression becomes, after substitution of equations 6 and 7 in equation 5,

$$r = k'K_2K_1^n[(nC_5)/(H_2)]^n = k[(nC_5)/(H_2)]^n \tag{8}$$

where $k = k'K_2K_1^n$. Equation 8 is identical with equation 4 derived from the data.

The rate equation for the isomerization of n-pentane according to the above analysis can be expressed in the form

$$r = k'K_2(nC_5^{=})^n = kK_1^{-n}(nC_5^{=})^n\,. \tag{9}$$

Identifying the parameters *k* and *n* of this equation with the parameters *k* and *n* in the experimental rate law (equation 4), and noting that K_1 is the equilibrium constant of equation 6, it was observed that the rate of isomerization of 1-pentene over a sample of catalyst containing only the acid function (no platinum present) was in good agreement with the rate calculated from equation 9. Whether one starts with 1-pentene or 2-pentene in this type of experiment would be expected to make little difference, since the rate of double bond migration is fast compared to the rate of skeletal isomerization of olefins. These results provide good supporting evidence for the kinetic analysis presented here.

In the example of n-pentane isomerization just considered, the rate of reaction is determined by the catalytic activity of the acid sites in the catalyst, since the rate limiting step in the reaction occurs on these sites. This situation is the usual one in the isomerization of alkanes on platinum-alumina catalysts at the conditions employed in catalytic reforming. Increasing the dehydrogenation activity of the catalyst by increasing the platinum content has no significant effect on the rate of isomerization [37]. Data on the isomerization of n-heptane to 2-methylhexane and 3-methylhexane in Table 2 show no change in the reaction rate when the platinum content is increased from 0.1 to 0.6 wt. percent. In this range of composition, the dehydrogenation activity is proportional to platinum content [37]. When the platinum content is as high as 0.1 wt. percent, the dehydrogenation activity is already high enough to maintain the equilibrium concentration of alkene (olefin) intermediates in the gas phase. Hence there is no effect of increasing the dehydrogenation activity further. The same result is found for the isomerization-dehydroisomerization of alkylcyclopentanes, as typified by the conversion of methylcyclopentane to cyclohexane and benzene (mostly the latter at typical reforming conditions). In this reaction, methylcyclopentenes are intermediates. As shown in Table 2, increasing platinum content from 0.3 to 0.6 wt. percent has very little effect on the rate of reaction.

Table 2. Effect of Pt content on reactivities of n-heptane (nC_7) and methylcyclopentane (MCP) over $Pt{-}Al_2O_3$ [37]

Pt Content/Wt. %	0.10	0.30	0.60
Reaction Rates[a]			
Isomerization of nC_7			
744 K	0.035	0.035	0.038
800 K	0.12	0.13	0.12
Dehydrocyclization of nC_7			
744 K	0.0022	0.0027	0.0045
800 K	0.020	0.025	0.035
Isomerization-Dehydroisomerization of MCP			
744 K	—	0.019	0.021
772 K	—	0.039	0.043

[a] Gram moles per hour per gram of catalyst at 21 atm. and H_2/hydrocarbon mole ratio = 5.

In the range of hydrogen pressures commonly used in reforming, the rates of isomerization of alkanes and isomerization-dehydroisomerization of alkylcyclopentanes decrease with increasing hydrogen pressure. However, when a much wider range of hydrogen pressures is considered, the directional effect of this variable changes. In the complete absence of hydrogen, the reactions are not observed at all. In the presence of hydrogen, the rates of reaction increase with increasing hydrogen pressure up to a given point, beyond which they decrease with further increase in hydrogen pressure. The change in the nature of the hydrogen pressure effect on the rates has been interpreted as follows: In the absence of hydrogen, or at low hydrogen pressures, the platinum sites become heavily covered with hydrogen-deficient hydrocarbon residues. As a result, the reaction rates are limited by the dehydrogenation activity of the catalyst, i.e., the rate of formation of intermediate olefins. The role of hydrogen is then one of keeping the platinum sites free of surface residues. The extent of coverage of the surface by such residues decreases with increasing hydrogen pressure, thus accounting for the beneficial effect of hydrogen pressure on the over-all rate of isomerization. At sufficiently high hydrogen pressures, however, coverage of platinum sites by surface residues ceases to be a limiting factor, and the rate does not continue to increase with increasing hydrogen pressure. This means that the formation of intermediate olefins is no longer the rate controlling step, and the overall reaction becomes limited by the isomerization of olefins on the acidic centers of the catalyst. The decrease in the rate of isomerization with further increase in hydrogen pressure is attributed to the effect of hydrogen pressure in limiting the concentration of olefin intermediates attainable at equilibrium, as can be seen from our discussion of n-pentane isomerization kinetics.

3. Dehydrocyclization of Alkanes

The dehydrocyclization of alkanes is a reaction of major importance in reforming. The mode of action of bifunctional reforming catalysts for this reaction is less clear than it is for the dehydrogenation of cyclohexanes, the isomerization of alkanes, or the isomerization-dehydroisomerization of alkylcyclopentanes. Consequently, the interpretation of kinetic data from a mechanistic viewpoint has been more difficult.

An early investigation of n-heptane dehydrocyclization kinetics on a platinum reforming catalyst was that of Hettinger et al. [16]. These investigators found that the rate of dehydrocyclization decreased with increasing total pressure at a constant hydrogen to hydrocarbon mole ratio. They also reported that the rate was substantially greater for n-nonane than for n-heptane, which is reasonable purely on the basis of the greater number of ways in which ring closure may occur for the former. In a subsequent paper from our laboratories [38], data on the kinetics of n-heptane dehydrocyclization on a platinum on alumina catalyst were reported over a wide range of hydrogen partial pressures. The n-heptane partial pressure was constant in this work. In the complete absence of hydrogen, no dehydrocyclization was detected.

In the presence of hydrogen the rate was found to increase with increasing hydrogen pressure up to a given point, beyond which the rate decreased with further increase in hydrogen pressure. This behavior is similar to that observed for isomerization and is interpreted in much the same way: The dehydrocyclization reaction involves a preliminary dehydrogenation step prior to ring closure. At sufficiently high hydrogen pressures, where coverage of platinum by carbonaceous residues is not a limiting factor, increasing hydrogen pressure serves to suppress the dehydrogenation step and hence decrease the rate of dehydrocyclization. However, at low hydrogen pressures (below about 6 atm at 744–800 K.) background reactions involving extensive dehydrogenation and polymerization become important. These reactions lead to extensive coverage of the platinum by carbonaceous residues. The limiting factor in the reaction then becomes the removal of the residues by reaction with hydrogen, thus accounting for the beneficial effect of hydrogen pressure on the rate.

In studies of the dehydrocyclization of various alkanes on a platinum-alumina catalyst containing 0.3 percent platinum [39], it has been observed that the rate of reaction increases in the sequence: n-hexane < n-heptane < n-octane. These results parallel the results of Hettinger et al. [16] mentioned earlier. Furthermore, the rates of dehydrocyclization of branched alkanes are lower than those of n-alkanes of the same carbon number. Thus, the rate of dehydrocyclization of 3-methylhexane is slightly lower than that of n-heptane, while the rate of dehydrocyclization of 2,2,4-trimethylpentane is much lower than that of n-octane. In the case of 2,2,4-trimethylpentane, the molecule does not readily rearrange to a structure which can satisfactorily undergo ring closure. The difficulty of rearrangement is also evident from the fact that the reaction products contain essentially no n-octane, methylheptanes, or dimethylhexanes. The major reaction of the 2,2,4-trimethylpentane is hydrocracking to lower carbon number alkanes, mostly isobutane. In the case of n-hexane dehydrocyclization, the major product at low conversion is methylcyclopentane rather than benzene or cyclohexane, which suggests that the reaction proceeds preferentially by formation of a five-membered ring structure, which must then undergo further reaction to yield six-membered ring products.

In contrast to the results which we have already considered on the effect of the platinum content of platinum-alumina catalysts on the rates of isomerization of n-heptane and isomerization-dehydroisomerization of methylcyclopentane, the rate of dehydrocyclization of alkanes increases significantly with increasing platinum content. Data illustrating this dependence are shown in Table 2 for the dehydrocyclization of n-heptane at 744 and 800 K. The rate is not directly proportional to platinum content, but does increase by about twofold over the range of platinum content from 0.1 to 0.6 wt. percent [37]. These results indicate that the rate of dehydrocyclization of alkanes is not determined simply by a step which occurs on the acid sites of the catalyst. The type of kinetic analysis which we have considered for the isomerization of n-pentane is not adequate for the dehydrocyclization reaction, since the factors controlling the reaction rate appear to be more complex.

4. Hydrogenolysis Reactions

Hydrogenolysis and hydrocracking reactions are a major concern in reforming, since they compete with reactions which lead to the formation of aromatic hydrocarbons. The investigation of these reactions is important from the standpoint of learning how to control them and hence achieve better selectivity of production of aromatics. The kinetics of hydrogenolysis of hydrocarbons have been studied on the various metals within Group VIII of the periodic table [28, 40, 41]. In particular, the kinetics of hydrogenolysis of ethane to methane, $C_2H_6 + H_2 \rightarrow 2\,CH_4$, have been studied in some detail [28].

In the catalytic hydrogenolysis of ethane on metals, the available evidence indicates that ethane is chemisorbed with dissociation of carbon-hydrogen bonds [28, 40–42]. This process yields a hydrogen deficient surface species which undergoes carbon-carbon bond scission. The following sequence of reaction steps leading to formation of monocarbon surface fragments may be visualized:

$$C_2H_6 \overset{1}{\rightleftharpoons} C_2H_5(\text{ads}) + H(\text{ads})$$

$$C_2H_5(\text{ads}) + H(\text{ads}) \overset{2}{\rightleftharpoons} C_2H_x(\text{ads}) + aH_2$$

$$C_2H_x(\text{ads}) \overset{3}{\rightarrow} \text{adsorbed } C_1 \text{ fragments}$$

The symbol (ads) signifies an adsorbed species, and the quantity a is equal to $(6 - x)/2$. The initial ethane chemisorption step involves dissociation of a carbon-hydrogen bond to form adsorbed C_2H_5. The latter then dehydrogenates to the species C_2H_x, and we assume in this discussion that equilibrium is effectively established between C_2H_5 and C_2H_x on the surface. The chemisorbed C_2H_x then undergoes scission of the carbon-carbon bond to yield monocarbon fragments which are hydrogenated to methane. We proceed with a steady state kinetic analysis in which the net rate of chemisorption of ethane in step 1 is equal to the rate of disappearance of C_2H_x via step 3. Assuming low surface coverage by all adsorbed species, we can write the equation

$$k_1 p_E - k_1' \theta_E \theta_H = k_3 \theta_x \tag{10}$$

where k_1 and k_1' are the forward and reverse rate constants for step 1, and k_3 is the rate constant for step 3. The symbols θ_E, θ_H, and θ_x represent the fractions of the surface covered by C_2H_5, H, and C_2H_x, respectively, and p_E is the ethane partial pressure. Assuming that equilibrium is effectively maintained between adsorbed C_2H_5 and adsorbed C_2H_x in step 2, we can write the equilibrium relation

$$K_2 \theta_E \theta_H = \theta_x p_H^a \tag{11}$$

where K_2 is an equilibrium constant and p_H is the hydrogen partial pressure. The rate of hydrogenolysis is simply the rate of step 3, and is given by the expression

$$r = k_3\theta_x \,. \tag{12}$$

Combining equations 10, 11, and 12, we obtain the rate equation

$$r = k_1 p_E/(1 + bp_H^a) \tag{13}$$

where $b = k_1'/k_3K_2$. Since the parameter b is temperature dependent, the dependence of the rate on hydrogen pressure is a function of temperature. If b decreases with increasing temperature, as found for cobalt [43], the rate equation at a sufficiently high temperature reduces to the expression, $r = k_1p_E$. This corresponds to the situation in which the chemisorption of ethane is completely irreversible. Under such conditions, step 2 of the reaction sequence may also become irreversible, so that the overall reaction may be considered simply as a sequence of irreversible steps. On the other hand, at a sufficiently low temperature, the term bp_H^a in equation 13 is large compared to unity, and the rate equation becomes

$$r = k_1p_E/bp_H^a = k_3K_1K_2p_E/p_H^a \tag{14}$$

where K_1 is the equilibrium constant of step 1 of the reaction sequence. Equation 14 corresponds to adsorption equilibrium being established between C_2H_6 and adsorbed C_2H_x. The equilibrium is characterized by an adsorption equilibrium constant K which is equal to the product of K_1 and K_2.

When the assumption of low surface coverage by the species C_2H_x is not satisfactory, equation 14 can be expressed in the more general form [40, 41]

$$r = k_3K^n(p_E/p_H^a)^n = kp_E^np_H^{-na} \,. \tag{15}$$

The parameter k is equal to k_3K^n. The negative exponent $(-na)$ on hydrogen partial pressure is consistent with kinetic data on many metals. For a given metal the experimentally determined exponent n on ethane partial pressure can be used to determine the value of a which gives the best agreement between the exponent $(-na)$ and the experimental exponent m on hydrogen partial pressure. If permissible values of a are limited to integers from 1 to 3, corresponding to a variation of x in C_2H_x from 4 to 0, the "best fit" values given in Table 3 are obtained for a number of silica-supported Group VIII metal catalysts. It is noteworthy that excellent agreement in Table 3 between the quantities m and $(-na)$ is obtained by simply using integral values of a. A high value of a indicates a high probability that a primary dicarbon surface species will undergo further dehydrogenation as opposed to carbon-carbon scission. Interestingly, the metals with a value of a equal to 3 very likely have the highest ratios of dehydrogenation to hydrogenolysis activities. Of the

Table 3. Kinetics of ethane hydrogenolysis — analysis of hydrogen pressure dependencies [40, 41]

Catalyst[a]	Temp./K	x[b]	a[c]	Exponents in Rate Expression		
				Experimental[d]		Calculated[e]
				Ethane (n)	Hydrogen (m)	Hydrogen ($-na$)
Co	492	4	1	1.0	−0.8	−1.0
Ni	450	2	2	1.0	−2.4	−2.0
Ru	461	2	2	0.8	−1.3	−1.6
Rh	487	0	3	0.8	−2.2	−2.4
Pd	627	0	3	0.9	−2.5	−2.7
Os	425	2	2	0.6	−1.2	−1.2
Ir	483	2	2	0.7	−1.6	−1.4
Pt	630	0	3	0.9	−2.5	−2.7

[a] All metals supported on silica.
[b] Specifies the composition of the surface species C_2H_x.
[c] Defined by the expression, $a = (6 - x)/2$.
[d] Exponents in the experimental power rate law, $r = kp_E^n p_H^m$.
[e] Based on equation 15 in the text.

metals in Table 3, platinum and palladium have by far the lowest hydrogenolysis activities [28, 41]. In the case of rhodium, the dehydrogenation activity may be unusually high, since it has very high activity relative to other metals for hydrogenation [44, 45] and ethane-deuterium exchange [46].

The kinetic analysis leading to equation 15 is not applicable to silica-supported iron or rhenium catalysts, for which the exponent on hydrogen partial pressure in the experimental power rate law is positive [28]. In these cases the rate of ethane hydrogenolysis may be determined by the hydrogenation-desorption of monocarbon fragments, rather than carbon-carbon bond scission [40].

Equation 15 ignores the possible competition between hydrogen and hydrocarbon for surface sites. If such an effect is important, it would alter the detailed form of equation 15 and could affect conclusions about the composition of the surface species C_2H_x [28, 40]. In any case, the simple kinetic analysis reveals an interesting trend in the variation of the composition of C_2H_x from metal to metal.

The assumption of adsorption equilibrium between ethane and the surface species C_2H_x, on which equations 14 and 15 are based, would appear to be a reasonable approximation at the lowest temperatures at which hydrogenolysis is observed. However, as the temperature is increased over a sufficiently wide range, the assumption of adsorption equilibrium becomes a poorer approximation. At sufficiently high temperatures, the reaction is probably represented best as a sequence of irreversible steps.

In comparing the catalytic activities of metals for ethane hydrogenolysis, it is instructive to consider the variation in activity as a function of the position

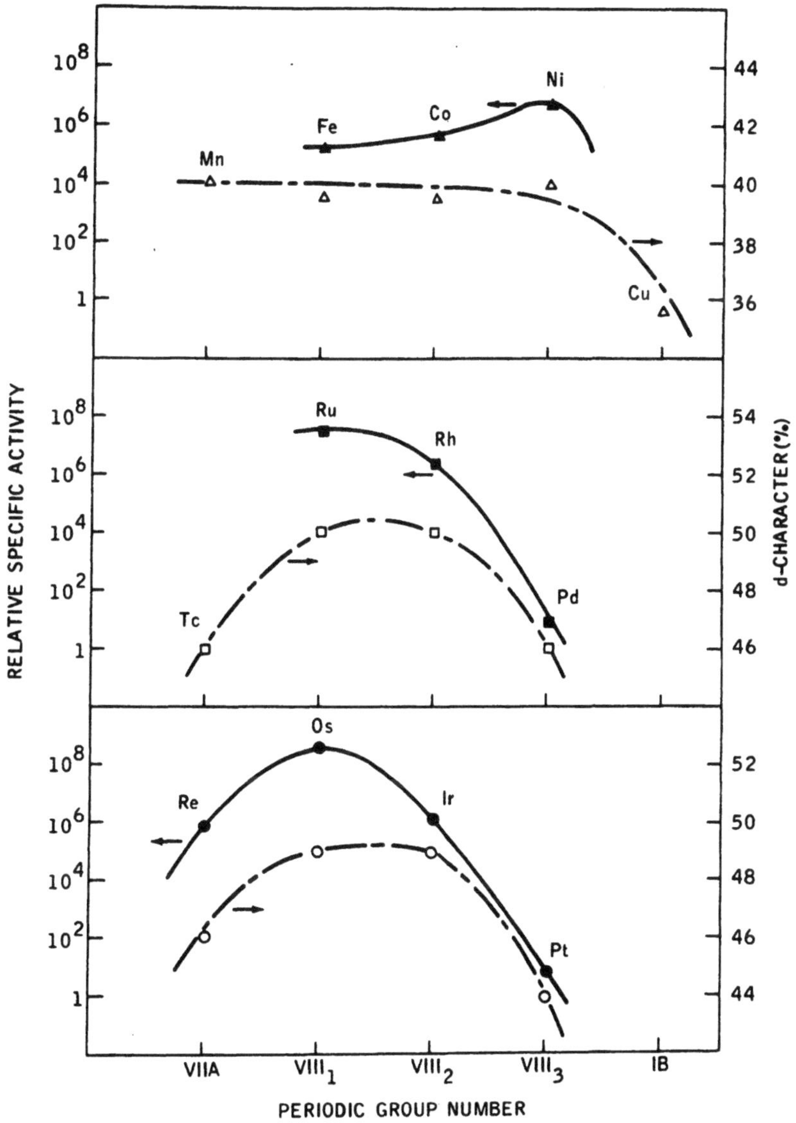

Figure 2. Catalytic activities of metals for ethane hydrogenolysis in relation to the percentage *d*-character of the metallic bond. The closed points represent activities compared at a temperature of 478 K and ethane and hydrogen pressures of 0.030 and 0.20 atm., respectively, and the open points represent percentage *d*-character. Three separate fields are shown in the figure to distinguish the metals in the different long periods of the Periodic Table [28, 41]

of the metal within a given period of the periodic table (Figure 2). The most complete data are available for the metals of the third transition series. Beginning with rhenium in Group VIIA, and proceeding in the direction of increasing atomic number to osmium, iridium, and platinum in Group VIII, we see that the hydrogenolysis activity attains a maximum value at osmium.

From osmium to platinum, the activity decreases by more than seven orders of magnitude. A similar variation is observed from ruthenium to palladium in the second transition series.

In the case of the Group VIII metals of the first transition series (iron, cobalt, and nickel), maximum catalytic activity is observed for the metal in the third sub-group, rather than for the metal in the first sub group. Thus, the pattern of variation of hydrogenolysis activity among the Group VIII metals of the first transition series is somewhat different from that observed among the Group VIII metals of the second and third transition series. This difference tends to parallel known chemical differences between elements of the first transition series on the one hand and the corresponding elements of the second and third transition series on the other [47].

Included with the plots of hydrogenolysis activities of the metals in Figure 2 are plots of the percentage *d*-character of the metallic bond, a quantity introduced by Pauling [48] in his valence bond theory of metals to represent the extent of participation of *d*-orbitals in the metallic bond. For the metals within a given long period or transition series, the patterns of variation of hydrogenolysis activity and percentage *d*-character from one metal to another are similar. Further consideration of the relation between hydrogenolysis activity and percentage *d*-character distinguishes again the metals of the first transition series from the metals of the second and third transition series. The hydrogenolysis activities of iron, cobalt, and nickel in the first transition series are comparable to those of metals with significantly higher *d*-character in the second and third transition series. Thus, while there is a degree of correlation between hydrogenolysis activity and percentage *d*-character, this parameter alone is not adequate for characterizing the catalytic activity of transition metals for hydrogenolysis [28].

From the standpoint of catalytic reforming, the low hydrogenolysis activity of platinum relative to most of the other Group VIII metals is of interest. A combination of low hydrogenolysis activity and high activity for reactions yielding aromatic hydrocarbons makes it an excellent component of a reforming catalyst. The high hydrogenolysis activities of metals such as ruthenium, rhodium, rhenium, osmium, and iridium tend to cast doubt on the feasibility of using these metals in place of platinum in reforming. However, it has been found that catalysts containing platinum in combination with certain of these metals have marked advantages over catalysts consisting of platinum alone on alumina [6, 7].

3. Nature of Platinum Reforming Catalysts

The platinum-alumina catalyst system was the first to be used widely in petroleum reforming. In commercial catalysts, the amount of platinum present is commonly in the range of 0.3 to 0.6 wt. percent. The catalysts also contain chlorine in amounts typically in the range of 0.3 to 1.0 wt. percent. One common method of preparing such a catalyst involves impregnation of alumina with chloroplatinic acid, followed by calcination in air at temperatures in the range 820 to 870 K [49]. The alumina generally has a surface area in the range of 150–300 $m^2\,g^{-1}$. In commercial reforming units the catalysts are commonly used in the form of pellets or extrudates with dimensions in the approximate range of 1.5–4 mm. The choice of size is based on considerations of pressure drop and diffusional limitations. For fundamental studies in small laboratory reactors, where pressure drop considerations are not a major issue, it may be preferable to use the catalyst in the form of small granules to minimize diffusional limitations.

A. Characterization of Dispersed Platinum

1. Chemisorption Studies

Chemisorption measurements have shown that freshly prepared platinum on alumina reforming catalysts are characterized by extremely high dispersion of platinum on the surface of the alumina carrier [50, 51]. In hydrogen chemisorption studies on such catalysts, the amount of hydrogen taken up is frequently close to one hydrogen atom per atom of platinum. Typical data on the chemisorption of hydrogen at room temperature on a platinum on alumina catalyst [52] are shown in Figure 3. The isotherm labelled A is the original isotherm determined on the "bare" catalyst surface. The "bare" surface was prepared by evacuation of the adsorption cell at high temperature (723 K) subsequent to the reduction of the catalyst in flowing hydrogen at 773 K. After the catalyst was cooled to room temperature in vacuum, it was contacted with hydrogen in the measurement of the isotherm. After isotherm A was completed, the adsorption cell was evacuated at room temperature for ten minutes (to approximately 10^{-4} Pa) and a second isotherm labelled B was measured. Isotherm A represents the total chemisorption. Isotherm B represents the weakly chemisorbed fraction, since it is removed by simple evacuation at room temperature. This isotherm includes adsorption on the alumina carrier. The difference isotherm, labelled A–B, is obtained by subtracting isotherm B from isotherm A and is independent of pressure over the range of pressures used in obtaining the isotherm. It represents the strongly chemisorbed fraction, i.e., the amount which cannot be removed by evacuation at room temperature. The quantity H/M in the right hand ordinate of Figure 3 represents the ratio of the number H of hydrogen atoms adsorbed to the number M of platinum atoms in the catalyst. If we assume a stoichiometry of one hydrogen atom per surface platinum atom in the case of the strongly chemisorbed fraction, the value of H/M determined from the difference isotherm A–B corresponds to the ratio of surface platinum atoms to total platinum atoms in the catalyst. This ratio is commonly called the metal dispersion, which is about 0.9 for the catalyst in Figure 3.

High resolution electron microscopy studies provide independent evidence of the highly dispersed nature of platinum in platinum-alumina reforming catalysts [53]. Such studies have shown that the platinum exists as very small

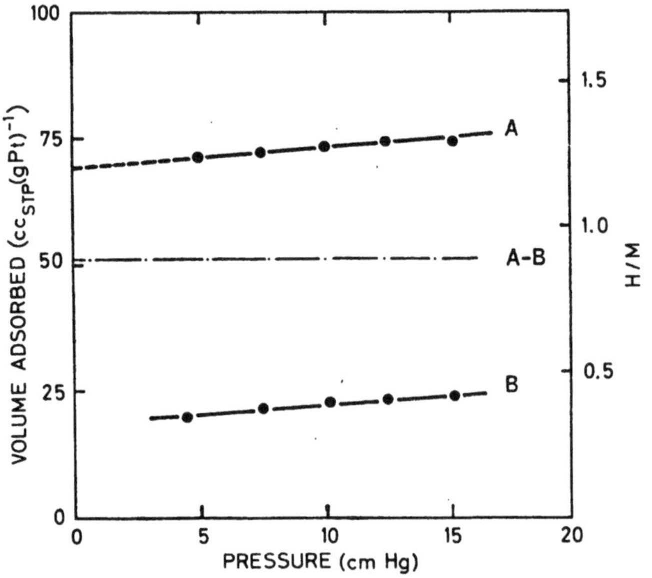

Figure 3. Typical hydrogen chemisorption data at room temperature on a platinum on alumina catalyst containing 1 wt. percent platinum. Isotherm A is the original isotherm, while isotherm B is a second isotherm determined after evacuation of the adsorption cell for 10 min (to a pressure of approximately 10^{-4} Pa) subsequent to the completion of isotherm A. The difference isotherm A–B, obtained by subtracting isotherm B from isotherm A, represents the strongly chemisorbed fraction [52]

crystallites or clusters of the order of 10 Å in size. The good agreement between conclusions derived from chemisorption and electron microscopy data is important, since a reliable characterization of metal dispersion is essential in a reforming catalyst.

2. Extended X-Ray Absorption Fine Structure (EXAFS) of Dispersed Platinum

Recent studies from our laboratory [54–55] on the extended X-ray absorption fine structure (EXAFS) of dispersed metal catalysts look very promising for the characterization of such systems. In a plot of X-ray absorption coefficient vs. the energy of the X-ray photons, the fine structure refers to fluctuations in the absorption coefficient on the high energy side of an absorption edge. The fluctuations of interest in EXAFS begin at an energy of approximately 30 eV beyond an absorption edge and extend over an additional range of 1000–1500 eV. The fine structure is observed in the absorption of X-rays by all forms of matter other than monatomic gases, and was first considered theoretically by Kronig [56–58]. The possibilities of EXAFS as a tool for investigating the structures of non-crystalline materials, however, have been realized only recently, and have emerged as a result of advances in methods of data analysis [59, 60] and experimental techniques, the latter being due primarily to the application of high intensity synchrotron radiation as an X-ray source.

In EXAFS we are concerned with the ejection of an inner core electron from an atom as a result of X-ray absorption. The ejected electron (photoelectron) is characterized by a wave vector K, which is given by the equation

$$K = (2\,mE)^{1/2}/\hbar \tag{16}$$

where m is the mass of the electron, $\hbar$ is Planck's constant/2π, and E is the kinetic energy of the photoelectron. The energy E is the difference between the X-ray energy and a threshold energy associated with the ejection of the electron. At the threshold energy, an X-ray absorption spectrum exhibits an abrupt change in absorption coefficient, i.e., the absorption edge. A typical spectrum [54] of platinum metal at 100 K is shown in Figure 4. The absorption coefficient is shown as a function of the energy E relative to the L_{III} edge. The data cover a wide enough range of energy to include all three of the characteristic L absorption edges, L_{III}, L_{II}, and L_I, corresponding respectively, to ejection of photoelectrons from $2p_{3/2}$, $2p_{1/2}$, and $2s$ states. At energies higher than the threshold value corresponding to a particular absorption edge, we note the fluctuations in absorption coefficient which constitute the extended fine structure.

In the analysis of EXAFS data, it is useful to consider a function $\chi(K)$ defined by the equation

$$\chi(K) = (\mu - \mu_0)/\mu_0 \tag{17}$$

where μ and μ_0 are atomic absorption coefficients characteristic of the absorption associated with the ejection of an inner core electron from an atom.

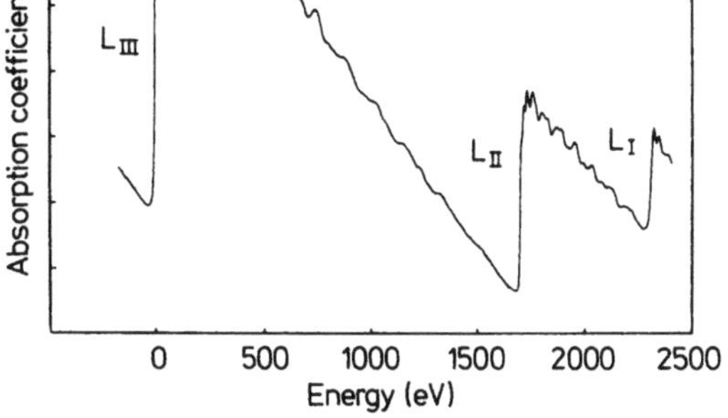

Figure 4. Typical L-absorption spectrum at 100 K of a 2.5 μ platinum foil [54]. The energy scale is that of a photoelectron emitted from the $2p_{3/2}$ state (L_{III} edge)

The coefficient μ refers to absorption by an atom in the material of interest while μ_0 refers to absorption by an atom in the free state, both of which are functions of K. The difference between the experimental absorption coefficient at a given energy and a value at the same energy determined by an average background line through the data, i.e., the fluctuation, is proportional to the numerator of equation 17. The increase in absorption coefficient at the edge (i.e., the difference between the minimum value at the energy corresponding to the onset of the edge and a value at the same energy obtained by a slight extrapolation of the average background line from the high energy side of the edge), which we call the step height, is proportional to the denominator μ_0 of equation 17. To account for the dependence of μ_0 on K, we determine a hypothetical step height for energies beyond the edge by employing empirical corrections based on the known energy dependence of the absorption coefficient of the element on either side of the edge. A value of $\chi(K)$ at a particular value of K is then determined by dividing the fluctuation by the step height.

Theories of EXAFS [61–65] based on the scattering of ejected photoelectrons by atoms in the coordination shells surrounding the central absorbing atom give an expression for $\chi(K)$ of the following form

$$\chi(K) = \sum_j A_j(K) \sin\left[2KR_j + 2\delta_j(K)\right] \tag{18}$$

where the summation extends over j coordination shells. In this expression, R_j is the distance from the central absorbing atom to atoms in the jth coordination shell and $2\delta_j(K)$ is the phase shift. The factor $A_j(K)$ is an amplitude function for the jth shell, and is defined by the expression

$$A_j(K) = (N_j/KR_j^2)F_j(K) \exp\left(-2\,K^2\sigma_j^2\right) \tag{19}$$

where N_j is the number of atoms in the jth shell, σ_j is the root mean square deviation of the interatomic distance about R_j, and $F_j(K)$ is a factor accounting for electron backscattering and inelastic scattering.

Fourier transformation of EXAFS data yields a radial structure function [61, 66] which exhibits a series of peaks at $R = R_j' = R_j - a_j$, the variable R representing the distance from the absorber atom. The difference a_j from a value of R_j corresponding to a particular coordination shell arises from the phase shift in equation 18. The EXAFS function $\chi(K)$ is commonly multiplied by a weighting factor K^n with $n = 1$ or $n = 3$ before the transform is taken [67]. Data on $K^3 \cdot \chi(K)$ vs. K at 100 K for platinum metal and dispersed platinum catalysts, and Fourier transforms of the data [52], are shown in Figure 5. The EXAFS fluctuations for the dispersed platinum catalysts are substantially smaller than those for platinum metal. Correspondingly, the magnitudes of the peaks in the Fourier transforms are also smaller (Note that the scales in the figures are not the same for the dispersed platinum catalysts and platinum metal). These features are a consequence of a lower average coordination number and/or a higher degree of disorder of the platinum atoms in the dispersed catalysts. The degree of disorder is characterized

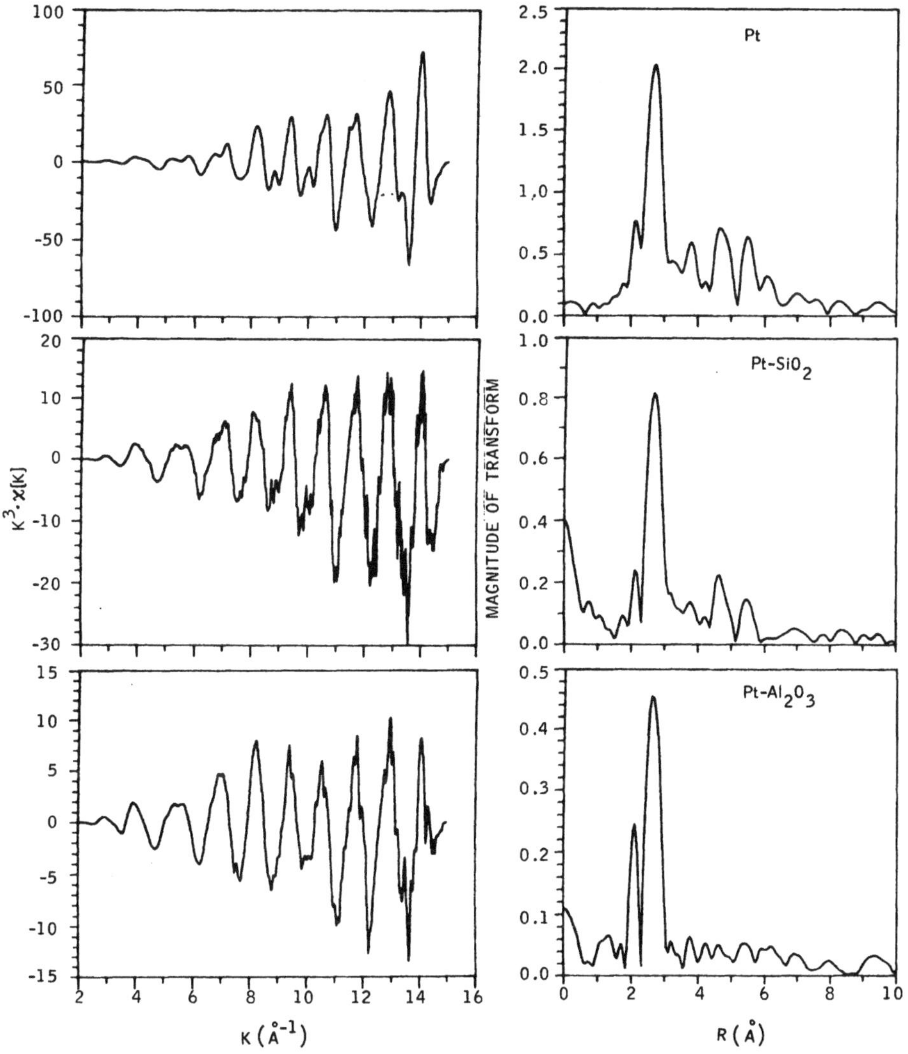

Figure 5. Normalized EXAFS data at 100 K and associated Fourier transforms for platinum metal and for dispersed platinum catalysts containing 1 wt. percent platinum [52]

by the parameter σ_j in equation 19. We note that the peaks in the Fourier transforms are located at values of R slightly lower than the true interatomic distances of platinum metal (about 0.12–0.15 Å lower), as a consequence of the phase shift.

A summary of properties of the platinum catalysts derived from the EXAFS data in Figure 5 is given in Table 4. Chemisorption measurements on the catalysts indicated platinum dispersions in the range of 0.7–0.9, where dispersion is defined as the ratio of surface platinum atoms to total platinum atoms [52]. Details of the procedure employed in the analysis of the EXAFS data are given elsewhere [52]. Included in the table are values of

Table 4. Properties of dispersed platinum catalysts derived from EXAFS data-comparison with bulk platinum [52]

	Pt Catalysts[a]		Bulk Pt [b]
	$Pt-SiO_2$	$Pt-Al_2O_3$	
N_1 [c]	8.0	7.2	12
R_1, Å [d]	2.774	2.758	2.775
$\Delta\sigma_1^2$, $Å^2$ [e]	0.0018	0.0037	—
σ_1, Å [f]	0.061	0.075	0.044

[a] The concentration of platinum in the catalysts was 1.0 wt. percent.

[b] The properties shown for bulk platinum, except for σ_1, were obtained from the literature. The value of σ_1 was calculated in a manner described elsewhere [52]. The EXAFS data on bulk platinum, in conjunction with the known values of N_1 and R_1 for bulk platinum, were used in the analysis of the EXAFS data on the dispersed platinum catalysts.

[c] Coordination number of the platinum atoms (nearest platinum atom neighbors). In the case of the dispersed platinum catalysts, the value is an average coordination number, with an estimated uncertainty of ±20 percent.

[d] Nearest neighbor Pt—Pt distances. The estimated uncertainty in the R_1 values for the dispersed platinum is 0.01–0.02 Å.

[e] The difference between the values of σ_1^2 for the dispersed platinum clusters and the corresponding value for bulk platinum at 100 K, as determined from EXAFS data.

[f] Root mean square deviation of interatomic distance about the equilibrium value. The values of σ_1 for the dispersed platinum clusters were determined from the values of $\Delta\sigma_1^2$ from the EXAFS data and the calculated value of σ_1 for bulk platinum.

the average coordination number N_1, which refers to the average number of nearest metal atom neighbors (i.e., atoms in the first coordination shell) about an atom in a platinum crystallite or cluster. While the estimated uncertainly in N_1 is high (±20%), the values are significantly lower than the value of 12 for bulk platinum. This result is expected, since most of the platinum atoms in the clusters are surface atoms with lower coordination numbers than the atoms in the interior of a crystal. Also, atoms at corners and edges have lower coordination numbers than the interior atoms in surface planes of crystals and become increasingly important as the size of a metal crystal decreases.

Also included in Table 4 are values of the interatomic distances R_1. The values of R_1 for the platinum catalysts differ from the value for bulk platinum by less than 0.02 Å, which is close to the estimated uncertainty in the determination of R_1 [52]. The quantity $\Delta\sigma_1^2$ in Table 4 is the difference between σ_1^2 for the dispersed platinum clusters and the corresponding quantity for bulk platinum at 100 K, where σ_1^2 is defined as the mean square deviation of interatomic distance about the equilibrium value. For both catalysts the value of $\Delta\sigma_1^2$ is positive, i.e., the values of σ_1 are higher for the dispersed platinum catalysts than for bulk platinum. The value of σ_1 for bulk platinum was calculated using a Debye temperature of 240 K, as discussed elsewhere [52].

The values for the dispersed platinum catalysts were then determined using the calculated value of σ_1 for bulk platinum and the experimentally determined values of $\Delta\sigma_1^2$ in Table 4. The values of σ_1 for the dispersed platinum are greater by a factor of 1.4 to 1.7 than the corresponding value for bulk platinum. The findings on σ_1 for the atoms in the highly dispersed platinum clusters are perhaps the most significant results of the EXAFS experiments described here. Such information on very highly dispersed metal catalysts has not previously been available.

B. Acidic Sites of Alumina

The surfaces of alumina carriers used in reforming catalysts have been shown to be acidic in nature. The acidic properties are readily demonstrated by the affinity of the alumina surface for such basic molecules as ammonia, trimethylamine, n-butylamine, pyridine, and quinoline [68, 69]. Furthermore, adsorption of certain acid-base indicators such as butter yellow gives a coloration similar to that observed in acid media [70, 71]. Another indication of the acidic properties of the alumina carriers is their ability to catalyze reactions commonly identified with acid catalysis, i.e., reactions which are typical of carbonium ion chemistry. A good example of such a reaction is the skeletal isomerization of an olefin, as typified by the conversion of 1-butene to isobutene.

The nature of the acidic sites on alumina has been the subject of much discussion. The surface is characterized by the presence of hydroxyl groups which could conceivably be a source of Bronsted (protonic) acidity. However, it has been concluded that the hydroxyl groups do not contribute significantly as a source of protonic acid sites [72], on the basis that infrared spectra obtained after adsorption of ammonia or pyridine on the surface do not exhibit bands due to ammonium or pyridinium ions. However, the surface of alumina in a reforming catalyst normally contains chloride ions which can interact with the hydroxyl groups to enhance their acidity [73]. Since the aluminas employed in reforming catalysts are normally heated to high temperatures (770–870 K) in their preparation, there are Lewis acid sites present as a result of dehydroxylation reactions [74]. A Lewis acid is defined as a species that can accept a pair of electrons from a base [75]. At the surface of dehydroxylated alumina there are incompletely coordinated aluminum atoms (i.e., aluminum atoms coordinated to three oxygen atoms instead of four) which could serve as electron acceptors. The interaction of these aluminum sites with a Lewis base such as ammonia, via bonding involving the lone electron pair on the nitrogen atom, is readily visualized.

The brief discussion of acidity in the previous paragraphs is intended to give only a very general picture of the acidic nature of the alumina surface. Questions about the detailed nature of the acidic sites on alumina are still of interest in research on bifunctional reforming catalysts. From the standpoint of reforming applications, the chlorine content of a platinum-alumina catalyst plays an important role in influencing the acidic character and hence the reforming properties. Studies of the acidity of the alumina surface with chlorine present are therefore of particular interest.

4. Bimetallic Catalysts

The term "bimetallic clusters" has been proposed by the writer in referring to highly dispersed metallic entities comprising atoms of two metallic elements [76, 77]. Such clusters may be so small, of the order of 10 Å in size, that virtually every atom in the cluster is a surface atom. In general, the clusters are supported on a high surface area carrier such as silica or alumina, and the total metal concentration may be of the order of 1 percent or lower. These novel systems introduce much flexibility in the design of catalysts. Virtually any property of a dispersed metal catalyst (including activity, selectivity, or stability) may be influenced by addition of a second metallic element to form bimetallic clusters. Systems of interest are not limited to combinations of metals which form bulk alloys. Two types of bimetallic clusters have been investigated extensively by the writer. One is a combination of a Group VIII and a Group IB metal, e.g., ruthenium-copper and osmium-copper [76]. The other is a combination of two Group VIII metals, e.g., platinum-iridium [7, 8]. Since the latter is of interest in reforming, we will discuss it in more detail. Another system which could conceivably be considered in the bimetallic cluster category is supported platinum-rhenium [6, 78] which is a combination of a Group VIII metal and a Group VIIA metal. This system will be considered after the discussion of platinum-iridium catalysts.

A. Platinum-Iridium

Bimetallic clusters of platinum and iridium can be prepared by coimpregnating a carrier such as silica or alumina with an aqueous solution of chloroplatinic and chloroiridic acids [7]. After the impregnated carrier is dried and possibly calcined at mild conditions (520–570 K), it is reduced in flowing hydrogen at elevated temperature (570–770 K). Some idea of the variation of metal dispersion with variation of the total metal content of such catalysts can be obtained from the data on hydrogen chemisorption [8] shown in Figure 6. All of the catalysts in the figure contain equal fractions by weight of platinum and iridium (corresponding to a Pt/Ir atomic ratio very close to unity). Data are shown for platinum-iridium clusters supported on both alumina and silica, the surface areas of which are, respectively, about 200 and 300 $m^2 \times g^{-1}$. The data represent strongly chemisorbed hydrogen, i.e., the amount of chemisorbed hydrogen which can not be removed by evacuation at room temperature to approximately 10^{-4} Pa. The amount adsorbed is expressed in terms of the quantity H/M, which represents the ratio of the number of hydrogen atoms adsorbed to the number of metal atoms (platinum and

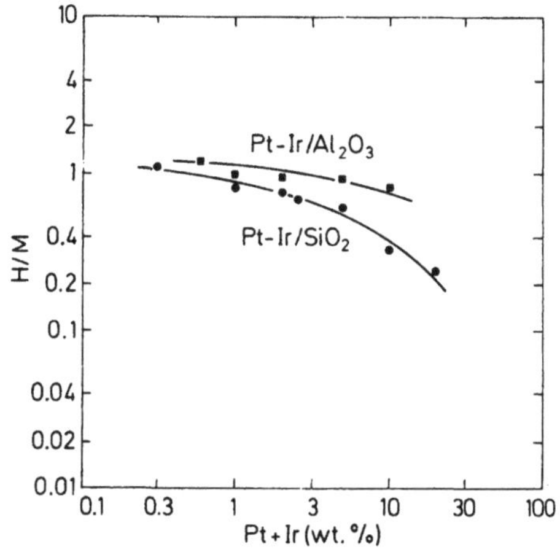

Figure 6. Hydrogen chemisorption at room temperature on platinum-iridium catalysts as a function of the platinum plus iridium content. All the catalysts contain equal percentages by weight of platinum and iridium. The quantity H/M represents the ratio of the number H of hydrogen atoms adsorbed to the number M of metal atoms (platinum and iridium) in the catalyst, and refers to the amount of hydrogen retained by the catalyst when the adsorption cell is evacuated at room temperature for 10 min (to a pressure of approximately 10^{-4} Pa) after completion of the adsorption isotherm; i.e., the H/M values represent the strongly chemisorbed fraction. The upper curve is for alumina supported platinum-iridium clusters while the lower curve is for silica supported clusters [8]

iridium) in the catalyst. Values of H/M are consistently higher for alumina supported platinum-iridium than for silica supported platinum-iridium, indicating higher metal dispersion on alumina. We see that H/M appears to approach a limiting value near unity as the metal concentration is decreased below about 1 wt. percent. Electron microscopy data on such catalysts show the average diameters of the metal clusters to be of the order of 10 Å or lower. Clusters of this size necessarily consist almost exclusively of surface atoms. The stoichiometry of the strongly chemisorbed fraction therefore corresponds closely to one hydrogen atom per surface metal atom. Thus the value of H/M for the strongly chemisorbed fraction is taken as a direct measure of the dispersion, which is defined as the ratio of surface atoms to total atoms in the metal clusters.

1. X-Ray Diffraction Studies

X-ray diffraction data are useful in characterizing the platinum-iridium system. The dependence of the lattice parameter of bulk platinum-iridium alloys on composition is shown in Figure 7 [8, 79]. Measurements of lattice parameters provide a way of demonstrating the presence of bimetallic clusters of platinum and iridium in the catalysts. X-ray diffraction data on a silica supported platinum-iridium catalyst containing 10 wt. percent platinum and 10 wt. percent iridium [8] are shown in the middle field of Figure 8. The metal dispersion of the catalyst as determined by hydrogen

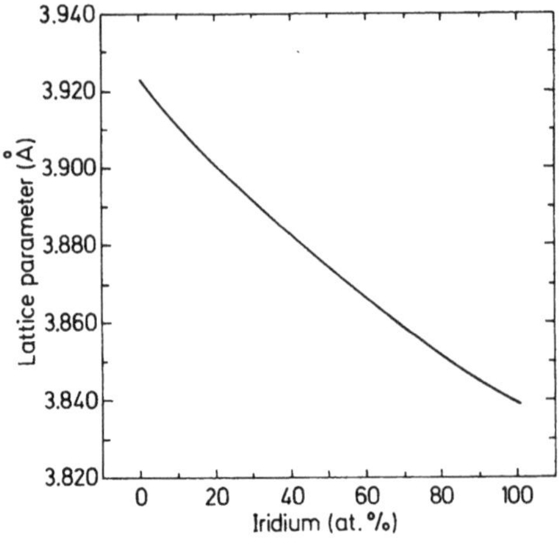

Figure 7. Lattice parameter of platinum-iridium solid solutions as a function of composition [8, 79]

chemisorption is 0.24. A portion of the diffraction pattern is shown which includes the (220) reflection. The upper field of the figure shows the corresponding part of the diffraction pattern of a physical mixture of bulk platinum and bulk iridium diluted with silica. In the lower field of the figure is shown the same region of the diffraction pattern for a physical mixture of equal weights of silica supported platinum and silica supported iridium, the mixture containing the same absolute amounts of platinum and iridium as the platinum-iridium bimetallic catalyst sample. It should be noted that the platinum-iridium bimetallic catalyst was diluted with an equal weight of silica. The background scattering arising from the silica carrier was subtracted from the total scattering in the diffraction patterns for the catalyst.

In the diffraction patterns in Figure 8, the platinum-iridium bimetallic clusters exhibit a single diffraction line (middle field of figure) about midway between the lines (upper field of figure) for bulk platinum and bulk iridium. The line for the clusters is broader than the lines for the bulk metals because of the small size of the clusters. An average platinum-iridium cluster size of 49 Å is calculated from the line width by means of the Scherrer equation [80]. The lattice parameter of the platinum-iridium clusters is 3.875 Å, as determined from the positions of a number of lines in the diffraction pattern. This value corresponds to a composition of 50 percent iridium in Figure 7,

▶

Figure 8. X-ray diffraction study of a platinum-iridium bimetallic cluster catalyst. The upper field is a reference pattern for a physical mixture of platinum and iridium metals diluted with silica. The middle field is a pattern for a silica supported platinum-iridium bimetallic cluster catalyst containing 10 wt. percent Pt and 10 wt. percent Ir, mixed with an equal weight of silica. The lower field is a pattern for a physical mixture of silica supported Pt clusters and silica supported Ir clusters, the catalysts containing 10 wt. percent Pt and 10 wt. percent Ir, respectively. The mixture was formulated to match the platinum and iridium contents of the sample in the middle field of the figure [8]

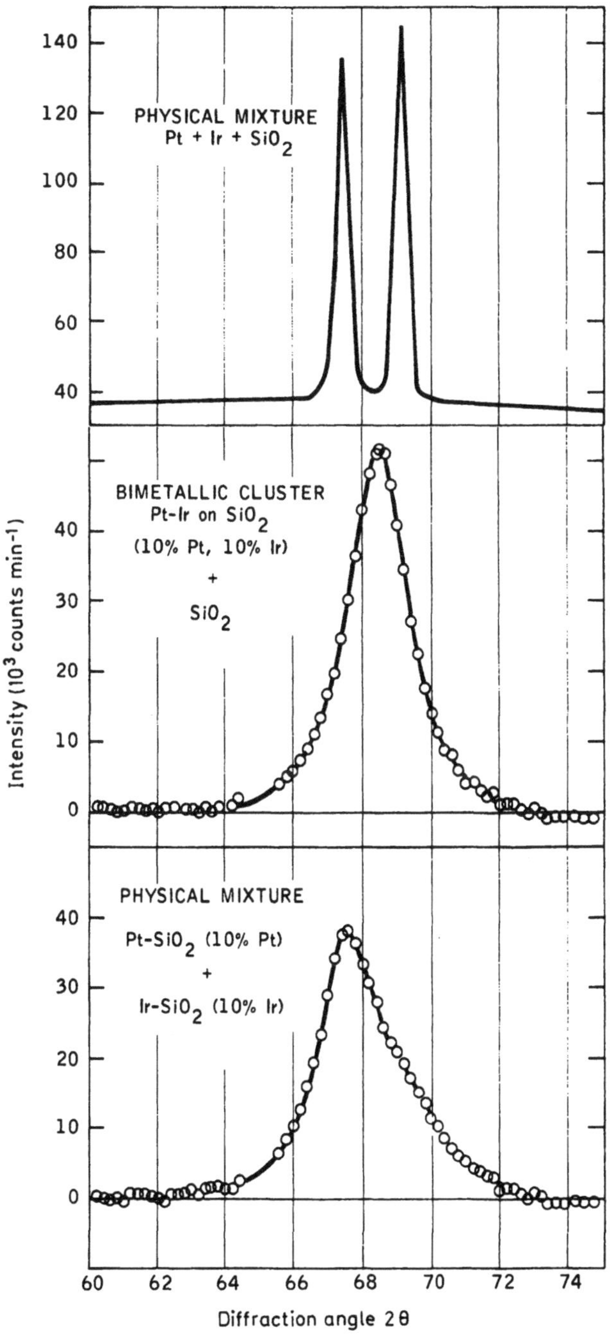
PHYSICAL MIXTURE
Pt + Ir + SiO2
BIMETALLIC CLUSTER
Pt-Ir on SiO2
(10% Pt, 10% Ir)
+
SiO2
PHYSICAL MIXTURE
Pt-SiO2 (10% Pt)
+
Ir-SiO2 (10% Ir)
Intensity (10^3 counts min^-1)
Diffraction angle 2θ
140
120
100
80
60
40
30
20
10
0
60
62
64
66
68
70
72
74

in agreement with the overall composition. The diffraction pattern of the physical mixture of silica supported platinum clusters and iridium clusters in the lower field of the figure consists of overlapping lines for the two individual types of clusters, and is clearly different from that of the platinum-iridium bimetallic clusters. The overlapping is due to the broadened nature of the individual lines resulting from the small sizes of the individual clusters of platinum and iridium.

X-ray diffraction has proved to be useful for the investigation of silica supported platinum-iridium clusters with metal dispersions as high as 0.60 [7, 8]. For platinum-iridium clusters of higher dispersion, the application of X-ray diffraction becomes very difficult, as it does for supported metal catalysts of any kind having such high dispersions. The lines in the diffraction patterns are then extremely broad and weak, becoming indistinguishable from the background. Furthermore, in the case of alumina supported platinum-iridium catalysts, there is the additional complication that the diffraction pattern of the alumina interferes with that of the platinum-iridium clusters.

In a typical preparation of highly dispersed platinum-iridium bimetallic clusters on a silica or alumina carrier, the carrier is first coimpregnated with chloroplatinic and chloroiridic acids and then dried at a temperature of 383 K prior to reduction in hydrogen at 773 K. If the catalyst is heated in air (calcined) at too high a temperature, say 773 K [7, 8], the iridium undergoes oxidation and agglomeration to form large crystallites of IrO_2 (>200 Å in size). On subsequent treatment in hydrogen at 773 K, the IrO_2 crystallites are reduced to metallic iridium crystallites. At this point, the material consists of a mixture of highly dispersed platinum or platinum-rich clusters and large crystallites of iridium. The diffraction profile for the (220) region for a silica supported platinum-iridium catalyst heated in air at 773 K prior to reduction in hydrogen is given in the upper field of Figure 9. The contrast between this pattern and the pattern for platinum-iridium bimetallic clusters in the lower field of Figure 9 is clear. These results demonstrate the importance of preparative conditions in the formation of bimetallic clusters in the platinum-iridium system. In general, exposure to air at temperatures below about 650 K does not appear to be harmful, but contact with air at temperatures above about 720 K should be avoided [7, 8].

2. Mössbauer Spectroscopy Studies

Mössbauer spectroscopy has been found to be useful for the characterization of platinum-iridium catalysts [7, 81]. It has the feature that it can be applied to catalysts in which virtually all of the metal atoms are surface atoms. In conducting Mössbauer spectroscopy experiments on platinum-iridium catalysts, one might incorporate the Mössbauer nuclides ^{195}Pt and/or ^{193}Ir in the catalysts. However, experiments with these nuclides are extremely difficult because of short lived sources and the requirement for measurements at liquid helium temperature. An alternative approach involves the addition of ^{57}Fe to platinum-iridium catalysts as a sensitive Mössbauer probe of

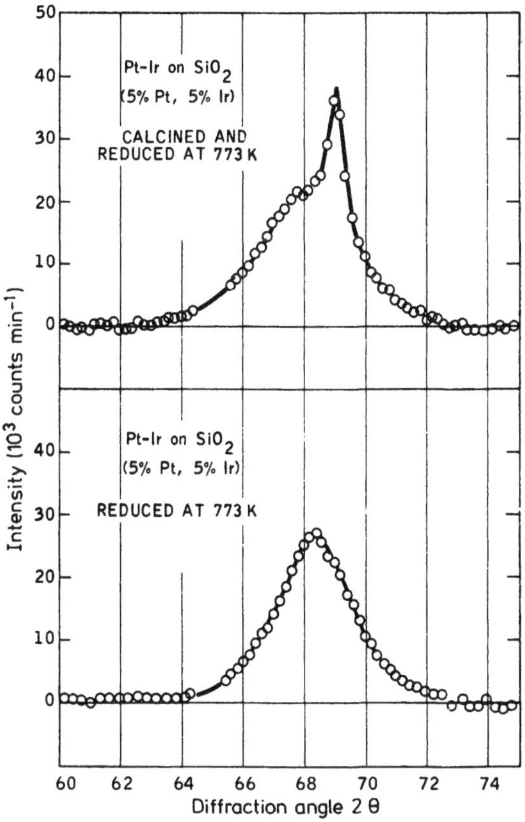

Figure 9. X-ray diffraction pattern on a silica supported platinum-iridium sample heated in air at 773 K prior to reduction in hydrogen at 773 K (upper field of figure), indicating a mixture of large iridium crystallites and much smaller platinum clusters. A diffraction pattern on a silica supported platinum-iridium bimetallic cluster catalyst of the same composition is shown in the lower field for comparison. The catalyst in the lower field had not been exposed to air at high temperature prior to reduction, which illustrates the importance of the preparative conditions in the formation of highly dispersed platinum-iridium bimetallic clusters. The data are for samples containing 5 wt. percent each of platinum and iridium, corresponding to a platinum-iridium atomic ratio very close to 1 [8]

the interaction between platinum and iridium [7, 81]. The Mössbauer resonance is sensitive to the environment of the ^{57}Fe nucleii in the catalyst, and hence it is possible to obtain information on the nature of the catalyst.

Mössbauer spectra at 298 K are shown in Figure 10 for alumina supported platinum, iridium, and bimetallic platinum-iridium catalysts containing ^{57}Fe (samples B, C, and D, respectively). All of the catalysts had metal dispersions (as determined by chemisorption) in the range 0.7–1 [81]. A Mössbauer spectrum is also shown for alumina containing only the iron (sample A). A spectrum (not shown in Figure 10) was also obtained on a physical mixture of samples B and C which we will designate as sample B–C. The two line (quadrupole split) spectrum for sample A and the corresponding Mössbauer

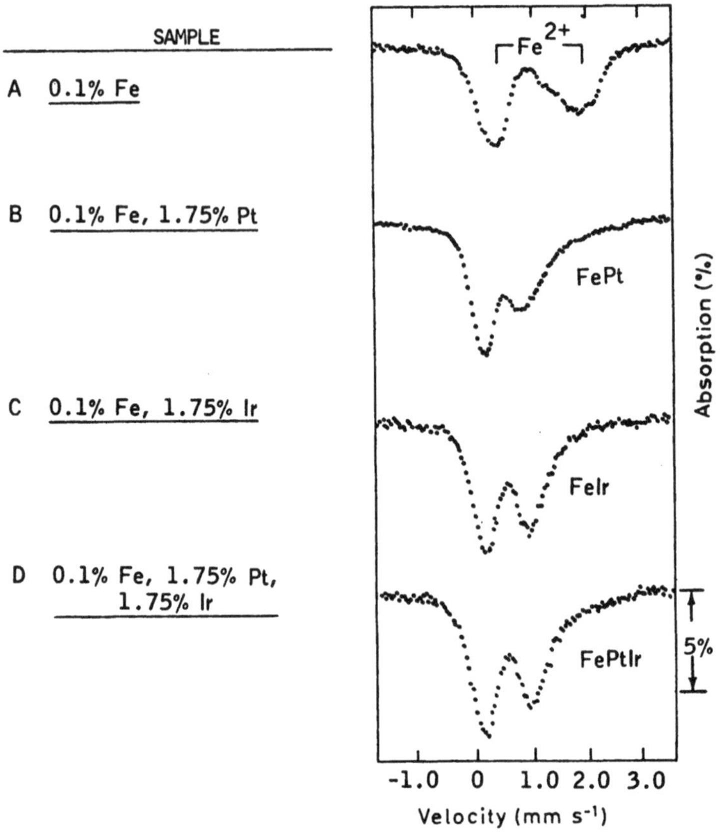

Figure 10. Mössbauer spectra at 298 K on alumina supported platinum-iridium, iridium, and platinum catalysts (samples D, C, and B, respectively) containing a small amount of Fe (0.1 wt. percent) enriched with ^{57}Fe as a Mössbauer probe. The platinum-iridium catalyst contained 1.75 wt. percent each of platinum and iridium, while the other two catalysts contained 1.75 wt. percent of either platinum or iridium. The Mössbauer spectrum on the alumina supported Fe sample (sample A) containing 0.1 wt. percent Fe enriched with ^{57}Fe serves as a reference for comparison with the spectra on the catalysts [7, 81]

parameters for this sample in Table 5 are characteristic of high spin ferrous ions in oxides [82], which indicates that the iron in the sample is reduced only to the ferrous state with hydrogen at 773 K. There is no evidence for the presence of iron in the metallic state. The large linewidths and the differences in intensity of the lines suggest that the iron has a number of different environments. The spectra for samples B, C, and D in Figure 10 are very different from that for sample A. Correspondingly, the Mössbauer parameters in Table 5 for samples B, C, D, and B–C contrast markedly with those for sample A. These results are indicative of differences in the chemical environment of the iron. The presence of platinum and/or iridium leads to a marked decrease in the amount of iron which exists as ferrous ions in the samples. The iron is therefore present in a different chemical state which

Table 5. Mössbauer spectroscopy data on Pt, Ir, and PtIr catalysts containing ^{57}Fe [7, 81]

Composition/ wt. %[a]	Sample[b]	Isomer Shift δ/mm sec^{-1}	Quadrupole Splitting/ mm sec^{-1}	A_2/A_1 [c]	W_2/W_1 [c]
0.1 Fe	A	1.20	1.69	1.01	1.32
0.1 Fe, 1.75 Pt	B	0.53	0.82	1.71	2.41
	B-600	0.55	0.86	1.90	2.73
	B-H_2O[d]	0.53	0.83	1.71	2.38
0.1 Fe, 1.75 Ir	C	0.58	0.97	1.18	1.58
	C-600	1.22	1.86	0.98	1.38
0.1 Fe, 1.75 Pt, 1.75 Ir	D	0.56	0.89	1.16	1.38
	D-600	0.57	0.92	1.50	2.16
	B-Ir[d]	0.57	0.86	1.13	1.40
0.1 Fe, 1.75 Pt mixed with 0.1 Fe, 1.75 Ir	B-C[e]	0.57	0.92	1.28	1.79

[a] The concentrations of metallic elements of interest, expressed as wt. percent of the total mass of material, which includes the metallic elements shown and the alumina carrier. The ^{57}Fe fraction of the 0.1 wt. percent Fe present in all cases serves as a Mössbauer probe of the catalyst. The composition listed as 0.1 Fe (sample A) is a reference material in which Fe is dispersed on alumina with no platinum or iridium present.

[b] Samples A, B, B-H_2O, C, D, and B-Ir were all calcined in air at 533 K in their preparation, while samples B-600, C-600, and D-600 were calcined at 873 K (600 °C).

[c] The ratios of the areas (A_2/A_1) and widths (W_2/W_1) of the two lines in the Mössbauer spectra in Figures 10, 12, and 13. The subscript 1 refers to the line corresponding to the lower Doppler velocity.

[d] Catalyst samples prepared by contacting sample B in its finally reduced form with water (designated B-H_2O) or with a chloroiridic acid solution (designated B-Ir), followed by drying and reduction steps identical to those employed in the preparation of the original sample B.

[e] Physical mixture of equal portions of samples B and C.

is characterized by an intimate association with highly dispersed clusters of platinum and/or iridium. The evidence for association of the iron atoms with platinum in samples similar to B has been discussed elsewhere in some detail [83]. Briefly, incorporation of iron atoms into the platinum clusters is indicated by the agreement of the isomer shift for iron in the catalyst with that for iron in a PtFe bulk alloy of similar composition. In addition, it was found that the iron in sample B, after reduction at 773 K, exhibited a unique chemical behavior in that it could be reversibly oxidized and reduced at room temperature. This property appears to be characteristic of iron in association with metals such as platinum and palladium [82–85]. It is not observed when the iron is present by itself on alumina. Sample C containing iridium and iron exhibited the same reversible oxidation-reduction behavior indicating the incorporation of the iron into the iridium clusters. The isomer shift for sample C, however, was not in good agreement with the value of 0.38 mm sec^{-1} [86] expected for dilute iron in iridium alloys. This difference may reflect an unusual chemical state for iron which is associated with surface iridium atoms in clusters. A similar situation has recently been reported and discussed for iron-ruthenium catalysts [87].

The Mössbauer parameters for sample D differ from those for samples B and C, which indicates that the iron atoms in sample D are not associated exclusively with either the platinum or the iridium in the catalyst. Furthermore, the area and linewidth ratios (A_2/A_1 and W_2/W_1) for sample D differ from those for the physical mixture B–C, which indicates that the iron atoms in sample D are not simply distributed between platinum clusters and iridium clusters with the platinum and iridium being totally isolated from each other. The data on Mössbauer parameters are consistent with the view that catalyst D consists of bimetallic platinum-iridium clusters with iron atoms incorporated therein.

Mössbauer spectra obtained at very low temperatures (18–23.5 K) on samples B, C, and D are shown in Figure 11. At these temperatures the samples become ferromagnetic and the spectra exhibit magnetic hyperfine splitting. As a result, the spectra consist of six lines instead of two. While

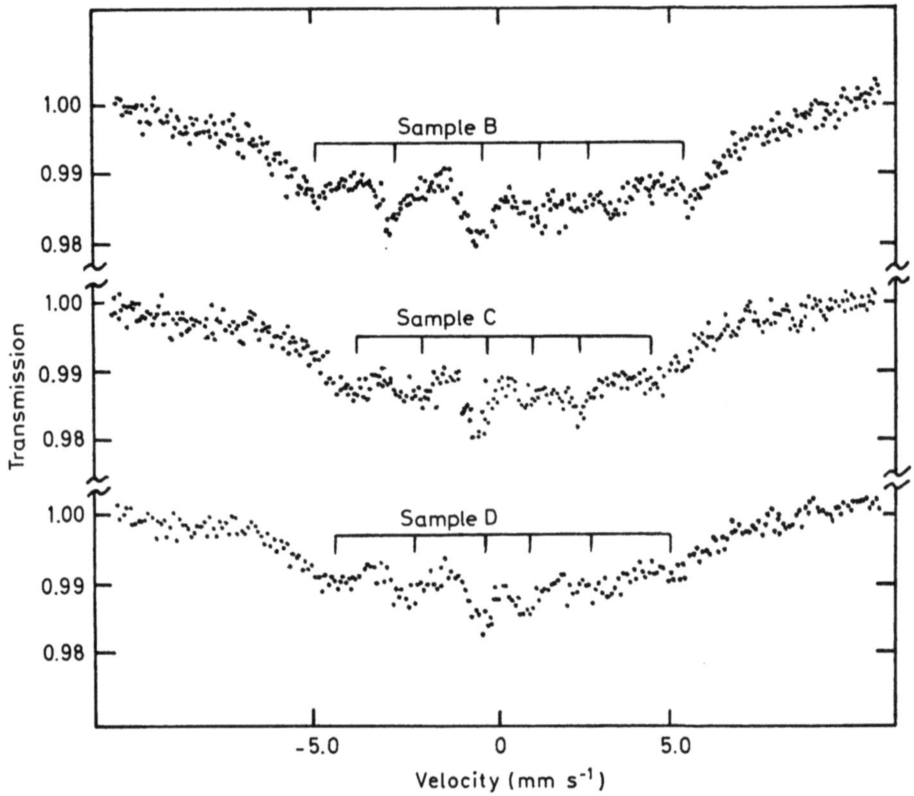

Figure 11. Low temperature (18–23.5 K) Mössbauer spectroscopy data on the same catalysts (samples B, C, D) used in obtaining the room temperature Mössbauer spectroscopy data in Figure 10. The data exhibit magnetic hyperfine splitting because of the ferromagnetic character of the samples at the low temperatures. The splitting leads to spectra consisting of six lines. The magnitude of the splitting varies for the different catalysts, indicating differences in the local magnetic fields at the ^{57}Fe nucleii. The positions of the six lines determined by computer analysis of the data are shown above each of the spectra [7, 81]

the six lines in the spectra of Figure 11 are not very pronounced, they are nonetheless real, as demonstrated by the computer analysis of the data. The positions of the lines from this analysis are shown above the spectra in Figure 11. The magnitude of the magnetic field experienced by the ^{57}Fe nucleii can be determined from the separation of lines in the spectrum [88, 89]. Magnetic fields derived from the separation of the outermost of the six lines of the hyperfine spectra in Figure 11 are given in Table 6. Since the internal magnetic field is different for each of the three samples, which differ with regard to the metal or metals (platinum and/or iridium) present, the results provide additional evidence that the iron atoms are associated with these metals. In studies on bulk PtFe, IrFe, and PtIrFe alloys, it has been observed [90, 91] that the magnetic fields at the iron nucleii are much smaller for IrFe than for PtFe. In addition, these studies showed that the magnetic fields for PtIrFe alloys were intermediate between those for PtFe and IrFe alloys. The same trend is found with the catalysts in Table 6, which is consistent with the view that the iron atoms in sample D are incorporated in bimetallic clusters of platinum and iridium. It is of interest to note that the magnetic field at the ^{57}Fe nucleii in the platinum catalyst (sample B) is similar to that reported for highly dispersed PtFe alloys on carbon [85] and for bulk PtFe alloys [91]. However, the fields at the ^{57}Fe nucleii in the iridium and platinum-iridium catalysts (samples C and D) are much larger than those observed with bulk IrFe and PtIrFe alloys [90, 91]. The data thus indicate a strong effect of the degree of metal dispersion on the magnetic properties of the IrFe and PtIrFe systems which is not observed with the PtFe system.

In the preparation of iridium-containing catalysts, the temperature of calcination in air is critical. Mössbauer spectra at 298 K are shown in Figure 12 for platinum, iridium, and bimetallic platinum-iridium catalysts (samples B-600, C-600, and D-600, respectively) which were prepared by calcination in air at 873 K (600 °C) prior to the final reduction in hydrogen at 773 K (500 °C). Sample B-600 gives a spectrum similar to that of sample B in Figure 10, and the Mössbauer parameters for these samples in Table 5 are also similar. These results indicate that sample B-600 also consists of platinum clusters with iron atoms incorporated therein. However, sample C-600

Table 6. Mössbauer spectroscopy data on Pt, Ir, and PtIr catalysts containing ^{57}Fe. Magnetic hyperfine structure results on the local magnetic fields at the ^{57}Fe nucleii [7, 81]

Catalyst Composition/wt. %[a]	Temperature/K	Magnetic Field/koe
0.1 Fe, 1.75 Pt	21	325
0.1 Fe, 1.75 Ir	23.5	260
0.1 Fe, 1.75 Pt, 1.75 Ir	18	295

[a] The concentrations of metallic elements of interest in the catalysts, including the ^{57}Fe enriched iron present as a Mössbauer probe. The concentrations are expressed as wt. percent of the total mass of material, which includes the metallic elements shown and the alumina carrier.

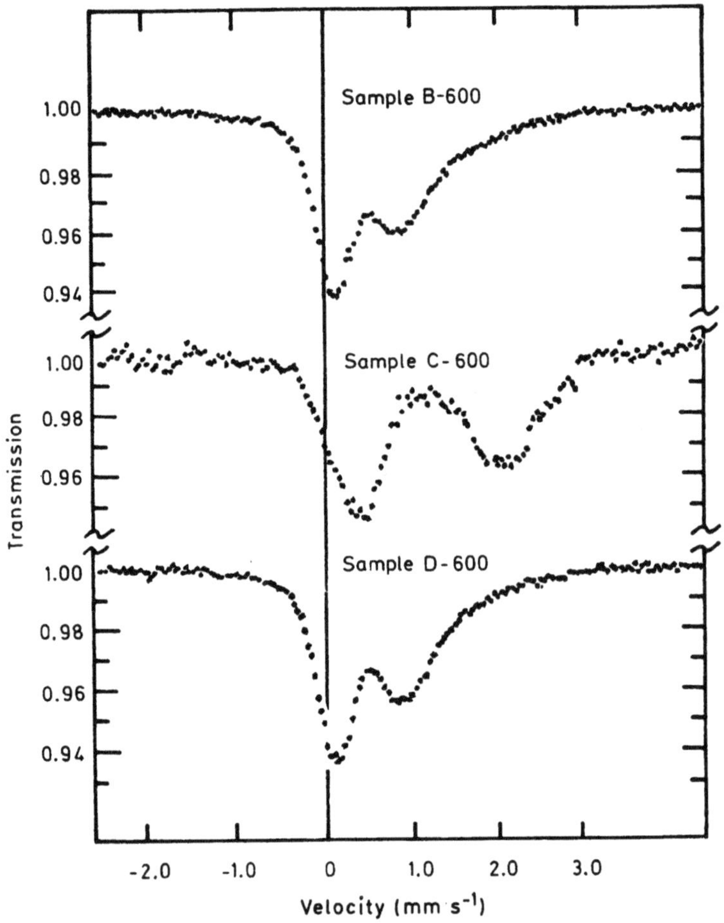

Figure 12. Mössbauer spectra at 298 K on alumina supported platinum-iridium, iridium, and platinum catalysts which were calcined in air at a temperature of 873 K (600 °C) in their preparation (samples D-600, C-600, and B-600, respectively). The catalysts have the same elemental compositions as the catalysts for which Mössbauer spectra are shown in Figure 10, but exhibit structural differences due to the higher calcination temperature (873 vs. 533 K) employed in their preparation [7, 81]

gives a spectrum which is very different from that presented for sample C in Figure 10. The spectrum for sample C-600 is nearly identical to that of sample A in Figure 10, in which the iron exists in the ferrous state. The iridium in sample C-600 is not present as highly dispersed clusters, as can be seen from chemisorption data [81]. It appears that the poorly dispersed iridium crystallites (dispersion < 0.1) are not associated to a significant extent with the iron in the sample. Hence sample C-600 behaves like a sample containing no iridium. The Mössbauer parameters in Table 5 are consistent with this statement. The spectrum of sample D-600 resembles that of sample B in Figure 10 more closely than it resembles the spectrum of sample D. The

Mössbauer parameters in Table 5, especially the ratios A_2/A_1 and W_2/W_1, show substantial differences between samples D and D-600. The iridium in sample D-600 is largely present in the same poorly dispersed form as the iridium in sample C-600, as has been found from x-ray diffraction studies of similar samples. Sample D-600 thus can be characterized approximately as consisting of highly dispersed platinum clusters incorporating iron atoms and separate iridium crystallites of much lower dispersion which are not associated with iron atoms to a significant extent. This characterization is consistent with chemisorption data [81]. The Mössbauer data on sample D-600, coupled with chemisorption data on this sample, show that calcination of alumina supported platinum-iridium in air at 600 °C is unsatisfactory for the formation of highly dispersed bimetallic clusters of platinum and iridium.

Mössbauer spectroscopy data on an alumina supported platinum-iridium sample prepared by a sequential impregnation procedure are of interest in providing information on the tendency of platinum and iridium to form

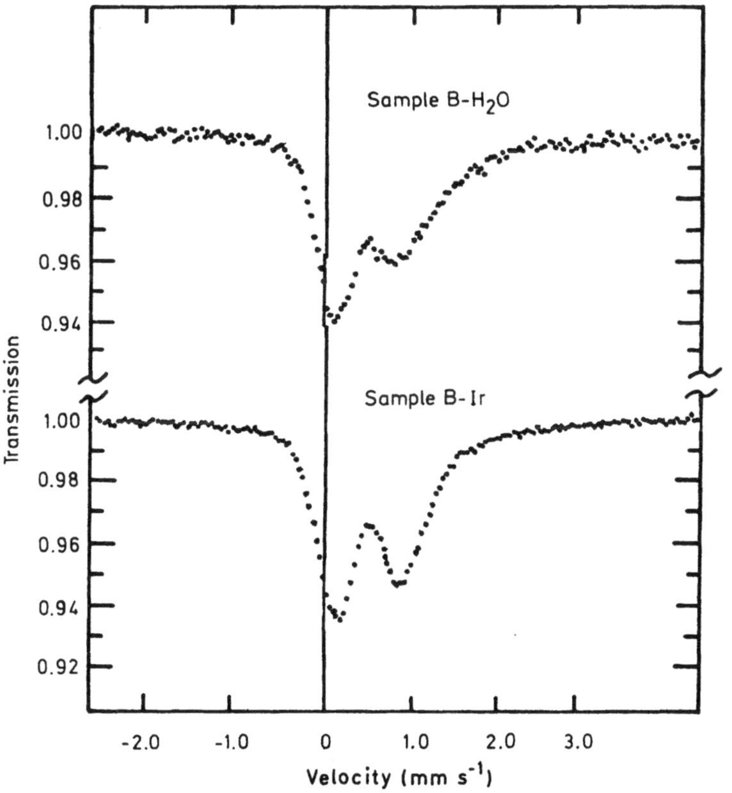

Figure 13. Mössbauer spectra at 298 K on catalyst samples B-H_2O and B-Ir prepared by contacting platinum-alumina catalyst (sample B of Figures 10 and 11) in its final reduced form with, respectively, water and a chloroiridic acid solution. The subsequent drying and reduction steps in the preparation of samples B-H_2O and B-Ir were identical to those employed in the preparation of sample B. The platinum and iridium contents of sample B-Ir are the same as those of sample D of Figure 10 [7, 81]

bimetallic clusters. The sample was prepared by contacting the platinum-alumina catalyst containing ^{57}Fe (sample B) with a chloroiridic acid solution after the hydrogen reduction step. On the basis of the evidence presented earlier, the sample, prior to contact with the chloroiridic acid solution, may be characterized as containing highly dispersed platinum clusters with iron atoms incorporated therein. After drying and reduction steps indentical to those employed in the preparation of samples B, C, and D, the sequentially impregnated material (sample B-Ir) had the same elemental composition as sample D. Another sample prepared in exactly the same manner, except that water was substituted for the chloroiridic acid solution, served as a reference material (designated as sample B-H_2O). Mössbauer spectra at 298 K on these samples are presented in Figure 13. The spectrum and Mössbauer parameters of sample B-H_2O are similar to those of sample B (Figure 10 and Table 5). Sample B is therefore unaffected by its treatment with water. The spectrum of sample B-Ir is similar to that of sample D in Figure 10, and the Mössbauer parameters in Table 5 are also similar. The data indicate that platinum clusters associated with iron are altered by addition of iridium to the sample. We conclude that the added iridium is incorporated in the platinum clusters to give PtIr clusters containing the iron probe atoms. If the added iridium was present as separate iridium clusters in sample B-Ir, the Mössbauer spectrum should have been the same as that of sample B-H_2O or sample B. These results provide a good illustration of the propensity of platinum and iridium atoms to mix and form bimetallic clusters, rather than to form monometallic clusters consisting exclusively of atoms of one or the other of the two metals.

B. Platinum-Rhenium

Platinum-rhenium-alumina catalysts generally contain an amount of rhenium comparable to the amount of platinum present (often about 0.3 wt. percent of each). The physical and chemical nature of the rhenium in the catalyst has been considered by various investigators [92–100]. Whether or not a highly dispersed alloy or bimetallic cluster of platinum and rhenium is present on the alumina is a matter of some interest. Although platinum and rhenium have different crystal structures and do not exhibit complete miscibility in the bulk, these factors may have limited significance in the case of highly dispersed metals.

Johnson and Leroy [92] concluded that the rhenium is present as a highly dispersed oxide at typical reforming conditions. They studied a series of alumina supported platinum-rhenium catalysts with platinum contents ranging from 0.31 to 0.66 wt. percent and rhenium contents ranging from 0.20 to 1.18 wt. percent. Their conclusions were based on measurements of hydrogen consumption during reduction of the catalysts at 755 K and on x-ray diffraction studies of the metal component of the catalyst after the alumina had been leached from the catalyst by treatment with a solution of fluoboric acid. The catalysts were prepared by contacting alumina with aqueous solutions of chloroplatinic acid and ammonium perrhenate. The consumption of hydrogen during reduction corresponded to complete reduction of platinum from the +4 oxidation state to the metal and of rhenium from the +7 to the +4 state. The x-ray diffraction data on the metal residue from the leached catalysts showed no evidence for the presence of rhenium metal or a platinum-rhenium alloy. Most of the rhenium was found in the solution employed in the leaching. Finally, the authors stated that data from an electron spin resonance experiment on one of the reduced platinum-rhenium catalysts were consistent with their conclusion that the rhenium was present in the +4 state.

In contrast to the results of Johnson and Leroy, studies by Webb [99] on rhenium-alumina catalysts containing 3.42 and 3.64 wt. % rhenium indicated that the rhenium was completely reduced from the +7 state to the metal by hydrogen at temperatures of 673–723 K. The catalysts were prepared by contacting alumina with an aqueous solution of perrhenic acid. Measurements of hydrogen consumption during reduction and of oxygen consumption during a subsequent oxidation were both consistent with changes of oxidation state between +7 and 0. Two possible explanations were considered in accounting for the differences in results of these investigations. First, the authors [99, 100] considered the possibility that the difference in the rhenium contents of the catalysts could play a role, since interaction of the rhenium with the alumina might be very strong at low concentrations and therefore inhibit reduction. Second, they noted that the water content of the hydrogen during reduction was lower in the experiments of Webb, due to his use of a system employing recirculating hydrogen from which water was continuously removed by means of a trap maintained at −196 °C. The experiments of Johnson and Leroy were conducted with a static system with product water remaining in the system, the water persumably being absorbed by the alumina in the catalyst. Johnson and Leroy estimated that the residual water pressure in their system was about 10 Pa. It was implied [100] that the water level in the experiments of Johnson and Leroy was typical of that encountered in actual catalytic reforming reactors.

Later experiments by McNicol [101] and by a group of French workers [93–97, 102], in which the water formed during reduction was removed by means of a trap at 78 K, were consistent with the results of Webb in showing a change in oxidation state of rhenium from +7 to 0. These workers indicated also that the properties of alumina supported platinum-rhenium catalysts depend on the method of preparation. In the case of catalysts which were simply dried in air at 383 K after impregnation of the alumina with H_2PtCl_6 and Re_2O_7, it was concluded that a platinum-rhenium alloy formed on reduction. This conclusion was based on the observation that the presence of platinum accelerated the reduction of oxygen chemisorbed on the rhenium and on results showing that the frequencies of the infrared absorption bands of carbon monoxide adsorbed on platinum and rhenium sites in platinum-rhenium catalysts were different from those found with catalysts containing only platinum or rhenium. However, for catalysts which were calcined in air at 773 K prior to reduction in hydrogen, it was concluded that the platinum exhibited much less interaction with the rhenium [101, 102].

On the basis of the work reported to date, it appears that the state of platinum-rhenium-alumina catalysts is sensitive to the conditions employed in their preparation. Under some conditions there appears to be evidence of significant interaction between the platinum and rhenium, and it has been concluded that highly dispersed alloys or bimetallic clusters of platinum and rhenium are then present. With other conditions the interaction between the two elements appears to be greatly decreased. Finally, the sensitivity of the catalyst to conditions of preparation and activation introduces considerable uncertainty in projections about the nature of the catalyst under actual reforming conditions.

5. Technological Aspects

During the 1950s and 1960s platinum-alumina catalysts dominated catalytic reforming. In the past decade, however, bimetallic (or more generally, polymetallic) catalyst systems have been widely introduced in reforming. The new catalysts include systems with higher activity in reforming and with much improved activity maintenance. Other advantages include higher reformate yields and the ability to operate at more demanding conditions (e.g., higher temperatures and much lower pressures).

One of the new catalysts is the platinum-rhenium-alumina catalyst first announced by Chevron [6]. As indicated in the previous section of this article, the catalyst commonly contains about 0.3 wt. percent each of platinum and rhenium. A major advantage of this catalyst over catalysts containing only platinum on alumina is its lower rate of activity decline during a reforming operation. The better activity maintenance makes it possible to process petroleum naphtha feedstocks for a longer period of time before regeneration of the catalyst is required; i.e., longer reforming cycle lengths can be obtained with platinum-rhenium than with platinum. Alternatively, for a given cycle length, the platinum-rhenium catalyst may be operated at a lower pressure. Such operation has the advantage of increasing the yields of liquid reformate for inclusion in gasoline. The choice of the best way to take advantage of the improved activity maintenance will in general depend on the particular refinery situation.

The new reforming catalysts also include platinum-iridium catalysts, some features of which were considered earlier in this article. Alumina supported platinum-iridium bimetallic cluster catalysts are substantially more active than platinum-alumina catalysts and are characterized by much better activity maintenance, as shown by the data in Figure 14, which are taken from a patent by the writer [7]. In Figure 14 the temperature required to produce 100 research clear octane number reformate is shown as a function of time on stream in the processing of a 372–444 K boiling range naphtha containing approximately 43, 45, and 12 percent by volume, respectively, of

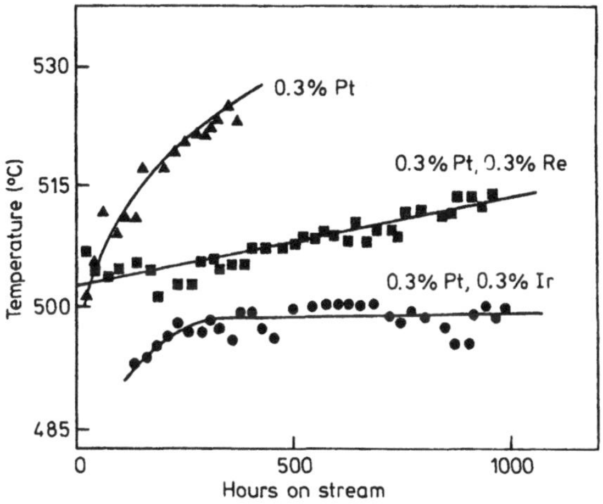

Figure 14. Data on the reforming of a 99–171 °C boiling range naphtha to 100 research clear octane number product over alumina supported platinum, platinum-rhenium, and platinum-iridium catalysts at 14.3 atm. pressure. The catalysts all contain 0.3 wt. percent of platinum and 0.3 wt. percent of either rhenium or iridium in the case of the bimetallics. The temperature required to produce 100 octane number product is shown as a function of time on stream. The temperatures shown are those of the fluidized solids bath (sand bath) surrounding the reactor. Hydrogen at a rate of 5000 standard cubic feet per barrel of naphtha was introduced with the naphtha feed. The weight hourly space velocity of the naphtha feed (grams of naphtha per hour per gram of catalyst) was 3 [7]

naphthenes, paraffins, and aromatics (The higher the catalytic activity, the lower the temperature required to produce the desired octane number reformate). The reactor pressure was 14.3 atm. and the weight hourly space velocity (grams of naphtha per hour per gram of catalyst) was 3. Hydrogen was introduced with the naphtha at a rate of 5000 standard cubic feet per barrel of naphtha. The reactor was surrounded by a fluidized solids bath (sand bath) for temperature control. The temperatures in the figure are the temperatures of the sand bath. As shown in the figure, the temperatures required to maintain 100 research clear octane number product are lower for the platinum-iridium catalyst (0.3 wt. percent Pt, 0.3 wt. percent Ir on alumina) than for either the platinum catalyst (0.3 wt. percent Pt on alumina) or the platinum-rhenium catalyst (0.3 wt. percent Pt, 0.3 wt. percent Re on alumina). Furthermore, the extent of increase of temperature required over a given period of time on stream to compensate for activity loss is lower for the platinum-iridium catalyst than for either of the other catalysts. The better activity maintenance of the platinum-iridium catalyst corresponds to a lower rate of deposition of carbonaceous residues on the catalyst. As already indicated, better activity maintenance makes it possible for the catalyst to operate for a longer period of time before regeneration is required. Regeneration of a reforming catalyst involves the combustion of carbonaceous residues from the catalyst surface using a gas of low oxygen content with close control of temperature and other operating conditions.

In the evolution of a major process such as catalytic reforming, the extensive information accumulated over a period of time ideally leads to a better understanding of the process. As a result, it may be possible to adopt a more sophisticated approach to the process. In the case of catalytic reforming, it is well known that various types of hydrocarbon reactions contribute to the end result. It would seem unlikely that a single catalyst would be optimum for all of the reactions. The possibility of using different catalysts for the different reactions thus suggests itself. A catalytic reformer can be divided into two zones. The first zone is concerned mainly with aromatization of naphthenic hydrocarbons while the second zone is concerned largely with aromatization of paraffinic hydrocarbons. The first two reactors of a catalytic reforming unit might be considered to constitute the first zone, and the last two reactors the tail zone, although there could be some flexibility here. The possibility of using different catalysts in these zones has been considered by the writer [103]. For example, a platinum-alumina or platinum-rhenium-alumina catalyst may be employed in the first zone and a platinum-iridium-alumina catalyst in the tail zone. Dual catalyst systems of this type provide additional possibilities for optimizing the performance of a catalytic reformer.

References

1. Sinfelt, J. H.: Advan. Chem. Engng. *5*, 37 (1964)
2. Sinfelt, J. H.: Heterogeneous catalysis in fossil fuel conversion. In: Materials science in energy technology. (Libowitz, G. G.; Whittingham, M. S., ed.) New York: Academic Press 1979, pp. 1—30
3. Haensel, V.: US Pat. 2,479,109 (1949)
4. Haensel, V.: US Pat. 2,479,110 (1949)
5. Haensel, V.; Donaldson, G. R.: Ind. Engng. Chem. *43*, 2102 (1951).
6. Jacobson, R. L. et al.: Proc. Amer. Petrol. Inst. Div. Refining *49*, 504 (1969)
7. Sinfelt, J. H.: U.S. Pat. 3,953,368 (1976)
8. Sinfelt, J. H.; Via, G. H.: J. Catal. *56*, 1 (1979)
9. Sinfelt, J. H.; Rohrer, J. C.: J. Phys. Chem. *65*, 978 (1961)
10. Rohrer, J. C.; Sinfelt, J. H.: J. Phys. Chem. *66*, 1193 (1962)
11. Sinfelt, J. H.: Chem. Engng. Sci. *23*, 1181 (1968)
12. Ciapetta, F. G.; Dobres, R. M.; Baker, R. W.: Catalytic reforming of pure hydrocarbons and petroleum naphthas. In: Catalysis (Emmett, P. H., ed.) New York: Reinhold 1958, Vol. VI, pp. 495—692.
13. Ciapetta, F. G.; Hunter, J. B.: Ind. Engng. Chem. *45*, 147 (1953)
14. Ciapetta, F. G.; Hunter, J. B.: ibid. *45*, 155 (1953)
15. Heinemann, H.; Mills, G. A.; Hattman, J. B. and Kirsch, F. W.: ibid. *45*, 130 (1953)
16. Hettinger, W. P.; Keith, C. D.; Gring, J. L. and Teter, J. W.: ibid. *47*, 719 (1955)
17. Rossini, F. D.; Pitzer, K. S.; Arnett, R. L.; Braum, R. M. and Pimentel, G. C.: Selected values of physical and thermodynamic properties of hydrocarbons and related compounds. API Res. Project 44. Pittsburgh: Carnegie Press, Inc. 1953
18. Mills, G. A.; Heinemann, H.; Milliken, T. H. and Oblad, A. G.: Ind. Engng. Chem. *45*, 134 (1953)
19. Weisz, P. B.; Swegler, E. W.: Science *126*, 31 (1957)
20. Hindin, S. G.; Weller, S. W.; Mills, G. A.: J. Phys. Chem. *62*, 244 (1958)
21. Sinfelt, J. H.; Hurwitz, H.; Rohrer, J. C.: ibid. *64*, 892 (1960)

22. Anderson, J. R.; Baker, B. G.: Proc. Roy. Soc. (London) Ser. A *271*, 402 (1963)
23. Carter, J. L.; Cusumano, J. A.; Sinfelt, J. H.: J. Catal. *20*, 223 (1971)
24. Boudart, M.; Ptak, L. D.: ibid. *16*, 90 (1970)
25. Matsumoto, H.; Saito, Y.; Yoneda, Y.: ibid. *19*, 101 (1970)
26. Anderson, J. R.; Avery, N. R.: ibid. *5*, 446 (1966)
27. Dautzenberg, F. M.; Platteeuw, J. C.: ibid. *19*, 41 (1970)
28. Sinfelt, J. H.: Advan. Catal. *23*, 91 (1973)
29. Sinfelt, J. H.; Rohrer, J. C.: J. Phys. Chem. *65*, 2272 (1961)
30. Sinfelt, J. H.; Rohrer, J. C.: ibid. *66*, 1559 (1962)
31. Coonradt, H. L. et al.: Ind. Eng. Chem. *53*, 727 (1961)
32. Anderson, J. R.; Kemball, C.: Proc. Roy. Soc. (London) Ser. A *226*, 472 (1954)
33. Andreev, A. A.; Gudkov, B. S.; Kiperman, S. L. and Shopov, D. M.: Kinet. Katal. (USSR) *9*, 35 (1968)
34. Sinfelt, J. H.; Hurwitz, H.; Shulman, R. A.: J. Phys. Chem. *64*, 1559 (1960)
35. Laidler, K. J.: Absolute rates of surface reactions. In: Catalysis (Emmett, P. H., ed.) New York: Reinhold 1954, Vol. I, p. 213
36. Keulemans, A. I. M.; Voge, H. H.: J. Phys. Chem. *63*, 476 (1959)
37. Sinfelt, J. H.; Hurwitz, H.; Rohrer, J. C.: J. Catal. *1*, 481 (1962)
38. Rohrer, J. C.; Hurwitz, H.; Sinfelt, J. H.: J. Phys. Chem. *65*, 1458 (1961)
39. Sinfelt, J. H.; Rohrer, J. C.: J. Chem. Eng. Data *8* (1), 109 (1963)
40. Sinfelt, J. H.: J. Catal. *27*, 468 (1972)
41. Sinfelt, J. H.: Catal. Rev. *9* (1), 147 (1974)
42. Cimino, A.; Boudart, M.; Taylor, H. S.: J. Phys. Chem. *58*, 796 (1954)
43. Sinfelt, J. H.; Taylor, W. F.: Trans. Faraday Soc. *64*, 3086 (1968)
44. Beeck, O.: Discuss. Faraday Soc. *8*, 118 (1950)
45. Schuit, G. C. A.; van Reijen, L. L.: Advan. Catal. *10*, 242 (1958)
46. Anderson, J. R.; Kemball, C.: Proc. Roy. Soc. (London) Ser. A *223*, 361 (1954)
47. Cotton, F. A.; Wilkinson, G.: Advanced Inorganic Chemistry. New York: Interscience 1962, pp. 661, 760
48. Pauling, L.: Proc. Roy. Soc. (London) Ser. A *196*, 343 (1949)
49. Sinfelt, J. H.: Ann. Rev. Mat. Sci. *2*, 641 (1972)
50. Spenadel, L.; Boudart, M.: J. Phys. Chem. *64*, 204 (1960)
51. Keavney, J. J.; Adler, S. F.: ibid. *64*, 208 (1960)
52. Via, G. H.; Sinfelt, J. H.; Lytle, F. W.: J. Chem. Phys. *71*, 690 (1979)
53. Wilson, G. R.; Hall, W. K.: J. Catal. *17*, 190 (1970)
54. Sinfelt, J. H.; Via, G. H.; Lytle, F. W.: J. Chem. Phys. *68*, 2009 (1978)
55. Lytle, F. W.; Via, G. H.; Sinfelt, J. H.: ibid. *67*, 3831 (1977)
56. Kronig, R. deL.: Z. Phys. *70*, 317 (1931)
57. Kronig, R. deL.: ibid. *75*, 191 (1932)
58. Kronig, R. deL.: ibid. *75*, 468 (1932)
59. Sayers, D. E.; Lytle, F. W.; Stern, E. A.: Phys. Rev. Lett. *27*, 1204 (1971)
60. Lytle, F. W.; Sayers, D.; Stern, E.: Phys. Rev. B *11*, 4825 (1975)
61. Sayers, D. E.; Lytle, F. W.; Stern, E. A.: Adv. X-Ray Analysis *13*, 248 (1970)
62. Stern, E. A.: Phys. Rev. B *10*, 3027 (1974)
63. Ashley, C. A.; Doniach, S.: ibid. B *11*, 1279 (1975)
64. Lee, P. A.; Pendry, J. B.: ibid. B *11*, 2795 (1975)
65. Lee, P. A.; Beni, G.: ibid. B *15*, 2862 (1977)
66. Sayers, D. E.: Ph.D. dissertation, Univer. Washington, 1971
67. Stern, E. A.; Sayers, D. E.; Lytle, F. W.: Phys. Rev. B *11*, 4836 (1975)
68. Oblad, A. G.; Milliken, T. H.; Mills, G. A.: Advan. Catal. *3*, 199 (1951)
69. Tamele, M. W.: Discuss. Faraday Soc. *8*, 270 (1950)
70. Benesi, H. A.: J. Amer. Chem. Soc. *78*, 5490 (1956)
71. Benesi, H. A.: J. Phys. Chem. *67*, 970 (1957)
72. Parry, E. P.: J. Catal. *2*, 371 (1963)
73. Gates, B. C.; Katzer, J. R.; Schuit, G. C. A.: Chemistry of catalytic processes. New York: McGraw-Hill 1979, pp. 259, 260
74. Pines, H.; Haag, W. O.: J. Amer. Chem. Soc. *82*, 2471 (1960)

75. Lewis, G. N.: Valence and the structure of atoms and molecules. New York: Chemical Catalog Co. 1923, p. 141
76. Sinfelt, J. H.: J. Catal. *29*, 308 (1973)
77. Sinfelt, J. H.: Accts. Chem. Res. *10*, 15 (1977)
78. Kluksdahl, H. E.: U.S. Pat. 3,415,737 (1968)
79. Pearson, W. B.: A handbook of lattice spacings and structures of metals and alloys. New York: Pergamon Press 1964, p. 704
80. Klug, H. P.; Alexander, L. E.: X-ray diffraction procedures for polycrystalline and amorphous materials (2nd ed.). New York: Wiley 1974, p. 656
81. Garten, R. L.; Sinfelt, J. H.: J. Catal. *62*, 127 (1980)
82. Garten, R. L.: J. Catal. *43*, 18 (1976)
83. Vannice, M. A.; Garten, R. L.: J. Molec. Catal. *1*, 201 (1975/76)
84. Garten, R. L.; Ollis, D. F.: J. Catal. *35*, 232 (1974)
85. Bartholomew, C. H.; Boudart, M.: ibid. *29*, 278 (1973)
86. Stevens, J. G.; Stevens, V. E.: Mössbauer effect data index. New York: Plenum 1973, p. 156
87. Vannice, M. A.; Lam, Y. L.; Garten, R. L.: Prepr. Amer. Chem. Soc. Div. Petrol. Chem. *23* (2), 495 (1978)
88. Frauenfelder, H.: The Mössbauer effect. New York: W. A. Benjamin 1962, pp. 51—53, 82, 83
89. Wertheim, G. K.: J. Appl. Phys. (Suppl.) *32*, 110S (1961)
90. Mizoguchi, T.; Sasaki, T.; Chikazumi, S.: Amer. Inst. Phys. Conf. Proc. *5*, 445/(1972)
91. Kanashiro, M.; Kunitomi, N.: J. Phys. Soc. Jap. *43* (5), 1559 (1977)
92. Johnson, M. F. L.; Leroy, V. M.: J. Catal. *35*, 434 (1974)
93. Bolivar, C.; Charcosset, H.; Frety, R.; Primet, M.; Tournayan, L.; Betizeau, C.; Leclercq, G. and Maurel, R.: J. Catal. *39*, 249 (1975)
94. Bolivar, C.; Charcosset, H.; Frety, R.; Primet, M.; Tournayan, L.; Betizeau, C.; Leclercq, G. and Maurel, R.: ibid. *45*, 163 (1976)
95. Betizeau, C.; Leclercq, G.; Maurel, R.; Bolivar, C.; Charcosset, H.; Frety, R. and Tournayan, L.: ibid. *45*, 179 (1976)
96. Tournayan, L.; Charcosset, H.; Frety, R.; Leclercq, C.; Turlier, P.; Barbier, J. and Leclercq, G.: Thermochimica Acta *27*, 95 (1978)
97. Charcosset, H.: Plat. Mets. Rev. *23* (1), 18 (1979)
98. Burch, R.: ibid. *22* (2), 57 (1978)
99. Webb, A. N.: J. Catal. *39*, 485 (1975)
100. Johnson, M. F. L.: ibid. *39*, 487 (1975)
101. McNicol, B. D.: ibid. *46*, 438 (1977)
102. Charcosset, H.; Frety, R.; Leclercq, G.; Mendes, E.; Primet, M. and Tournayan, L.: ibid. *56*, 468 (1979)
103. Sinfelt, J. H.: US Pat. 3,791,961 (1974)

Subject Index

Catalysis –
Science and Technology
•
Volume 1

EDITORS:

PROF. DR. J. R. ANDERSON (Melbourne/Australien)
PROF. DR. M. BOUDART (Stanfort/USA)

CONTRIBUTORS:

PROF. DR. K. AIKA, DR. M. E. DRY,
DR. H. HEINEMANN, PROF. DR. A. OZAKI, DR. J. H. SINFELT,
PROF. DR. J. CH. R. TURNER

Bei Fragen zur Produktsicherheit wenden Sie sich bitte an:
If you have any questions regarding product safety,
please contact:

Walter de Gruyter GmbH
Genthiner Straße 13
10785 Berlin
productsafety@degruyterbrill.com